J. Delisle

# The Physical Chemistry and Mineralogy of Soils

VOLUME II: SOILS IN PLACE

# The Physical Chemistry and Mineralogy of Soils

VOLUME II: SOILS IN PLACE

## C. Edmund Marshall
Professor Emeritus, University of Missouri

A Wiley-Interscience Publication

John Wiley & Sons, New York • London • Sydney • Toronto

Copyright © 1977 by John Wiley & Sons, Inc.

All rights reserved. Published simultaneously in Canada.

No part of this book may be reproduced by any means, nor transmitted, nor translated into a machine language without the written permission of the publisher.

**Library of Congress Cataloging in Publication Data** (Revised)

Marshall, Charles Edmund, 1903–
   The physical chemistry and mineralogy of soils.

   Includes bibliographies.
   CONTENTS: v. 1. Soil materials.—v. 2. Soils in place.
   1. Soil chemistry. 2. Soil mineralogy.

S592.5.M37 1964     631.4′1     64-20074
ISBN 0-471-02957-2

Printed in the United States of America

10 9 8 7 6 5 4 3 2 1

# *Preface*

Any preface to guide the reader into this book will certainly be personal. This was the way I came to pedological problems. By the time I took my Ph.D. at Rothamsted, where I worked on the chemistry of humus, I had a grasp of colloid chemistry and of some applications in soil science; but the subjects of soil formation and development were practically unknown to me. With this background I traveled to Zürich. There I found colloid chemistry to my heart's content and, withal, a magical bonus, the teaching of soil science by Georg Wiegner. Thus through Wiegner I came to know of Dokuchaev and the other great Russians. Soil geography now had meaning. The exchange complex was clearly the chief link between the geographical and the chemical aspects of soil. It has remained so to this day. The present volume attempts a modern expression of this linkage.

After the year in Zürich I began my teaching career at Leeds, where my colleagues were N. M. Comber, H. T. Jones, and J. S. Willcox; all were interested in the acidic soils of the north of England and especially in Podzols. So from time to time I cheerfully interrupted my research on clays and clay minerals to join them in hunting for and sampling Podzols. We were a very congenial group, and our association bore fruit in Jones and Willcox's classic paper on complexing mechanisms in the formation of Podzols. One summer, during my years at Leeds, Geoffrey Milne, returning on leave from East Africa, joined us in the laboratory. He brought stories of vastly different perspectives in soil genesis, and of soils that reacted with common reagents in strange ways. Initially through him, I began to take an interest in tropical soils.

This was intensified when F. Hardy published Harrison's great monograph "The katamorphism of igneous rocks under humid tropical conditions." Over the years I have read and reread this work, examining the data through various chemical calculations. A few of the most significant results are presented here, using the perspective of the modern geochemical approach to pedology.

In the same years I began to explore the uses and improvement of microscopic techniques in connection with the relation of soils to parent materials. One could employ procedures using thin sections of coherent rocks on the one hand, or the mounted sand grains of the sedimentary petrologist on the other. I remember how I was taken aback one day when a colleague in geology commented that most surface soils were probably contaminated with wind-blown material. Hence calculations relating them to the rocks beneath would be vitiated. I thereupon began to consider how the methods of the sedimentary petrographer could be extended to wind-blown materials, which fall mainly in the silt and clay fractions. Continuation of this work in Missouri finally led to the development of quantitative methods based on resistant heavy minerals as described in Chapter 5.

Although I have never been a soil surveyor, the experience of sampling and describing soil profiles in detail has always fascinated me. Many opportunities have been afforded through my colleagues in Missouri since 1936. Thus it was no accident that the first profile studied in detail was of the Putnam silt loam, and that the particular site was the virgin Tucker prairie, east of Columbia, which had been used by my predecessor Hans Jenny as the base in assessing the effects of cultivation on organic matter and nitrogen content. Data from several highly detailed studies of Missouri profiles are used in this book. These owe much to the field experience of my senior colleague H. H. Krusekopf, and, most of all, to the devotion, originality, and industry of the graduate students who carried out these laborious investigations: E. P. Whiteside, R. P. Humbert, J. F. Haseman, Yi Hseung, J. E. Brydon, and C. L. Scrivner.

The highly detailed sampling of a soil profile was undertaken for another purpose by my colleague W. J. Upchurch. The exchange complex, original and modified, was examined for evidence of cationic movement and of mineralogical change, using modern geochemical methods. These studies form a considerable part of Chapters 5 and 10. They provide the experimental side of the geochemical considerations that permeate most of this book and originated, as far as I was concerned, with the publication of *Solutions, Minerals, and Equilibria*, by Garrels and Christ, in 1965. I feel that such geochemical evidence will now become exceedingly important, indeed crucial, in the development of pedology. Future experiments in mineral weathering and synthesis, interpreted in this manner, should contribute greatly by defining limiting factors in the inherently complex systems presented by natural soils.

I have avoided the subject of soil classification, relying mainly on authors' nomenclatures. I feel that the comparisons of different systems now available in *Soil Genesis and Classification*, by Buol, Hole, and McCracken, excuse this omission.

I indicated in the Preface to Volume I that Volume II would include consideration of plant growth in relation to soil colloids. This topic has been omitted here because the literature has ramified so greatly that I have been unable to cope with it.

Finally I thank the authors and publishers who have allowed me to use figures and tables from their publications. They are: Professor Hans Jenny, the Iowa State University Press, the Soil Science Society of America, the American Society of Agronomy, *Soil Science, American Journal of Science*, the Clay Minerals Society, Freeman Cooper and Company, Butterworth and Company, Springer Verlag, and *Annals of the Agricultural College of Sweden*. I am especially indebted to Professor S. Buol of North Carolina State University for slides of thin sections of a soil profile, from which Figures 43, 44, and 45 were made.

C. EDMUND MARSHALL

*Columbia, Missouri*
*April 1977*

# Contents

**1 Pedology in Relation to Physicochemical and Mineralogical Processes**    1

    The Development of Pedological Concepts    1
    Soil Formation and Soil Development    4

    *Weathering, Synthesis, and Soil Development,* 8
    *Quantitative Changes in Soils,* 9
    *Pedological Processes,* 9

    *References,* 10

**2 Early Stages in Rock Weathering**    12

    Soil Minerals in Relation to Their Chemical Environment    21

    *Standard Free Energies in the Formulation of Equilibria,* 21
    *Fundamental Equations of Weathering Reactions,* 25
    *Pedological Considerations,* 37
    *Concept of the "Weathering Sequence,"* 38

    *References,* 42

**3 Hydrothermal Synthesis of Minerals in Relation to Pedology**    44

    General Considerations    44
    The Feldspar Group    47
    The Mica Groups    49
    The Clay Groups    49

    *References,* 54

## 4 The Chemical Expression of Climatic Factors 57

Thermodynamic Factors 57
*Changes in the Chemical Potential of Soil Water,* 57

Cyclic Climatic Changes and the Soil Solution 58
*Temperature Variations,* 58
*Moisture Variations,* 61
*Hysteresis Curves,* 61

Mineral Equilibria 63
Kinetic Factors 65
*Mechanisms,* 66
*Capillary Conductivity,* 72
*Diffusion,* 74

Solubility Differences 75
Differences in States of Oxidation Leading to Chemical Depositions 76
*General,* 76
*The Production of Concretionary Material,* 79

The Complexing of Metallic Elements 82
Colloid-Chemical Considerations 86
*Peptization,* 86
*Colloid-Chemical Aspects of Podzolization,* 87
*Humus in Nonpodzolic Soils,* 90

*References,* 91

## 5 Quantitative Aspects of Soil Profiles 94

Lysimeter Studies 94
*Missouri Results,* 95
*Soil Characterization,* 95
*The Moisture Regime,* 99
*Chemical Composition,* 100
*Discussion,* 111

Index Mineral Studies 114
*General,* 114
*Choice of Index Minerals,* 118

*Method of Calculation*, 122
*Soil Formation and Soil Development*, 124
*The Constant Volume Hypothesis*, 126
*Barshad's Method; A Critique*, 126
*Clay Formation in Relation to Factors Affecting Soil Profiles*, 131
*Soil Profiles Evaluated by Index Methods*, 132
*Changes in Total Mass*, 132
*Changes in Nonclay Fractions*, 133
*Changes in Clay Content*, 137
*Movement of Individual Elements*, 144

*References*, 146

## 6 Characterization of the Pedological Environment and Its Products: Physical and Chemical Methods as Applied to Soil Profiles    149

| | |
|---|---|
| Climatic Factors | 150 |

*Temperature*, 150
*Precipitation*, 151

| | |
|---|---|
| Geological Factors | 153 |

*Parent Materials*, 154

| | |
|---|---|
| Biotic Factors | 158 |

*Vegetation*, 158
*Microbiological and Other Regimes*, 163

| | |
|---|---|
| Topographic Factors | 163 |

*Slope and Aspect*, 164
*Drainage and Depth of Water Table*, 164

| | |
|---|---|
| Duration | 164 |

*References*, 167

## 7 Physical Description of Soil Profiles    170

| | |
|---|---|
| Field Observations and Descriptions | 170 |
| Microscopic Observations and Descriptions | 170 |
| Particle Size Distribution | 176 |
| Aggregate Analysis | 178 |
| Behavior Toward Water | 178 |

*Moisture in Relation to Depth*, 178
*Pore Size Distribution*, 179
*Behavior at High pF Values (Adsorption)*, 181
*Permeability; Infiltration Capacity*, 182

*References*, 183

## 8  Mineralogical Analysis — 185

Minerals of the Clay Fraction — 185

*General Considerations*, 185
*Specific Groups of Soils*, 187

Silts and Sands — 216

*References*, 218

## 9  Chemical Analyses of Soils and Soil Fractions — 222

The Mineral Portion of Soils — 222
The Organic Portion of Soils — 228
Individual Elements and Compounds as Indicators of Pedological Processes — 238

*Calcium*, 239
*Other Cationic Elements*, 243
*Phosphorus*, 243

*References*, 244

## 10  Ionic Properties of the Exchange Complex and the Soil Solution — 247

Relationships Between Exchangeable Cations — 247

*Single Cation Functions*, 250
*Detailed Study of a Missouri Soil*, 252

Exchangeable Cations in Relation to Soil Classification and Characteristics — 255

*Alfisols and Mollisols*, 256
*Ultisols*, 256
*Oxisols*, 256
*Histosols*, 258

*Podzols and Brown Earths,* 259
*Selectivity Curves,* 260

The Soil Profile in Relation to the Soil Solution     263

*Other Ionic Ratios,* 272

Oxidation-Reduction Conditions in Soil Profiles     273

*References,* 276

## 11   Obstacles and Vistas     278

| | |
|---|---|
| Definition of Solid Phases | 278 |
| Exchange Reactions | 279 |
| Electrokinetics | 279 |
| Index Methods | 280 |
| Soil Organic Matter | 280 |
| Cationic Equilibria and Moisture Changes | 281 |
| Silica | 281 |
| Aluminum | 282 |
| Iron in Soil Profiles | 283 |
| The Clay Minerals | 284 |
| Instrumentation in the Service of Pedology | 285 |

**Author Index**     **289**

**Subject Index**     **295**

# The Physical Chemistry and Mineralogy of Soils

**VOLUME II: SOILS IN PLACE**

# 1 *Pedology in relation to physicochemical and mineralogical processes*

**THE DEVELOPMENT OF PEDOLOGICAL CONCEPTS**

The study of soils as a distinct discipline (pedology) was begun by Dokuchaev in the period 1880–1900 (4). The initial grand generalizations were arrived at by correlating observable features of soil profiles with geographic and climatological facts. In due course chemical and mineralogical processes were considered. Dokuchaev clearly stated that soils were formed by the simultaneous operation of a small group of soil-forming factors—namely, climate, relief or topography, vegetation, and parent material—all to be considered in relation to duration or time. These factors operate through physical, chemical, and biological processes. In some cases the results were so distinctive that the same root name was used both for the characteristic soil profile and for the sum of the processes that gave rise to it. Thus soil scientists began to speak of podzolization, salinization, or lateritization, at the same time that they attempted to improve their chemical, physical, or biological definitions of these processes. The interplay of Dokuchaev's ideas with the researches of two succeeding generations of Russian soil scientists is well described in *Fundamentals of Soil Science and Soil Geography* by Gerasimov and Glazovskaya (5).

Through the work of many investigators we have inherited four interrelated groups of problems. First, the detailed study of processes of soil formation and development, which is steadily becoming more quantitative and physicochemical; second, attempts at the comprehensive classification of soils as natural objects; third, the isolation and individual study of agronomic factors in the use of soils for crop growth; fourth, connections between measurable properties and applications in engineering and construction. Much of the emphasis in this book is placed on the first group.

Dokuchaev's list of the soil-forming factors was critically examined in the late 1930's by Jenny (9), who came to the conclusion that logically the influence of vegetation or the biotic factor should not be represented simply by the dominant observable vegetation but rather by the potential for growth of the whole ecological assemblage. His treatment of time as a soil-forming factor also differed from that of Dokuchaev. Jenny gave it the same independent rank as the other factors; thus his generalized functional equation was as follows:

$$s = f(cl, o, r, p, t, \ldots)$$

where $cl$ is climate, $o$ is the biotic factor, $r$ is topography, $p$ is parent material, and $t$ is duration.

The equation of Dokuchaev placed time or duration in a separate category from the other factors. It represented, so to speak, the integration interval for the other factors, and was written $\pi = (K, O, Y)B$, where $\pi$ is soil, $K$ is climate, $O$ is organisms, $Y$ is the geologic substratum, and $B$ is the age of soil. It seems probable that when the soil-forming processes eventually are analyzed in physicochemical terms, the advantages of a separate category for duration will become apparent.

As Dokuchaev pointed out, the characteristic feature of soils that have been subjected appreciably to soil-forming factors is their horizonation—that is, the development of an anisotropic character along a line more or less at right angles to the surface (Jenny). This is to be distinguished from their general three-dimensional heterogeneity—a property that has become increasingly vivid in our minds through the microscopy of thin sections (Kubiena, 10, 11; Brewer, 1) and of vertical exposures of profiles (Kubiena).

These two kinds of heterogeneity are deeply disturbing to the physical chemist as he seeks to establish mechanisms and provide explanations. He faces the task of choosing the way in which he will simplify chemical and physical aspects to supply partial explanations, which when fitted together will give a coherent, true picture. It is worthwhile examining the nature of soil heterogeneity in some detail as a means of introducing the reader to considerations that arise again and again.

The physical description of a soil as composed of three interpenetrating phases—solid, liquid, and gaseous—refers ideally to the moist condition. Dry or frozen soils can be regarded physically as two-phase systems, although one realizes that thin adsorbed films of water are present on all the solid surfaces. From the viewpoint of the phase rule, however, the gaseous and liquid phases would each be taken as homogeneous and in contact with a number of solid phases, all tending to approach equilibrium through the medium of the liquid and gas. To take the gaseous phase first, the relatively rapid rate of self-diffusion implies that equilibration with regard to water vapor, $CO_2$, $O_2$, and $N_2$ will be chiefly a function of the interconnecting pore space. But in dry or moist soils the variations in this quantity are relatively restricted. The range of possible values could conceivably cover about one order of magnitude—say from 50 to 5%. Values smaller than 5% would be likely to arise only through flooding, which may not remove all air, yet effectively eliminates the gaseous interconnections of different parts of the soil. According to Buckingham's relationship (2) between the amount of a gas diffusing per unit time and the square of the porosity, gaseous diffusion could vary through 2 orders of magnitude in response to variations in pore space. Even so, equilibration through the gas phase over distances of the order of centimeters is relatively rapid when compared with the cyclical time interval involved in changes in moisture content.

Because diffusion in aqueous solution is slower by a factor of over $10^4$ than gaseous diffusion, the situation for solutes is quite otherwise. Even in the flooded condition, movement by diffusion alone over centimeter distances is very slow. The speed is reduced by several more orders of magnitude when the soil passes through the moisture range between field capacity and the wilting point. Thus inhomogeneities can persist, and through reactions at particular points in the moisture or vegetative cycle they can undergo accentuation with time. A further factor, of intermittent occurrence, is the downward leaching of rainwater under the influence of gravity.

These factors, which make for the persistence and even the accentuation of inhomogeneities, are almost absent from related granular mineral systems in nature. This is very strikingly brought out by comparisons of the thin sections of soils studied by Brewer with thin sections of sedimentary rocks. The latter, formed by deposition from suspension, remain in a relatively constant aqueous environment for a long time. For example, recent electron microscope studies have shown how sand grains can become cemented by crystalline clay particles very uniformly distributed at the points of contact. Here long periods of time were probably involved, and there was no intermittent leaching under gravity. Such a sandstone could be very different from the corresponding sandy soil in its heterogeneity.

The importance of the biotic factor in soil heterogeneity can hardly be overstressed. Its influence can be felt long after the original organic structures have disappeared. Thus root channels sometimes become coated or filled with deposits of clay or hydrous oxides (cutans, Brewer), or iron is mobilized by reduction or chelation by organic matter at discrete sites, then oxidized to hydrous ferric oxide, thus giving rise to particular kinds of glaebules (Brewer) or concretions. Clearly an alternation of oxidizing and reducing conditions favors such an involvement of iron or manganese. We examine the chemical conditions for the mobility and deposition of oxides of these elements in soil later.

## SOIL FORMATION AND SOIL DEVELOPMENT

Although the general idea of soil formation can be separated in the mind from that of soil profile development, it must be admitted that the historical sequence actually followed in a given case may not correspond at all to this mental distinction. If soils are to be distinguished from unconsolidated rocks or parent material through having been influenced by the biotic factor, the boundary between soil or solum and parent material becomes a matter of the sensitivity of the methods by which the influence of living matter can be detected.

There are therefore cogent reasons for treating all those processes together and taking that material whose properties do not change with depth as parent material, whether it be consolidated or unconsolidated. This of course involves assumptions about the original state of the visible horizons of the solum. Particularly in sedimentary materials it becomes necessary to determine by careful experiment whether the whole profile arose from uniform material. The techniques used are described in Chapter 5. Following upon the successful demonstration of uniformity, it becomes possible to assess the various changes that have occurred. There are many advantages to basing such calculations on original unchanged rock, since distinctions between soil development processes and rock weathering are difficult to make. Both Whiteside (19) and Brewer (1) advocate this. The former, in classifying rocks from the standpoint of the pedologist, concluded that unconsolidated materials could with advantage be treated as rocks for this purpose.

The clearly apparent changes involved in the development of soil profiles have been investigated by chemical methods for many years. Indeed almost every refinement in our views on the nature of soils has a pedological aspect. This extends back to the early Russian investigators. The first chemical method to be used was that of total analysis, exactly as for silicate rocks. Then came various acid extractions for the purpose of separating the inert sand, and so on, from the more reactive and agriculturally important components. A very important change in outlook followed the recognition of the exchange complex of soils as a crucial and sensitive indicator of pedological changes. This was effected by Gedroiz in Russia, von Sigmond in Hungary, and Hissink in Holland. Pedological changes were found to affect both the total exchange capacity and the proportions of cations that made up that capacity. Furthermore, the colloidal constituents of soil (clay, humus) were the chief components of the exchange complex. Hence Wiegner (20) in Switzerland, following earlier work by Van Bemmelen in Holland, was able to bring to bear a general colloid-chemical viewpoint on problems of soil formation and development. Briefly stated, the mobility or arrest of colloidal materials in soil profiles was explained by the properties of the electrical double layer and the phenomenon of protective action of one colloid on another. These considerations were further elaborated by Mattson (16), whose work we examine in some detail later.

In the decade 1930–1940 an important change of viewpoint in regard to mineral soils followed the demonstration of the crystalline character of many soil colloids and of the connection between the atomic structure of these minerals and the property of ionic exchange. Such minerals could be inherited, or they could have pedological origin. The author played an active part in linking together the structural and colloid-chemical aspects of clay minerals (13–15). The improvement of X-ray techniques, of differential thermal analysis, and of electron optical methods quickly led to quantitative estimates of the clay minerals in soils. Thus during the 1960's it became clear that in most mineral soils the crystalline components of the clay fraction were quantitatively dominant over the amorphous. However the latter came into prominence wherever the influence of acidic organic matter was high (as in the B horizon of Podzols) and also in soils from volcanic ash. Amorphous and poorly crystalline material was more evident in surface soils than in subsoils. Factors such as the presence of organic matter delay or slow down crystallization and increase the proportion of amorphous components.

This raises central problems for pedology. In the transformation of one identifiable mineral to another, is the production of an amorphous intermediate necessary? In other words, are competing mechanisms present, one direct and another through amorphous products? If so, what are the factors affecting the respective rates of reaction? These, for the most part, are questions for the future. This book shows that pedology is now very much concerned with initial and final states. As these problems are solved, however, more and more attention will naturally be given to mechanisms and rates of reaction.

It is evident also that in following searches for detailed mechanisms, pedology is approaching consideration of each of the soil-forming factors in mechanistic terms. This work attempts to summarize and clarify the present position. One must recognize however, that other approaches may also be enlightening.

The recognition that soil profile features reflected the operation of relatively stable soil forming factors carried with it certain other concepts which had their roots in biology. On a geographic basis it was clearly apparent that between the characteristic examples of related Great Soil Groups there lay zones of transition. In certain localities also evidence of temporal successions in vegetation could be found. Thus shorter and longer operations of the same climatic or vegetative factors were compared. To explain these phenomena the concepts of "stage of development" or "maturity" were used. Once these ideas were accepted, examples were also found suggesting retrogression or degradation. Pursued even further, biological analogies then suggested the use of such terms as "growth" and "senescence." The greater the influence of the biotic factor, the more this use of biological terms is found. For instance the Russian investigator Vil'yams thought of the long history of evolution on the earth in terms of the changing conditions of what he calls the "single soil-forming process." More recently Pallmann (17) and Kubiena (11) have placed great emphasis on the biotic factor. The former stressed the close connection between the ecological succession of plant life on a given site and the corresponding development of the soil. The latter concluded that the greater the dominance of the biotic factor, the greater would be the heterogeneity of the soil and the greater the applicability of microscopic as contrasted with chemical methods of investigation.

The concept of the "mature soil" as expounded by Marbut (12) was intended to facilitate clear-cut comparisons by the exclusion of local variations due to site factors and of soils that were in a youthful stage of development. He stated,

Maturity represents the full expression of the influence of the environment on soil material, yet I do not want to convey the impression that a soil may not change after maturity. The stage of maturity is not a stage of absolute fixedness of characteristics, but after maturity has been reached the changes that take place, take place much more slowly than previously, and differences do not become so great.

Marbut's views clearly point more directly to the study of rate processes by physicochemical means than do those of the Russian school or Kubiena. Marbut proceeds directly from the definition of maturity to a discussion of the possible bases for the initial subdivision of mature soils.

He concludes that the presence or absence of sparingly soluble carbonates derived in mature soils by pedogenic processes provides a first subdivision of world-wide application. This is frankly chemical. Complete analyses on successive layers of the soil profile furnish good criteria for this subdivision into the pedocals (with carbonates) and the pedalfers (without carbonates). It is simpler to apply than to attempt a primary subdivision into mature and immature soils, or as Glinka (6) termed them, ectodynamomorphic and endodynamomorphic soils.

Marbut's criterion can now be expressed in relation to the soil solution. Cole (3) showed that the chemical potential of calcium hydroxide throughout the soil system can be expressed in terms of the chemical potential and dissociation constant $K_w$ of water, the equilibrium constant for the hydrolysis of calcium carbonate, and the equilibrium partial pressure of $CO_2$ in presence of solid calcite. Since for very dilute systems the water can be taken as in its standard state, and the hydrolysis constant $k_1$ is known ($= 10^{-18.30}$) the following relationship was found at 25°C (see Vol. 1, Chap. 2):

$$\text{pH} - \tfrac{1}{2}\text{pCa} = \tfrac{1}{2} \log \frac{k_1}{(k_w)^2} - \tfrac{1}{2} \log P_{CO_2}$$

Since the $CO_2$ partial pressure in soils can rarely fall below that of the normal atmosphere, namely, $P = 0.0003$, the limiting chemical potential for calcium hydroxide at 25°C works out at $-20.160$ cal/mole, corresponding to an activity $a_{Ca(OH)_2}$ of $10^{-14.78}$, and a (pH $- \tfrac{1}{2}$pCa) value of 6.61, assuming that the activity of water is unity.

In soils, as soon as solid phase calcium carbonate is absent, pH $- \tfrac{1}{2}$pCa changes drastically to lower values. For instance, Schofield and Taylor (18) found that an acidic Rothamsted subsoil gave a value of 3.36, whereas a calcareous subsoil gave 6.18. The latter value is lower than 6.61, possibly indicating variations in $CO_2$ and in calcium ion activity.

In the light of these quantities, easily measurable for soils in bulk by the method of dilute small exchange (see Vol. I, Chap. 1), we must also consider the matter of soil heterogeneity. It is a common observation that many noncalcareous soils produce efflorescences and localized deposits of calcium carbonate upon drying. This arises through the operation of mineral hydrolysis as applied to primary mineral grains and of Donnan-type hydrolysis as applied to the colloidal exchange complex. Clearly, upon drying, when small amounts of moisture, insufficient to dissolve such localized carbonate are present, the value of pH $- \tfrac{1}{2}$ pCa in these spots will be much higher than for the soil as a whole. At any given moisture content and $CO_2$ pressure, only a limited amount of calcium carbonate can be present in solution, and quantities in excess of this will be deposited. Thus the complete definition of a pedocal involves both the soil-water ratio and the $CO_2$ pressure. Similar considerations reveal that the formation of a pedocal from noncalcareous material can occur progressively through cycles of drying and wetting, provided losses in drainage are less than the accretions from hydrolysis.

In considering Donnan hydrolysis, two factors are important: (1) a separate aqueous phase must exit (the outer solution), (2) hydrogen and aluminum ions must replace the calcium, leading to an acidic exchange complex. Thus the necessary counterpart of regions of carbonate accumulation are acidic regions in the exchange complex. Obviously when ample water is present and sufficient time available, the reverse reaction to the hydrolysis will occur. Thus it is highly unlikely that the major part of the original exchangeable calcium will find itself changed to carbonate. Considering that 1% $CaCO_3$ in the soil is the equivalent of 20 meq exchangeable calcium per 100 g of soil, it would seem that other sources of mobile calcium should be considered. As mentioned earlier, the hydrolysis of primary mineral grains can also give rise to calcium in solution and as solid $CaCO_3$. Plagioclose feldspars are the primary source and, as Graham (7) showed, they are actively decomposed by acidic clays. Thus we have a mechanism for the restocking of the exchange complex with calcium, as well as for the ultimate appearance of $CaCO_3$: 1% $CaCO_3$ in the soil corresponds to about 2% anorthite or 14% oligoclase. Thus the accumulation of $CaCO_3$ in amounts greater than 1% or so from noncalcareous parent material must generally arise from primary mineral decomposition. Such decompositions are reversible only under special conditions—alkaline systems with sufficiently high cation activities are required. Since there are no low temperature syntheses yet available for calcium or calcium-sodium feldspars, we are really in the dark on reversibility.

The example of Marbut's pedocals thus serves as a preliminary warning that one should not expect relatively simple field criteria to correspond to simple chemical criteria, and vice versa. Once the situation has been carefully examined, practical chemical procedures may be devised.

### *Weathering, Synthesis, and Soil Development*

Our view of mineral transformations in rocks and soils differs from the perspective of the past. For many years weathering was looked on as a degradation of crystalline minerals with reorganization possible only under special conditions. In recent times the process described as diagenesis has been defined in terms of changes under mild conditions at the earth's surface, to be contrasted with metamorphism, which involves elevated temperatures and pressures. Much more comprehensive understanding has resulted from experiments in mineral synthesis combined with thermodynamic evaluations of energy changes. By this route changes are viewed in their appropriate chemical environment, whatever it may be. The unequivocal definition of the chemical environment then acquires primary importance. It is implicit in the thermodynamic treatment of chemical changes that they can be represented, and indeed are frequently realized, as reversible reactions. Hence weathering, synthesis, diagenesis, and metamorphism are all considered together. One of the major tasks in this book is to apply these methods to the study of soils. Clearly a serious obstacle in such a discussion is soil heterogeneity, but by suitable choice of conditions it may sometimes be circumvented. It must also be recognized that the detailed physical chemistry of soil organic matter is insufficiently known for complete thermodynamic treatment to be applied. Thus it would be premature to expect all the major pedological processes to find expression in quantitative chemical thermodynamics or kinetics.

*Quantitative Changes in Soils*

From early times in the history of pedology, chemical and physical methods were used for the quantitative comparison of soil horizons. Mineralogical methods were at first mainly qualitative. They have become quantitative through increased understanding of the clay minerals and of their X-ray diffraction patterns and their thermochemical properties. Straightforward chemical analyses indicated that podzolic soils, for instance, showed depletion of iron and aluminum in the $A_2$ horizon with concentration of these elements and of organic matter in the B horizon. Each comparison was made with what was assumed to be unchanged parent material, the C horizon. By such comparison, carried out on Podzols originating from different parent materials, a general picture of podzolic processes emerged. For many years it remained descriptive rather than quantitative.

For firm quantitative conclusions to be drawn, three steps are required: first, for a given soil profile, uniformity of horizons with regard to their parent material must be demonstrated; second, an immobile constituent must be chosen, then used as the basis for quantitative comparisons; third, the kinetic study of processes needs to be undertaken. Over the past 30 years steady progress has been made in relation to the first two steps, but there has been little progress with the third. The evidence now available by these methods is reviewed in Chapter 5.

*Pedological Processes*

Although in a formal sense the processes of soil development can be divided into physical, chemical, and biological phases, there is much overlapping. For instance, the translocation downward of clay can initially be thought of in purely physical terms; but the exact conditions for peptization and later deposition involve considerations of surface chemistry and the properties of the electrical double layer. In some cases it is easy to trace chemical changes in soil profiles back to biological origins. Many soil organisms have qualitatively clearly apparent biological and chemical functions in the soil when they are alive, and a series of less well-defined functions when they are dead, until what remains of their effects is included in the standard physical and chemical description of the organic matter.

Hallsworth's comment (8) that in any given situation there is, superimposed on the visible horizonation, a distinctive depth variation for each property measured, is highly pertinent. Many pedological processes may not be apparent to the field observer and may show themselves in different quantitative variations with depth. In some cases more may be learned by working from the original rock upward than from the soil surface downward.

Another factor of great importance is the inheritance from past regimes of climate and vegetation. Present-day mechanisms may be operating on profile material that had already been affected by earlier and different combinations of soil-forming factors. This makes for difficult interpretations, which may depend in part on very approximate ideas on the rates of soil development under different conditions. Evidence that helps to date soil features is important, and often exceedingly hard to find. Almost all agricultural operations tend to obscure original profile features in the upper part of the soil—hence the pedologist's preoccupation with virgin sites.

## REFERENCES

1. Brewer, R., *Fabric and Mineral Analysis of Soils*, Wiley, New York (1964).
2. Buckingham, E., Contributions to our knowledge of the aeration of soils, U.S. Department of Agriculture, Bureau of Soils Bulletin No. 25 (1904).
3. Cole, C. V., Hydrogen and calcium relationships of calcareous soils, *Soil Sci.*, **83**, 141 (1956).
4. Dokuchaev, V. V., *Russian Chernozem*, Israel Program for Scientific Translations, Jerusalem (1965).
5. Gerasimov, I. P. and M. A. Glazovskaya, *Fundamentals of Soil Science and Soil Geography*, Israel Program for Scientific Translations, Jerusalem (1965).
6. Glinka, K., *Die Typen der Bodenbildung*, Borntraeger, Berlin (1914).
7. Graham, E. R., Acid clay—An agent in chemical weathering, *J. Geol.*, **49**, 392 (1941).
8. Hallsworth, E. G., In *Experimental Pedology*, E. G. Hallsworth and D. V. Crawford, Eds., Butterworth, London (1965).

9. Jenny, H., *Factors of Soil Formation*, McGraw-Hill, New York (1941).
10. Kubiena, W. L., *Micropedology*, Collegiate Press, Ames, Iowa (1938).
11. Kubiena, W. L., *Micromorphological Features of Soil Geography*, Rutgers University Press, New Brunswick, N.J. (1970).
12. Marbut, C. F., *Soils: Their Genesis and Classification*, U. S. Department of Agriculture Graduate School and Clark University (1928). Reprinted by the Soil Science Society of America (1951).
13. Marshall, C. E., Layer Lattices and the base exchange clays, *Z. Krystallogr.*, **91A**, 433 (1935).
14. Marshall, C. E., The electrochemistry of the clay minerals in relation to pedology, *Trans. 4th Int. Congr. Soil Sci. Amsterdam 1950*, **I**, 71 (1950).
15. Marshall, C. E. and C. A. Krinbill, The clays as colloidal electrolytes, *J. Phys. Chem.*, **46**, 1077 (1942).
16. Mattson, S., The laws of soil colloidal behavior. III and IV. Isoelectric precipitates, *Soil Sci.*, **30**, 459 (1930); **31**, 57 (1931).
17. Pallmann, H., Grundzüge der Bodenbildung, *Schweitz. Landwirtsch. Monatsh.*, **22**, 1 (1942).
18. Schofield, R. K. and A. M. Taylor, The measurement of soil pH, *Soil Sci. Soc. Am. Proc.*, **19**, 164 (1955).
19. Whiteside, E. P., Some relationships between the classification of rocks by geologists and the classification of soils by soil scientists, *Soil Sci. Soc. Am. Proc.*, **17**, 138 (1953).
20. Wiegner, G., *Boden und Bodenbildung in Kolloidchemischer Betrachtung*, Springer, Leipzig (1923).

# 2 Early stages in rock weathering

The chemical factors of significance in the early stages of rock breakdown are related to the chemical environment and its persistence or variation in time. It is universally agreed that moisture is essential to chemical weathering and that oxygen and carbon dioxide are important. However the relative ineffectiveness of oxygen and carbon dioxide as compared with water alone is not apparent from textbook descriptions. Accessibility of primary weathering sites to the atmosphere is widely assumed, without consideration of the hindrance to diffusion caused by a thick mantle of weathered rock. Secondary carbonates are much less commonly found in the weathering crusts of igneous rocks than is tacitly assumed to be the case, and the assumption, frequently made, that their absence is due to solution by leaching, seems out of harmony with the main experimental evidence. For instance, in J. B. Harrison's magnificent study of some 12 rocks of Guyana [then known as British Guiana] (7), carbonates were never found in the initial stages. Merrill's observations on rocks in the United States (17) reveal few cases of carbonates appearing. The reason becomes apparent when a model system is considered in which the rock is overlaid by a moist but not saturated porous sand. Utilizing the data provided by Buckingham (4), it is easy to calculate the diffusion of carbon dioxide through any given thickness of such material, assuming some definite air space. For instance, with an air porosity of 10% and a porous layer 1 m thick, the quantity of carbon dioxide that arrives at the rock surface is only 3 mg/cm$^2$ per year. Since the average mantel of unconsolidated material is considerably thicker, it is evident why other processes usually take precedence over the formation and deposition of alkaline earth carbonates.

Naturally more oxygen is able to reach any given depth because of its much greater proportion in the atmosphere. The factor for $O_2$ as compared with $CO_2$ is about 700. Hence oxidation may be expected to show itself in regions where carbonate formation is negligible. Its recognition is greatly aided by the intense color and low solubility of ferric hydroxide.

Variation in pore space of the disintegrated mantle of rock is probably of minor effect, since only a relatively narrow range of air porosities is likely to be encountered. In passing from moist but porous media to systems completely saturated with water, essentially an abrupt change occurs. With it comes a striking reduction in diffusion velocity of a given molecular species, such as oxygen or carbon dioxide. As already mentioned, gaseous diffusion exceeds diffusion in water at the same temperature by a factor of over $10^4$. Hence the presence of a saturated zone effectively blocks the movement of these gases over appreciable distances. In a similar way, diffusion of soluble products away from the decomposing rock surface must be very slow unless it is aided by mass movement of water. The latter may be of two kinds—steady flow under a uniform hydraulic gradient, and intermittent flow caused by the emptying and filling of pores, as imposed, for instance, by variations in the height of the free water table.

Thus the situation with regard to water supply and movement is highly critical, and we cannot expect any single model to serve to portray, even in qualitative terms, all the variations that may be encountered. Conditions of very common occurrence would seem to involve (*a*) diffusion in solution to and from the decomposing rock surfaces through a stationary water layer, with the effective thickness of this layer quantitatively an exceedingly important variable; (*b*) possible diffusion of gases through porous layers to within short distances of the rock surfaces (as pointed out earlier, this usually shows itself in a restricted way by oxidation of ferrous iron but often not by deposition of carbonates); (*c*) mass movement of water at some distance from the decomposing rock surfaces.

Certain chemical factors in this situation also call for comment. As demonstrated later, the evidence on the weathering of single silicate minerals under acid, neutral, and alkaline conditions clearly proves (*a*) that attack under alkaline conditions can be extremely vigorous, (*b*) that there is minimum attack near neutrality, (*c*) that poorly organized layers richer in silica than the original mineral remain attached to the weathering surfaces (19, 20). These layers vary in thickness and in composition with the pH. Normally such immobile, hydrated material would carry a negative charge and would be capable of cation exchange.

A more quantitative physicochemical approach to carbonate formation is now possible, as illustrated, for example, in Chapter 3 of the Garrels and Christ work (6). The problem is first to delimit the possible range of variation by the consideration of extreme cases in which one factor or another predominates. Within this range it may then be possible to see what information will be required to set up a quantitative model. Clearly our first concern must be with conditions for the production of solid phase calcium carbonate. As far as this species is concerned, four equilibria are to be investigated: carbonate–hydroxide, carbonate–bicarbonate, bicarbonate–carbonic acid, carbonic acid–gaseous $CO_2$. On the weathering mineral side, we must consider the pH of the solution phase, often assumed to be the same as the abrasion pH, in relation to the weathering reactions, as well as the calcium content and the ionic strength.

The first thing to note is that certain specified weathering reactions operating alone should be automatically buffered at a definite pH. All systems that produce a mixture of soluble silicate and silicic acid fall into this category through the applicability of the buffer equation pH = p$K$ + log salt/acid, when p$K$ represents the negative logarithm of the first dissociation constant of silicic acid. For instance, the sodium beidellite → kaolinite reaction can be written:

$$2NaAlSi_7O_{12}Al_4O_8(OH)_4 + 13H_2O \rightarrow 5Si_2O_3 \cdot Al_2O_2(OH)_4 + 2NaOSi(OH)_3 + 2Si(OH)_4$$

Here the salt-acid ratio is 1.0; thus the pH will be equal to p$K$ for the first dissociation of silicic acid, that is, about 9.8. Other examples, with their pH values, are as follows:

albite → sodium beidellite, pH 9.8
orthoclase → kaolinite, pH 9.8
orthoclase or albite → gibbsite, pH 9.5
orthoclase → muscovite, pH 9.5
muscovite → gibbsite, pH 9.5

Some equilibria correspond to the presence of a soluble salt in solution rather than a mixture of a weak acid and a salt. In this case the pH is a function of the salt concentration $c$ as follows, assuming $K_w = 10^{-14}$, and $K_A$ is the dissociation constant of the acid.

$$\text{pH} = 7 + \tfrac{1}{2} pK_A + \tfrac{1}{2} \log c$$

As an example, the change anorthite → gibbsite involves one equivalent of calcium in solution for each mole of silicon. If $c$ is taken as the total silicate in solution, then at $c = 10^{-3}$ mole/liter the pH will be 10.4 and at $c = 10^{-4}$ mole/liter it will be 9.9.

An example such as muscovite → kaolinite involves no release of silica into solution. Hence potassium hydroxide is theoretically the soluble product, and the pH will depend on the amount of potassium in solution.

All the cases above are greatly affected by Donnan-type hydrolysis, which of itself would produce the appropriate hydroxide alone in the outer solution. Its dominance is seen in the ratio of cations released to silica in solution, which is always considerably greater than unity. But this does not mean that the pH is always higher than indicated by the figures just given. We have also to take into account the practically universally present carbonic acid → bicarbonate → carbonate equilibrium in most cases as a second buffer system. Its effect on the Donnan hydrolysis is to increase the cations in solution while at the same time reducing the pH. Weathering in contact with the atmosphere is greatly affected by the conditions that control the access of $CO_2$ and oxygen to the active sites at the rock surface (see below).

Thus the internal environment for rock weathering commonly lies in the interaction of three systems—a Donnan hydrolysis system, a silicate buffer, and a carbonate buffer.

In relation to possible mineral equilibria, calcium carbonate formation can be included in the compositional diagram for the system $HCl-H_2O-Al_2O_3-CaO-CO_2-SiO_2$ as given by Helgeson, Brown, and Leeper (8) (Figure 1; this type of diagram is more fully explained later in the chapter). This scheme does not include anorthite, whose place is taken by lawsonite, which is compositionally a hydrated anorthite. The horizontal line for calcium carbonate lies much above the areas for montmorillonite and kaolinite in pure calcium systems. Thus calcium carbonate is produced only when log $\sqrt{Ca}/H^+$ is high, about 6.6 with normal atmospheric $CO_2$. With regard to silica, for values in solution considerably below the solubility of quartz (below $10^{-5.2}$) it appears that calcium carbonate and gibbsite are compatible with each other. Thus the general scarcity of calcium carbonate as a primary product in the weathering of igneous rocks can be laid to two factors: the $\sqrt{Ca}/H$ is rarely high enough, and this ratio is rarely found in combination with sufficiently low values of soluble silica. If leaching conditions are very severe, as in the basic igneous weathering investigated by Harrison, there is no chance of any such combination occurring. Indeed even with intermittent wetting and drying, cases offering a $\sqrt{Ca}/H$ sufficiently high that solid calcium carbonate is compatible with various aluminosilicate species are quite infrequent. It is not likely to occur in equilibrium with kaolinite, since high values of $\sqrt{Ca}/H$ imply high pH values, which in turn would impose high values of Na/H. These in combination with moderate to high silica, would produce such species as sodium beidellite and analcime as well as mixed calcium-sodium zeolites.

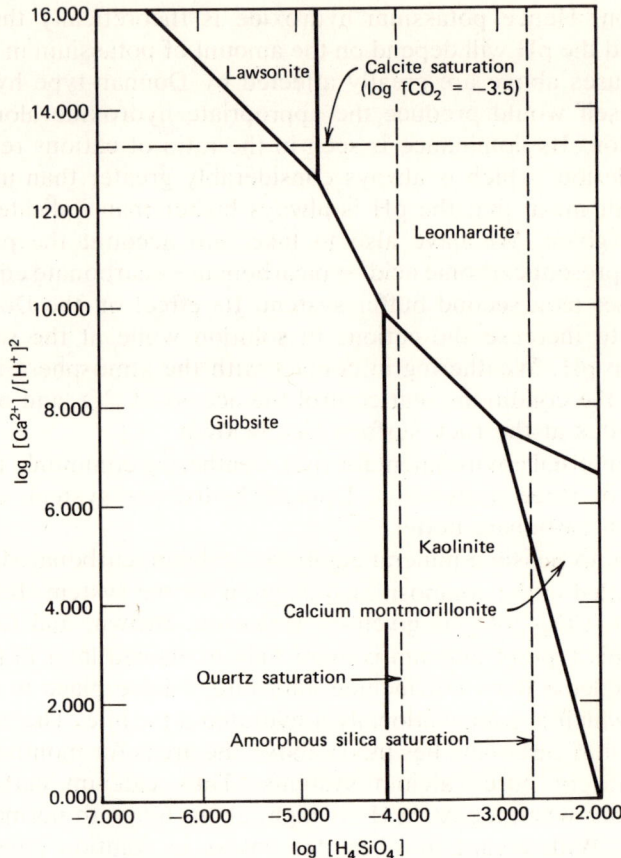

**Fig. 1** Garrels and Christ type of diagram for the system $HCl$-$H_2O$-$Al_2O_3$-$CaO$-$CO_2$-$SiO_2$ at 25°C (8). Used by permission, Freeman Cooper and Co.

It is not certain, however, that the conditions of alkalinity in such experiments as those of Correns and his associates (5, see Vol. I, Chap. 2) were as extreme as those prevailing close to decomposing basic igneous rocks. At depth, under tropical high water table conditions, within 3 mm of a dolerite rock, Harrison was able to recognize crusts of gibbsite (primary Laterite). Loss of silica was extreme. Weathering of this rock under ordinary atmospheric exposure was demonstrably quite different and gave rise to no gibbsite, but quartz, hydrated oxides of iron and titanium, and hydrated silicates of aluminum and iron. This example clearly shows the necessity for a much more complete definition of the chemical environment of rock weathering (and equally for soil formation) than has yet been attained.

Thus under the same overall climatic conditions weathering can proceed through different sequences of reactions, depending on what is sometimes referred to as the microclimate. As pointed out earlier, the differences between tropical and temperate weathering include large divergencies in products as well as quantitative differences in rate. Both the residual materials and those carried away in solution reflect these differences. It is well known that the streams of the tropics that drain areas of primary rocks carry much greater quantities of silica, both absolutely and relatively to the alumina and iron, than those of temperate regions. The high silica content of certain tropical plants (e.g., tabaschir deposited in the stems of bamboo) illustrates the mobility of silica in these regions.

The effect of temperature differences on the very earliest stages of rock breakdown has been examined by Blanck (3). He made complete chemical analyses of fresh and weathered rocks in arctic and high mountain regions and compared them with similar data under desert conditions. In each case the rock was under desiccation for the greater part of the year. The thaw or the occasional desert downpour provided the aqueous environment for limited chemical breakdown. Under arctic conditions the disintegrated rock and the soil derived from it showed (on a moisture- and humus-free basis) very little change in composition as compared with the original rock. Oxidation of ferrous iron and hydration were the most prominent changes. There were slight changes in calcium, magnesium, sodium, and potassium according to the type of rock under consideration. The ratio of $SiO_2$ to $Al_2O_3$ was extremely constant. Under desert conditions a more vigorous chemical attack was demonstrated. Silica and bases moved noticeably from weathering surfaces, the bases accumulating to some extent in the fine earth, since the rainfall was insufficient for extensive transportation. The $SiO_2/Al_2O_3$ ratio became markedly lower in the weathered rock and the fine earth derived from it. Blanck draws attention to this qualitative difference between arctic and desert weathering and sums up the situation in the following sentence: *"Es kommt demnach also der Wärme und nicht dem Wasser an sich die wichtigste Rolle beim chemischen Verwitterungsprozess zu."* (In the chemical weathering process it happens that warmth rather than water plays the predominant part.) There is, however, a factor in the situation besides temperature difference, namely, the presence of abundant organic matter in the arctic soils. Its possible influence on the weathering of the underlying rock should not be lost sight of, although Blanck believed that his conclusions were still valid in spite of it.

The material assembled by Blanck represents the most complete series of comparisons available of the early stages of the weathering of different rocks (primary, metamorphic, and secondary) under different climatic conditions. Complete chemical analysis, coupled in most uses with analyses of HCl extracts, were chiefly relied on. A little information about mineralogical changes can be deduced from such analyses; but much remains uncertain, owing to the lack of detailed optical work or X-ray determinations. Furthermore, little mention is made of drainage conditions. The work of Harrison, in the limited tropical area of Guyana, is much more complete.

Following the comparison of arctic and desert conditions, Blanck takes up the weathering of acid and basic igneous rocks, gneisses, sandstones, and limestone rocks under humid, temperate conditions, chiefly through German data. The weathering of granites shows itself first in hydration and oxidation of ferrous iron; silica remains about constant, while alumina increases or decreases slightly in different cases. Calcium is lost in the early stages before potassium, sodium, or magnesium. All decrease when soil formation processes begin to operate in presence of humified organic matter. Under somewhat similar climatic conditions basalts were found to show much greater chemical changes. In the early stages silica decreased very slightly, whereas alumina increased, but in the overlying soil silica had increased markedly, and alumina had decreased. Losses of alkalis and alkaline earths were greater than in the weathered granites.

Blanck also discusses the weathering of granite under the influence of humus, such as is commonly the case in heath and moorland regions. Here again, in the early stages the silica and alumina change little, but considerable losses of alkalis and alkaline earths occur. In later stages (formation of *"Bleicherde"*), large losses of alumina and iron were found so that silica showed an increase. The sesquioxides may be deposited in illuvial zones nearer to the rock surface than the *Bleicherde* horizon. No evidence was found for a kaolinization of the granite.

In a discussion of soils of the Mediterranean regions Blanck devotes much attention to the genesis of the "Terra Rossa." The analytical data presented indicate clearly that this cannot be simply a colored residue after solution of the limestone; rather, soil-forming processes have caused differential downward movement of iron and aluminum. Much speculation has centered around colloid-chemical explanations of these facts (see Chapter 8).

Blanck's examination of decomposing rocks in subtropical regions (chiefly Chile) indicated a tendency for silica to decrease and for sesquioxides to increase in percentage. Losses of alkalis and alkaline earths were only moderate except in one instance of a Chilean andesite that lost the greater part of the calcium and sodium.

Blanck's examples contain numerous cases of small amounts of carbonates being found in the decomposed rock. These would be expected in the type of profile he was concerned with, in which the total depth to bedrock was no more than a few feet and the drainage conditions were good. Access of atmospheric $CO_2$ to the decomposing rock surfaces was thus appreciable.

Since the action of water on silicates produces a strongly alkaline environment, whereas atmospheric moisture with its content of $CO_2$ is acidic, it is evident that the depth of the transition zone in which the reaction shifts from the one to the other may be small or large, depending chiefly on conditions for the movement of $CO_2$. Thus one might expect to find the analyses of a decomposing rock at increasing distances upward from the rock surface to show early trends indicative of an alkaline environment and later signs of an acid environment. The latter should become greatly accentuated in presence of humus. A number of the examples discussed by Blanck show decreases in the $SiO_2/Al_2O_3$ ratio near the rock surface and increases in the overlying soil, suggesting that this situation has prevailed, although in each case there may be other complicating circumstances as well.

A broad review in the light of these considerations strongly suggests that one characteristic of typically tropical rock decomposition as compared with that in temperate regions is an alkaline environment, maintained against the acidic factors ($CO_2$, humus, and sulfuric acid from sulfides, etc.) by the greater rate of rock decomposition, with its consequent thicker mantle of decomposed rock. The effect of increasingly continuous rainfall would be a heightened water table immediately above impermeable rock, hence further restriction of access of carbon dioxide and oxygen in the regions of active rock decomposition. The removal of soluble products would be effected by mass movement of water, which, on the average, would not be vertical but rather a movement along directions somewhat inclined to the horizontal. Vertical leaching and an acidic environment would still be features of soil formation processes in the overlying material, near the soil surface.

Under tropical conditions both the day-to-day and the seasonal variations in soil moisture conditions can be very great. Hence a simple statement of annual rainfall may be quite misleading. This is well brought out in Chapter 1, "Atmospheric Climate and Soil Climate," of Mohr and van Baren's *Tropical Soils* (18). Both in the initial stages of rock breakdown and in subsequent soil development, the soil climate is the important factor. In the tropics these two processes are frequently well separated vertically in space, so that contrasts between them become apparent. In temperate regions they are more often intermingled. It then becomes difficult to separate primary mineral breakdown from soil development processes. In favorable cases, however, much can be accomplished by the quantitative mineralogical and chemical analysis, layer by layer, of complete soil profiles.

Considerable light can also be thrown on weathering processes by analyses of drainage waters. Those of Harrison in Guyana are extremely enlightening. He gives detailed figures for percolating waters in contact with the following rocks: (*a*) granite at Mazaruni Quarry, at least 5.5 m from soil surface; (*b*) epidiorite at Issorora apparently about 12.2 m from soil surface, (*c*) hornblende schist in mine tunnel (Peter's Mine) 55 m below rock surface, probably about 61 m below soil surface. As might be expected, these solutions consisted chiefly of soluble silica with carbonates and bicarbonates of alkalis and alkaline earths. From their composition they were clearly alkaline in reaction. The granite gave a solution corresponding to a mixture of carbonates with a little bicarbonate. The epidiorite and hornblende schist gave more alkaline solutions evidently consisting of carbonates with hydroxides.

The most significant analysis, however, was that of percolating water collected some distance above the granite surface at the base of a thick (7 m) layer of argillaceous granite sand. This was just above the zone of active kaolinization. This argillaceous granite sand varied relatively little in chemical composition with depth. Mineralogical examination showed much kaolinite throughout. Small residues of potash feldspars were found, decreasing in amount toward the surface. The water that had percolated this material was also alkaline, having the composition of a carbonate-bicarbonate buffer. Since the layer varied little in composition throughout its depth, somewhat alkaline weathering conditions evidently prevailed. The transition from acidic conditions imposed by carbonic acid must have occurred relatively close to the soil surface, after passage through the 36 cm soil and subsoil layers from which all potash feldspar had been removed.

Harrison's work thus affords decisive evidence that the most active zone of granitic weathering occurs close to the rock surface where the environment is alkaline. In the thick (4.3–7.0 m) argillaceous granite sand, conditions of mild alkalinity and slow weathering prevail. Only in the topmost 61 cm do acidic soil-forming factors show themselves in an increase in the $SiO_2/Al_2O_3$ ratio.

Under similar climatic conditions basic igneous rocks were found to give a crust of "primary Laterite," chiefly gibbsite, close to the rock surface. All the evidence indicates that here aluminum passes into solution and is redeposited over very short distances in a strongly alkaline environment. Depending on the moisture regime, this "primary Laterite" may either accumulate as such (high level laterite) or be resilicified to give ferruginous kaolinite earths (low level Laterite). Thus the final product depends on the exact leaching conditions; but in both cases it is attained in an alkaline environment. The lateritic or kaolinitic earth so formed is extremely stable. The slight acidity of rainwater containing carbonic acid would exert little effect on it. However surface decomposition of organic matter would produce small transitory quantities of complex-forming organic acids. These might be expected, in the course of time, to carry some iron and aluminum from the surface soil downward, to be rather easily redeposited on oxides and hydroxides already in the subsoil.

This interpretation of the weathering conditions of primary rocks in the tropics may be compared with the results of the weathering of volcanic ash. Here is a material of great reactivity, due partly to its content of volcanic glass and partly to its moderately high specific surface. In Japan, New Zealand, and Hawaii, the first colloidal weathering product is amorphous allophane. This should be much more labile than crystalline kaolinite or halloysite. As tropical weathering continues, soils from volcanic ash show a definite sequence of stages. In the Dutch East Indies, Mohr and Van Baren (18) have classified these according to the chemical and physical characteristics of the ash, the climatic conditions within the soil, and the vegetative cover. The end result under high rainfall and high temperature is a Laterite, dominated by hydrous oxides of iron and aluminum. Under much lower rainfall with intermittent periods of drought, black earths dominated by montmorillonite and organic matter are produced.

Knowing that the resultant soils may be gibbsitic, kaolinitic, or montmorillonitic, we must ask whether the crystalline clays minerals undergo an orderly sequence of transformations dependent mainly on climatic factors. Jackson (10) believes that a definite weathering sequence exists, and he has postulated a weathering sequence, discussed later in this chapter, which is essentially in harmony with more recent evidence from the geochemical approach of Garrels and Christ (6).

## SOIL MINERALS IN RELATION TO THEIR CHEMICAL ENVIRONMENT

### *Standard Free Energies in the Formulation of Equilibria*

In the past 20 years several complementary aspects of mineral transformations in soil systems have become clearer. Hydrothermal synthesis of clay minerals began essentially with the work of Noll (21) in the 1930's. The subject then remained almost stationary, until in the 1950's investigations of mineral stability over a broad front were organized and carried through (Osborne, Roy, Barrer). It then became possible to consider thermodynamic aspects of mineral stability (Garrels and Christ) through which the experiments on synthesis could fruitfully be interpreted. This came about initially through determinations of the standard free energies of well-characterized mineral species, starting from standard heats of formation. It would take us too far afield to discuss these methods, but the consequences of having such information available are very pertinent. The simplest example of a weathering reaction might be represented by the following equation:

$$\text{mineral A} + \text{water} \rightleftarrows \text{mineral B} + \text{soluble species C}$$

In terms of the phase rule, since A and B are solid phases, this system will only have one degree of freedom, thus will be invariant at constant temperature or constant pressure. If the activity or chemical potential of the soluble species can be defined by analytical techniques combined with calculations of the activity coefficient, and if the change in chemical potential of the water is inappreciable or calculable, knowledge of A leads to knowledge of B. This type of consideration applies very broadly. Its use does not depend on a congruent solubility existing either for A or B—merely that the procedures adopted lead to a reliable value for the activity or chemical potential of C as a well-defined chemical species in a true equilibrium. The slowness of weathering reactions is of more concern to the pedologist than to the geologist. For both, however, once the components in the reaction are correctly identified and equilibrium conditions are demonstrated, calculations can be undertaken. However great care must be exercised in the interpretation of analytical results. For instance, a procedure that measures total aluminum in true solution would need supplementation by other information if the equation used involved the ion $(AlOH)^{2+}$.

To use free energy changes and chemical potentials in a uniform way, all participating species are referred to their standard states. Thus a reaction is characterized by the free energy change in going from the reactants in their standard states to the products in their standard states. Of course under any actual equilibrium conditions the free energy change for the reaction is zero. We then have $(\Delta F)_r^0 = -RT \ln K$; $(\Delta F)_r^0$ being the free energy change for the reaction and $K$ the equilibrium constant.

These relationships are conveniently illustrated by considering reactions that kaolinite undergoes with water (6; pp. 353, 355). Consider the equation:

$$H_4Al_2Si_2O_9 + 5H_2O \rightleftarrows Al_2O_3 \cdot 3H_2O + 2H_4SiO_4$$
kaolinite + water ⇌ gibbsite + silicic acid

The standard free energy change can be calculated from the known free energies of formation as follows:

Products    Reactants

$$\Delta F_r^0 = [\Delta f_f^0 + 2\Delta_f^0] - [\Delta F_f^0 + 5\Delta F_f^0]$$
gibbsite  silicic acid kaolinite  water

$$= [-554.6 + 2(-300.3)] - [-884.5 + 5(-56.69)]$$
$$= +12.7 \text{ kcal}$$
$$= -RT \ln K = -1 \cdot 364 \log [H_4SiO]^2$$

where the square brackets indicate activity and the numerical values are taken from Garrels and Christ (6).

$$\log [H_4SiO_4]^2 = \frac{12.7}{-1.364} = -9.31$$

$$[H_4SiO_4] = 10^{-4.65} = 2.21 \times 10^{-5} M$$

Later values (8) of the thermochemical data give

$$[H_4SiO_4] = 10^{-4.25} = 5.59 \times 10^{-5} M$$

Thus the activity of silicic acid is fixed at this low value. The concentration of silica in solution as determined by the colorimetric method with molybdate will probably approximate this value at equilibrium, since at such a high dilution the activity coefficient is almost unity. The colorimetric method measures monomeric silicic acid, which is what the equation stipulates.

Now consider another aspect. Kaolinite can react with hydrogen ions to give aluminum ions and silicic acid in solution. For instance:

$$H_4Al_2Si_2O_9 + 6H^+ \rightleftarrows 2Al^{3+} + 2H_4SiO_4 + H_2O$$

for which we have

$$\sqrt{K'} = K_1 = \frac{[Al^{3+}][H_4SiO_4]}{[H^+]^3}$$

or $\log K_1 = \log [Al^{3+}] + \log [H_4SiO_4] + 3 \text{ pH}$

However other formulations are possible. Thus

$$H_4Al_2Si_2O_9 + 4H^+ + H_2O \rightleftarrows 2(AlOH^{2+}) + 2H_4SiO_4$$

$$K'' = \frac{[AlOH^{2+}]^2 \cdot [H_4SiO_4]^2}{(H^+)^4}$$

$$\sqrt{K''} = K_2 = \frac{[AlOH^{2+}][H_4SiO_4]}{(H^+)^2}$$

Also

$$H_4Al_2Si_2O_9 + 2H^+ + 3H_2O \rightleftarrows 2Al(OH)_2^+ + 2H_4SiO_4$$

$$K''' = \frac{[Al(OH)_2^+]^2 \cdot [H_4SiO_4]^2}{[H^+]^2}$$

$$\sqrt{K'''} = K_3 = \frac{[Al(OH)_2^+][H_4SiO_4]}{[H^+]}$$

or

$$\log K_3 = \log [Al(OH)_2^+] + \log [H_4SiO_4] + pH$$

In presence of solid phase silica, either crystalline or amorphous, $[H_4SiO_4]$ could be taken as constant, so that a plot of log [Al] against pH would give straight lines with different slopes in the three cases. The problem, of course, is to distinguish and quantitatively determine $Al^{3+}$, $AlOH^{2+}$, $Al(OH)_2^+$, and other possible species that form mixtures over a considerable pH range. This is an extremely important consideration in pedological processes: we return to it later.

In each of these three equilibria the equations involve individual ions, but a more important consideration is the way in which they occur together. Thus in the first we have $[Al^{3+}]/[H^+]^3$, in the second $[Al(OH)^{2+}]^2/[H^+]^4$, and in the third $[Al(OH)_2^+]^2/[H^+]^2$. In each case we can substitute $K_w/[OH^-]$ for $[H^+]$, which results in

$$\frac{[Al^{3+}][OH^-]^3}{(K_w)^3}, \quad \frac{[AlOH^{2+}]^2[OH^-]^4}{(K_w)^4}, \quad \text{and} \quad \frac{[Al(OH)_2^+]^2[OH^-]^2}{(K_w)^2}$$

where $K_w$ is the dissociation constant for water. All three reduce to an ionic activity product for aluminum hydroxide. Thus thermodynamically we deal with the chemical potential or free energy of this hydroxide, along with the corresponding quantity for silicic acid. This is not surprising when we recall that from the standpoint of strict thermodynamics, ions are simply mechanistic devices: thermodynamic conclusions do not depend on mechanisms, only on initial and final states. However it is useful to consider the role played by individual ions in the reactions of soil minerals. Many expressions and diagrams utilizing individual ionic activities are employed.

$$H_4Al_2Si_2O_9 + 5H_2O \rightleftharpoons Al_2O_3 \cdot 3H_2O + 2H_4SiO_4$$
$$\text{kaolinite + water} \rightleftharpoons \text{gibbsite + silicic acid}$$

The standard free energy change can be calculated from the known free energies of formation as follows:

$$\begin{array}{cc} \text{Products} & \text{Reactants} \\ \Delta F_r^0 = [\Delta f_f^0 + 2\Delta_f^0] & - [\Delta F_f^0 + 5\Delta F_f^0] \\ \text{gibbsite \ silicic acid} & \text{kaolinite \ water} \end{array}$$

$$= [-554.6 + 2(-300.3)] - [-884.5 + 5(-56.69)]$$
$$= +12.7 \text{kcal}$$
$$= -RT \ln K = -1 \cdot 364 \log [H_4SiO]^2$$

where the square brackets indicate activity and the numerical values are taken from Garrels and Christ (6).

$$\log [H_4SiO_4]^2 = \frac{12.7}{-1.364} = -9.31$$

$$[H_4SiO_4] = 10^{-4.65} = 2.21 \times 10^{-5} M$$

Later values (8) of the thermochemical data give

$$[H_4SiO_4] = 10^{-4.25} = 5.59 \times 10^{-5} M$$

Thus the activity of silicic acid is fixed at this low value. The concentration of silica in solution as determined by the colorimetric method with molybdate will probably approximate this value at equilibrium, since at such a high dilution the activity coefficient is almost unity. The colorimetric method measures monomeric silicic acid, which is what the equation stipulates.

Now consider another aspect. Kaolinite can react with hydrogen ions to give aluminum ions and silicic acid in solution. For instance:

$$H_4Al_2Si_2O_9 + 6H^+ \rightleftharpoons 2Al^{3+} + 2H_4SiO_4 + H_2O$$

for which we have

$$\sqrt{K'} = K_1 = \frac{[Al^{3+}][H_4SiO_4]}{[H^+]^3}$$

or $\log K_1 = \log [Al^{3+}] + \log [H_4SiO_4] + 3 \text{ pH}$

However other formulations are possible. Thus

$$H_4Al_2Si_2O_9 + 4H^+ + H_2O \rightleftharpoons 2(AlOH^{2+}) + 2H_4SiO_4$$

$$K'' = \frac{[AlOH^{2+}]^2 \cdot [H_4SiO_4]^2}{(H^+)^4}$$

$$\sqrt{K''} = K_2 = \frac{[AlOH^{2+}][H_4SiO_4]}{(H^+)^2}$$

Also

$$H_4Al_2Si_2O_9 + 2H^+ + 3H_2O \rightleftarrows 2Al(OH)_2^+ + 2H_4SiO_4$$

$$K''' = \frac{[Al(OH)_2^+]^2 \cdot [H_4SiO_4]^2}{[H^+]^2}$$

$$\sqrt{K'''} = K_3 = \frac{[Al(OH)_2^+][H_4SiO_4]}{[H^+]}$$

or

$$\log K_3 = \log [Al(OH)_2^+] + \log [H_4SiO_4] + pH$$

In presence of solid phase silica, either crystalline or amorphous, $[H_4SiO_4]$ could be taken as constant, so that a plot of log [Al] against pH would give straight lines with different slopes in the three cases. The problem, of course, is to distinguish and quantitatively determine $Al^{3+}$, $AlOH^{2+}$, $Al(OH)_2^+$, and other possible species that form mixtures over a considerable pH range. This is an extremely important consideration in pedological processes: we return to it later.

In each of these three equilibria the equations involve individual ions, but a more important consideration is the way in which they occur together. Thus in the first we have $[Al^{3+}]/[H^+]^3$, in the second $[Al(OH)^{2+}]^2/[H^+]^4$, and in the third $[Al(OH)_2^+]^2/[H^+]^2$. In each case we can substitute $K_w/[OH^-]$ for $[H^+]$, which results in

$$\frac{[Al^{3+}][OH^-]^3}{(K_w)^3}, \quad \frac{[AlOH^{2+}]^2[OH^-]^4}{(K_w)^4}, \quad \text{and} \quad \frac{[Al(OH)_2^+]^2[OH^-]^2}{(K_w)^2}$$

where $K_w$ is the dissociation constant for water. All three reduce to an ionic activity product for aluminum hydroxide. Thus thermodynamically we deal with the chemical potential or free energy of this hydroxide, along with the corresponding quantity for silicic acid. This is not surprising when we recall that from the standpoint of strict thermodynamics, ions are simply mechanistic devices: thermodynamic conclusions do not depend on mechanisms, only on initial and final states. However it is useful to consider the role played by individual ions in the reactions of soil minerals. Many expressions and diagrams utilizing individual ionic activities are employed.

## Fundamental Equations of Weathering Reactions

Once the initial reactants and the final products have been correctly identified, it should be possible to use both mass action and thermodynamic considerations in the expression of reversible equilibria connecting them. This has only been fully recognized in recent years. Garrels and Christ have laid the ground work for these developments.

First it is helpful to use a simple means of relating the compositions of aluminosilicates and their weathering products to each other. This is attempted in Figure 2, in which the vertical axis represents the atomic ratio of Al to Si + Al and the horizontal axis is tetrahedral Al to total Al, or what often amounts to the same thing, equivalents of charge balancing cations per Al total. Each mineral is represented by a point. The feldspars, feldspathoids, and zeolites all lie on the vertical line drawn through 1.0 on the horizontal axis; the actual values depend on the extent of proxying of aluminum for silicon in the tetrahedral structure. The vertical axis itself corresponds to quartz at the origin and gibbsite at Al/(Si + Al) = 1.0, with kaolinite midway. Minerals with both tetrahedral and octahedral groups lie between these vertical lines. Weathering involves three general types of change: (1) reduction of tetrahedral units and their replacement by octahedral, with accompanying loss of cations; (2) preferential loss of silica as compared with aluminum; (3) increasing hydration. Often the actual changes involve two or all three simultaneously. The first type involves movement to the left in Figure 2, the second movement upward and the third no change in position. The figure does not include the ferromagnesium minerals, since their reactions involve magnesium and iron with additional complexities. The dashed lines drawn to the left from anorthite and albite indicate by their slope the compositional changes found by Nash and Marshall (19) in the initial stages of their respective reactions with water. Later stages may, of course, involve other slopes.

The almost horizontal line for muscovite corresponds to the data of Marshall and Macdowell (16).

This diagram has the useful property of illustrating that in the change from calcic feldspar (anorthite) to gibbsite, no minerals of intermediate composition need appear. Albite and orthoclase would be much more likely to progress toward gibbsite through minerals of intermediate composition. Of course the actual dominant factor is the chemical environment for change and the associated free energy difference.

We now begin by considering equations for the production of various minerals from feldspars, in which hydrogen ions and water molecules participate on the left and silicic acid and alkali are soluble products on

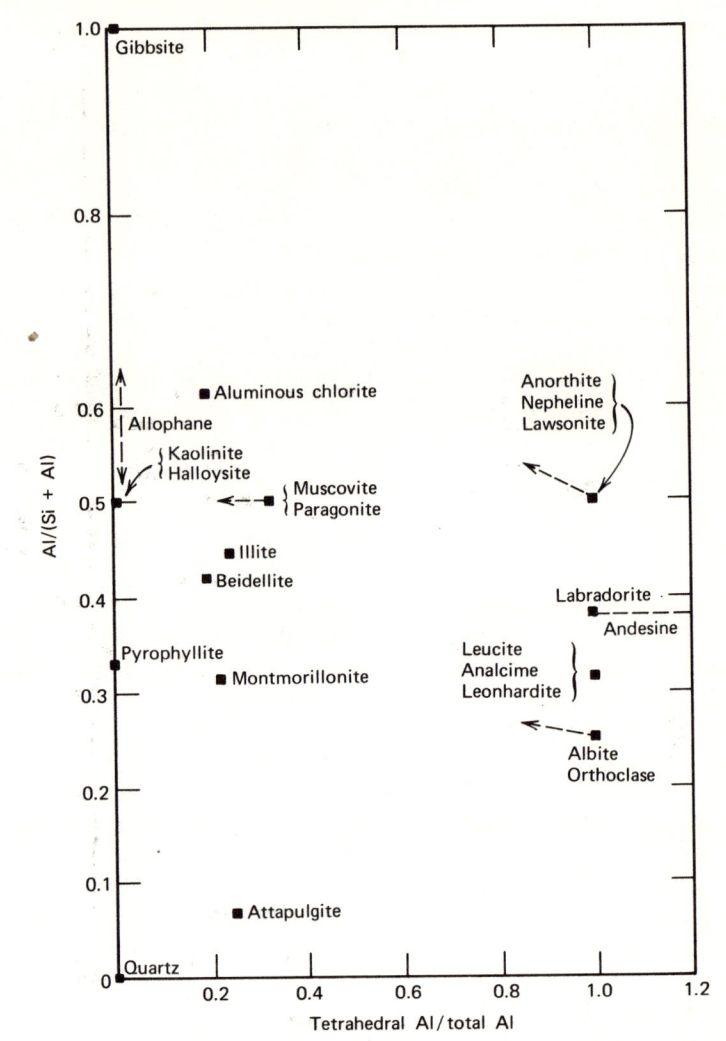

Fig. 2  The relationships of aluminosilicates to their weathering products.

the right. The same type of formulation can be used for orthoclase as for albite. The first example is discussed in some detail, the rest more briefly.

*Orthoclase* → *muscovite*.

$$3KAlSi_3O_8 + 11 H_2O \rightleftarrows KAlSi_3O_6 \cdot Al_2O_4(OH)_2 + K_2SiO_3 + 5Si(OH)_4$$

This formulation in terms of complete molecules can easily be transformed to involve ions, since $K_2SiO_3 \rightleftarrows 2K^+ + SiO_3^{2-}$ and $SiO_3^{2-} + 2H^+ + H_2O \rightleftarrows Si(OH)_4$. Addition of these three equations gives an equation involving $H^+$ and $K^+$, namely:

$$3KAlSi_3O_8 + 12H_2O + 2H^+ \rightleftarrows KAlSi_3O_6 \cdot Al_2O_4(OH)_2 + 2K^+ + 6Si(OH)_4$$

The solid phases and the water are each given unit activity, the solution being so dilute that the water remains in its standard state.

We can then write the mass action equation in terms of $H^+$, $K^+$, and soluble $Si(OH)_4$ as follows:

$$K_r = \frac{[K^+]^2}{[H^+]^2} \cdot [Si(OH)_4]^6$$

or in logarithmic form:

$$\log K_r = 2 \log [K^+] + 2\, pH + 6 \log [Si(OH)_4]$$

$$\frac{\log K_r}{2} = \log K_1 = \log [K^+] + pH + 3 \log [Si(OH)_4]$$

Thus there are three variables, and graphically they can be reduced to two, since $K^+$ and $H^+$ are related. Hence log $[K^+]/[H^+]$ plotted against log $[Si(OH)_4]$ will give a straight line of slope $\frac{6}{2} = 3$ (Figure 3) whose intercept is determined by the value of log $K_r$. This line can be regarded as the phase boundary or join between feldspar and mica for aqueous systems. Then any solution on which $[K^+]$, $[H^+]$, and $Si(OH)_4$ have been determined can be placed on such a diagram and its position compared with the line. If the system is of additional complexity with some unknown factors present, it may be possible to vary $K^+/H^+$, to establish new equilibria, and to determine whether they follow the straight line just described. A variety of possible reactions as described below give similar straight lines with different slopes.

In terms of sericite or illite, which have a somewhat lower charge and K content than ideal muscovite, modification of the equation can be made. It seems best to regard illite and sericite as separate species from muscovite, since there is not only a lower charge per unit cell but usually also some magnesium and iron in the octahedral layer. To regard these minerals as simply muscovite in which hydronium $H_3O^+$ has taken the place of $K^+$ is too great a simplification. It can be seen from Figure 2 that illite is closer in composition to beidellite than to ideal dioctahedral mica and that the Al/(Al + Si) ratio is distinctly lower in illite than in muscovite. Thus the actual weathering of orthoclase or microcline to sericite or illite may be defined by a line of Figure 3, which lies between those for muscovite and beidellite.

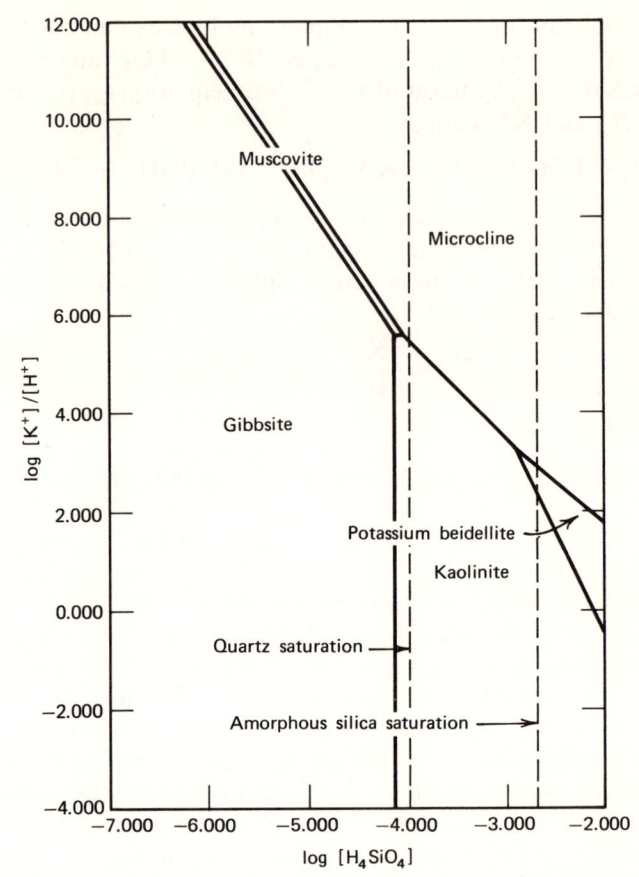

**Fig. 3.** Garrels and Christ type of diagram for the system HCl-H$_2$O-Al$_2$O$_3$-K$_2$O-SiO$_2$ at 25°C (8). Used by permission, Freeman Cooper and Co.

*Albite → sodium beidellite → hydrogen beidellite.* In dealing with the montmorillonite group (smectites), it is convenient to begin with members having little substitution of magnesium for aluminum in the octahedral layer. Since the high potassium compositions lead to illites rather than to smectites, we should consider sodium, or sodium, hydronium, and calcium as the balancing cations. Under natural conditions the exchange complex is often dominated by calcium, but we can get a broad view of the situation relating to the phase boundaries and their slopes by considering the primary change from albite to sodium beidellite. For this purpose beidellite is defined as the smectite whose charge arises by the proxying of one-eighth of the silica by aluminum, with sodium or other cations to balance the charge. This leads to the following equations:

$$5\text{NaAlSi}_3\text{O}_8 + 20\text{H}_2\text{O} \rightleftarrows \text{Na[AlSi}_7\text{O}_{12} \cdot \text{Al}_4\text{O}_8(\text{OH})_4]$$
$$+ 4\text{NaOH} + 8\text{Si(OH)}_4$$

$$5\text{NaAlSi}_3\text{O}_8 + 4\text{H}^+ + 16\text{H}_2\text{O} \rightleftarrows \text{Na[AlSi}_7\text{O}_{12} \cdot \text{Al}_4\text{O}_8(\text{OH})_4]$$
$$+ 4\text{Na}^+ + 8\text{Si(OH)}_4$$

Proceeding to give the two solid phases and the water unit activity, we have:

$$K_r = \frac{[\text{Na}^+]^4}{[\text{H}^+]^4} \cdot [\text{Si(OH)}_4]^8$$

or

$$K_2 = \frac{[\text{Na}^+]}{[\text{H}^+]} \cdot [\text{Si(OH)}_4]^2$$

or

$$\log K_2 = \log [\text{Na}] + \text{pH} + 2 \log [\text{Si(OH)}_4]$$

Thus there is a clear distinction between this line of slope 2 and that for the feldspar-mica equilibrium, slope 3.

For passage to the hydrogen beidellite, an extra hydrogen ion is involved. Hence the equation becomes

$$K'_r = \frac{[\text{Na}]^5}{[\text{H}]^5} [\text{Si(OH)}_4]^8$$

*Anorthite → calcium beidellite.* Basic igneous rocks often weather to give clays of the swelling group. Since the latter are more silicious than anorthite (see Figure 2), an external source of silica is required: but a feldspar with the composition of andesine (An 67:Ab 33%) would need no supplementation, and those richer in the albite component would liberate silica. This seems to have been the case in the soil from diabase studied by Humbert and Marshall (9). The feldspar was a labradorite, which would liberate silica upon weathering to beidellite. Both quartz and chalcedony were formed throughout the profile; the amount increased toward the surface. The clay was identified as montmorillonitic, with some mica.

*Feldspars → kaolinite.* The Al/Si + Al ratio is the same for kaolinite as for anorthite ($\text{CaAl}_2\text{Si}_2\text{O}_8$) and nepheline ($\text{NaAlSiO}_4$), hence no $\text{Si(OH)}_4$ enters into the equation; as follows:

$$\text{CaAl}_2\text{Si}_2\text{O}_8 + 3\text{H}_2\text{O} \rightleftarrows \text{Si}_2\text{O}_3 \cdot \text{Al}_2\text{O}_2(\text{OH})_4 + \text{Ca(OH)}_2$$

$$\text{CaAl}_2\text{Si}_2\text{O}_8 + 2\text{H}^+ + \text{H}_2\text{O} \rightleftarrows \text{Si}_2\text{O}_3 \cdot \text{Al}_2\text{O}_2(\text{OH})_4 + \text{Ca}^{2+}$$

$$K_4 = \frac{[\text{Ca}^{2+}]}{[\text{H}^+]^2}$$

Similarly for nepheline

$$K'_4 = \frac{[Na^+]^2}{[H^+]^2}$$

In this type of weathering cations would be lost, but no silicon or aluminum. Furthermore, since there are noncalcic micas with the same ratio Al/(Al + Si), it seems likely that in the presence of monovalent cations, paragonite or muscovite might be produced as an intermediate step. The calcic brittle mica margarite has a considerably higher ratio Al/(Si + Al) but is not commonly found under normal weathering.

Kaolinite has often been observed as a product of the weathering of granites under conditions that involve effective leaching. Orthoclase gives the following equations, which apply similarly to albite.

$$2KAlSi_3O_8 + 11H_2O \rightleftarrows Si_2O_3Al_2O_2(OH)_4 + 2KOH + 4Si(OH)_4$$

$$2KAlSi_3O_8 + 2H^+ + 9H_2O \rightleftarrows Si_2O_3Al_2O_2(OH)_4 + 2K^+ + 4Si(OH)_4$$

$$K_r = \frac{[K^+]^2 [H_4SiO_4]^4}{[H^+]^2}$$

or

$$K_5 = \frac{[K^+]}{[H^+]} [H_4SiO_4]^2$$

Figure 2 indicates that in the course of arriving at the kaolinite composition, the weathering feldspar is likely to encounter compositions ranging from montmorillonite to illite. With less effective removal of soluble products, such clays would be expected, and they are found to occur in many soils.

*Micas → kaolinite.* For muscovite (and similarly paragonite) we have:

$$2KAlSi_3O_6Al_2O_4(OH)_2 + 5H_2O \rightleftarrows 3Si_2O_3Al_2O_2(OH)_4 + 2KOH$$
$$2KAlSi_3O_6Al_2O_4(OH)_2 + 2H^+ + 3H_2O \rightleftarrows 3Si_2O_3Al_2O_2(OH)_4 + 2K^+$$

$$K_r = \frac{[K^+]^2}{[H^+]^2}$$

$$K_6 = \frac{[K^+]}{[H^+]}$$

No movement of silica is required. Nevertheless the apparently simple $K^+ \rightleftarrows H^+$ interchange is deceptive. A profound rearrangement is required to go from mica to kaolinite, probably involving hydration of the mica as well as movement of silicon and aluminum over molecular distances.

Secondary micas such as illite lose silica in weathering to kaolinite. The equations are quantitatively intermediate between those for muscovite and those for beidellite. Actual experiments on muscovite show that the ratio of cations to silica removed initially is very high (16), corresponding to the small slope of the arrow in Figure 2, which represents about 20 eq cations per atom of silicon released.

*Beidellite → kaolinite.* Figure 2 reveals that the compositions of beidellite and montmorillonite are sufficiently different to demand different equations. Soil clays tend to approximate more closely to beidellite than to typical montmorillonite. In working out the equations for beidellite, the possible presence of magnesium and trivalent iron is disregarded.

$$2NaAlSi_7O_{12} \cdot Al_4O_8(OH)_4 + 15H_2O \rightleftharpoons 5Si_2O_3Al_2O_2(OH)_4 + 2NaOH + 4Si(OH)_4$$

$$2NaAlSi_7O_{12} \cdot Al_4O_8(OH)_4 + 2H^+ + 13H_2O \rightleftharpoons 5Si_2O_3 \cdot Al_2O_2(OH)_4 + 2Na^+ + 4Si(OH)_4$$

$$K_r = \frac{[Na^+]^2 [Si(OH)_4]^4}{[H^+]^2}$$

$$K_6 = \frac{[Na^+]}{[H^+]} \cdot [Si(OH)_4]^2$$

$$\log K_6 = \log [Na] + pH + 2 \log [Si(OH)_4]$$

In considering the reactions of soil clays the fact that the compositions of montmorillonite, beidellite, and aluminous chlorite lie very close to the same vertical line in Figure 2 means that we should investigate not only losses in silica but possibly also interlayering of gibbsite.

*Feldspars → gibbsite*

1. Anorthite

$$CaAl_2Si_2O_8 + 8H_2O \rightleftharpoons Al_2(OH)_6 + Ca(OH)_2 + 2Si(OH)_4$$

$$CaAl_2Si_2O_8 + 2H^+ + 6H_2O \rightleftharpoons Al_2(OH)_6 + Ca^{2+} + 2Si(OH)_4$$

$$K_7 = K_r = \frac{[Ca^{2+}]}{[H^+]^2} \cdot [Si(OH)_4]^2$$

How calcic feldspar proceeds directly to gibbsite as demonstrated by Harrison (7) becomes a matter of particular interest. If the residual composition after loss of soluble products is essentially gibbsite, Figure 2 would suggest that loss of silica must be very high. In such a change the residue would attain high values of Al/(Si + Al) without ever passing through the characteristic compositions of micas or clays. In Figure 2 the arrow from anorthite indicates by its slope the relative change in Al/(Si + Al) and cationic equivalents per aluminum as found by reaction with water (19). If extended to zero cations, this line would intersect no stable compositions until it reached the vertical axis at about an Al/(Al + Si) ratio of 0.75. This would indicate that the earliest stages of the reaction differ from later stages in that the latter involve greater proportionate loss of silica. It is quite probable that such a change occurs as weathering proceeds. Similar changes have indeed been demonstrated with other minerals.

Geochemical data indicate that unusually large amounts of silica are found in stream waters in areas of basic igneous rocks. This is easily understood in terms of the very small fragments of the three dimensional –Si–O–Al lattice that are released upon decomposition and reorientation of the aluminum. A high proportion of monomeric $Si(OH)_4$ would be expected. Harrison's figures (7) show this mobility of silicon very clearly.

2. Albite, orthoclase, or microline → gibbsite

$$2NaAlSi_3O_8 + 13H_2O \rightleftarrows Al_2(OH)_6 + Na_2SiO_3 + 5Si(OH)_4$$
$$2NaAlSi_3O_8 + 2H^+ + 14H_2O \rightleftarrows Al_2(OH)_6 + 2Na^+ + 6Si(OH)_4$$

$$K_r = \frac{[Na^+]^2}{[H^+]^2} \cdot [Si(OH)_4]^6$$

or

$$K_8 = \frac{[Na^+]}{[H^+]} \cdot [Si(OH)_4]^3$$

$$\log K_8 = \log [Na^+] + pH + 3 \log [Si(OH)_4]$$

In the early stages of the reaction with water, experimental evidence (19, 20) indicates that loss of cations considerably exceeds that of silica as indicated by the arrow in Figure 2. This has a smaller slope than that for anorthite. The more silicious fragments remaining when these feldspars decompose are more readily held at the surface than the silicic acid monomer units liberated from anorthite. The decomposition of these feldspars is practically bound to lead through such compositions as beidellite and mica before kaolinite is reached. Thus gibbsite seems likely to be a secondary or tertiary product rather than a primary one.

Kaolinite → gibbsite

$$Si_2O_3 \cdot Al_2O_2(OH)_4 + 5H_2O = Al_2(OH)_6 + 2Si(OH)_4$$

$$K_r = [Si(OH)_4]^2$$
$$K_9 = [Si(OH)_4]$$

Garrels and Christ evaluated $K_9$ from thermochemical data to give log $K_9 = -4.7 = \log [Si(OH)_4]$. Thus when both crystalline gibbsite and kaolinite are present, the activity of Si(OH)$_4$ is fixed. In plotting log [K$^+$]/[H$^+$] against log [Si(OH)$_4$], the boundary between kaolinite and gibbsite can be represented as a vertical line at log [SiOH$_4$] = $-4.7$. Later values used by Helgeson et al. (8) give the value $-4.2$ at 25°C (Figure 3).

In systems that do not involve balancing cations, the relationships can be shown in other ways. Kittrick (14) uses pH $- \frac{1}{3}$pAl$^{3+}$ as the vertical axis and pSi(OH)$_4$ as the horizontal axis. By simplifying montmorillonite to a pyrophyllite formula, he was able to construct a diagram showing amorphous silica, montmorillonite, kaolinite, and gibbsite (Figure 4). From this it appears that montmorillonite can exist only over a narrow range of silicic acid activities. However since in actual cases the activity of the balancing cation is also a factor, there may be more room for variation than Kittrick's simplification implies. It is known from low temperature syntheses that the presence of salts greatly improves the production and crystallinity of montmorillonite. Kittrick has determined the intersection of the montmorillonite and kaolinite lines by determining the Si(OH)$_4$ in solution when both minerals are present (12). In a similar way the intersection of the horizontal gibbsite line with the sloping kaolinite line was established by measuring the Si(OH)$_4$ in solution when both minerals were present (13).

*Analcime → beidellite.* The importance of analcime as a constituent of alkali soils has been recognized only recently. This silicate has been found associated with smectite clays in a number of California soils (1). It was synthesized by Noll (21) at 300°C and was found to be associated with montmorillonite in some of the mixtures employed. The author has produced it at 200°C and has shown that in systems involving both Mg(OH)$_2$

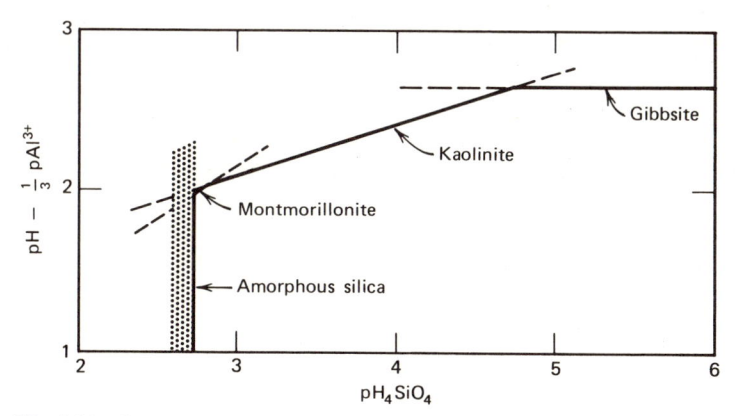

*Fig. 4* Kittrick's diagram for certain minerals in the Al$_2$O$_3$-SiO$_2$-H$_2$O system at 25°C (14). Used by permission, Pergamon Press.

and $Al_2(OH)_6$, the maintenance of an alkaline reaction together with sufficient sodium in relation to the aluminum, involves the simultaneous formation of a trioctahedral smectite and analcime. A slight reduction in pH causes a mixture of dioctahedral and trioctahedral smectites to be produced. Since the slope is defined by the equations below, to fix the analcime → sodium beidellite line definitely, only one point must be determined by the composition of the solution in presence of both minerals.

$$5NaAlSi_2O_6 \cdot H_2O \rightleftarrows NaAlSi_7O_{12}Al_4O_8(OH)_4$$
$$+ 2Na_2SiO_3 + Si(OH)_4 + H_2O$$
$$5NaAlSi_2O_6 \cdot H_2O + 4H^+ + H_2O \rightleftarrows NaAlSi_7O_{12}Al_4O_8(OH)_4$$
$$+ 4Na^+ + 3Si(OH)_4$$

$$K_{10} = K_r = \frac{[Na^+]^4}{[H^+]^4} \cdot [Si(OH)_4]^3$$

or

$$\log K_{10} = 4 \log [Na^+] + 4 \text{ pH} + 3 \log [Si(OH)_4]$$

*Albite → analcime*

$$NaAlSi_3O_8 + 3H_2O \rightleftarrows NaAlSi_2O_6 + H_2O + Si(OH)_4$$

$$\log K_{11} = \log [Si(OH)_4]$$

There is no interchange of $Na^+$ and $H^+$, hence the equilibrium determines the $Si(OH)_4$ in solution. In Figure 10.7 in the book by Garrels and Christ (6), a value for $\log [Si(OH)_4]$ of about $-2.4$ is used.

The triple point albite–sodium beidellite–analcime would then fall at a value of $\log [Na^+]/[H^+]$ (Figure 5) of about 5.5. Above this value mixtures of analcime and sodium beidellite could coexist up to 6.3. Analcime and kaolinite follow up to 7.0, and finally analcime and gibbsite. The silicic acid values, however, are very low for this last equilibrium, from $10^{-4.2}$ downward. The analcime–beidellite range in silicic acid is from $10^{-3.5}$ to $10^{-4.2}$. The higher concentrations are thus in the range of values found in soils (i.e., between quartz and amorphous silica). Thus analcime, sodium beidellite, and amorphous silica could coexist at a point on the analcime–beidellite line. These deductions are based on the correctness of the silicic acid concentration $10^{-2.4}$ taken from the diagram of Garrels and Christ. If this is in error, the analcite lines must be displaced in the diagram. In alkali soils the presence of calcium affects these equilibria because calcium beidellite and calcium carbonate are both possible secondary phases.

*Muscovite → potassium beidellite → hydrogen beidellite.* For muscovite or paragonite to produce beidellite, additional silica must be added.

### Early Stages in Rock Weathering

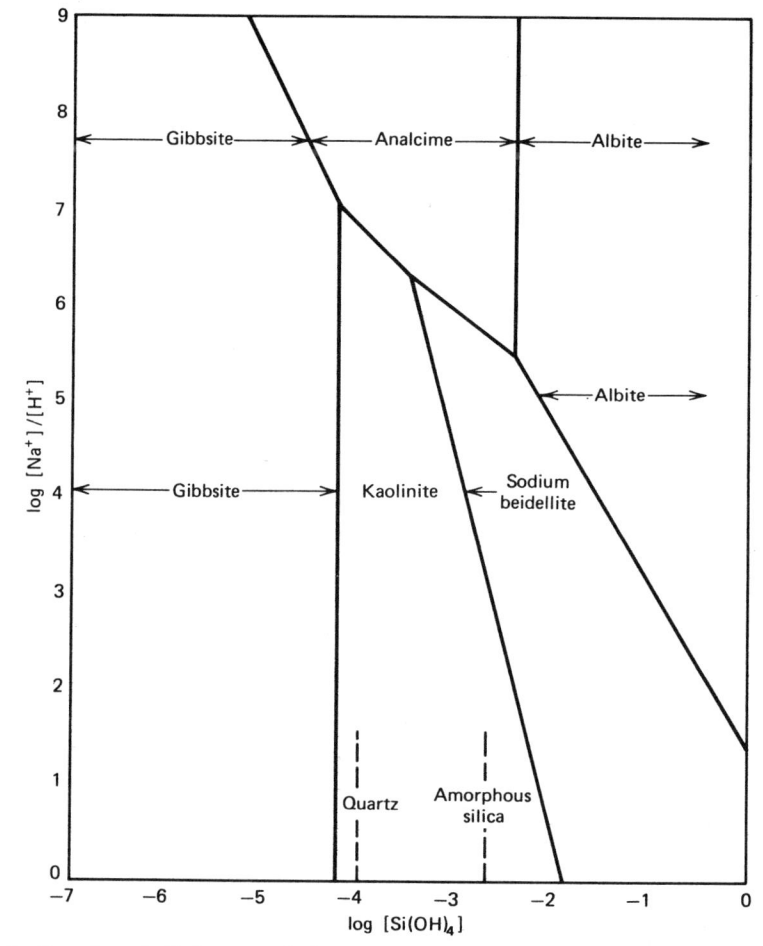

*Fig. 5* Garrels and Christ type of diagram for the system HCl-H$_2$O-Al$_2$O$_3$-Na$_2$O-SiO$_2$ at 25°C with inclusion of approximate data for analcime.

Thus we have the following equations:

$$5KAlSi_3O_6 \cdot Al_2O_4(OH)_2 + 6Si(OH)_4 = 3KAlSi_7O_{12}Al_4O_8(OH)_4 + 2KOH + 10H_2O$$

$$5KAlSi_3O_6 \cdot Al_2O_4(OH)_2 + 6Si(OH)_4 + 2H^+ = 3KAlSi_7O_{12}Al_4O_8(OH)_4 + 2K^+ + 12H_2O$$

$$K_r = \frac{[K^+]^2}{[H^+]^2} \cdot \frac{1}{[Si(OH)_4]^6}$$

or

$$K_{12} = \frac{[K^+]}{[H^+]} \cdot \frac{1}{[Si(OH)_4]^3}$$

$$\log K_{12} = \log [K^+] + pH - 3 \log [Si(OH)_4]$$

Alternatively, for hydrogen beidellite:

$5KAlSi_3O_6 \cdot Al_2O_4(OH)_2 + 6Si(OH)_4$
$\qquad = 3HAlSi_7O_{12} \cdot Al_4O_8(OH)_4 + 5KOH + 7H_2O$
$5KAlSi_3O_6 \cdot Al_2O_4(OH)_2 + 6Si(OH)_4 + 5H^+$
$\qquad = 3HAlSi_7O_{12} \cdot Al_4O_8(OH)_4 + 5K^+ + 12H_2O$

$$K'_{12} = K'_r = \frac{[K^+]^5}{[H^+]^5} \cdot \frac{1}{[Si(OH)_4]^6}$$

$$\log K'_{12} = 5 \log [K^+] + 5 pH - 6 \log [Si(OH)_4]$$

Clearly the production of silica in solution by weathering of other minerals such as albite or orthoclase will greatly influence the reaction. In this way beidellite would come from two sources simultaneously, and it would be possible to write an equation for the production of beidellite from orthoclase plus muscovite which does not involve silicic acid in solution. For this reaction $K_r$ is simply a function of $[K^+]/[H^+]$ raised to the appropriate power. Given the standard free energies of formation of orthoclase, muscovite, and beidellite, the value could be calculated. In this way quantitative expression can be given to one case of the general proposition that the weathering of a mineral is influenced by the environment provided by that of its neighbors. It is interesting to look at the final equation for the case of the formation of hydrogen beidellite from potash feldspar and muscovite (or, alternatively, albite and paragonite).

$3KAlSi_3O_8 + 4KAlSi_3O_6 \cdot Al_2O_4(OH)_2$
$\qquad + 7H^+ \rightleftarrows 3HAlSi_7O_{12} \cdot Al_4O_8(OH)_4 + 7K^+$

$$K_r = \frac{[K^+]^7}{[H^+]^7}$$

$$K_{13} = \frac{[K^+]}{[H^+]}$$

The activity of water does not enter into this equation. When all three solid phases are present $[K^+]/[H^+]$ is fixed and is independent of the activity of $Si(OH)_4$ and that of water. Perhaps the case of negligible loss of silica is not simply academic. Conditions in certain horizons of Prairie soils or Chernozems, where passage of water completely through the profile is a very rare event, may provide such cases.

*General.* Almost all the cases considered involve equivalent interchange between $H^+$ and a cation. Thus one variable in the equation is [cation$^+$]/[$H^+$] and the other is [Si(OH)$_4$]. This allows us to arrive at the Garrels and Christ diagram in which the vertical axis is log [cation]/[H] and the horizontal axis log [Si(OH)$_4$] (Figures 1, 3, 5). The slopes of the lines are defined by the various equations as formulated earlier; their actual positions can then be fixed by correctly identifying a given point in each line. A separate diagram is required for each cation considered.

In making calculations from thermochemical data, one limitation must be closely watched. The physical chemist expects mass action equations to be agencies of precision, and the activities that go into them are normally accurate to 1% or better. Suppose [$K^+$] carries a maximum error of 1%, then 1364 log $K$ can vary by 5.9 cal. Very few thermochemical results on minerals correspond to this accuracy. In working out equations involving combined uncertainties of, say, 59 cal or 0.059 kcal, the required activity of a given constitutent in solution could be defined only to within about 10%. Thus if data of only fair precision can be obtained on the solution phase, thermochemical quantities accurate to 0.05 kcal became available. In comparing thermochemical quantities obtained by different procedures, it is common to treat variations of 0.1 kcal/mole as small, whereas they actually correspond to relatively large uncertainties in the solution phase.

*Pedological Considerations*

Figures 2 and 4 suggest many conclusions regarding pedological systems. Kittrick's diagram corresponds to changes along the Al/(Si + Al) axis of Figure 2, and a similar representation could be used for all cases in which the balancing cations remain constant. Thus beidellite and aluminous chlorite, being on the same vertical line in Figure 2, would lend themselves to the Kittrick representation. Similarly changes up and down the feldspar–feldspathoid–zeolite vertical line could be shown. The main difficulty here is in establishing the values of $pH - \frac{1}{3}pAl^{3+}$, since in such systems total aluminum in solution is extremely low and its exact distribution among the various ionic forms is difficult to determine. As Kittrick points out, what can be done at present is to examine the compatibility of existing mixtures in the soil with predictions from thermochemical data including, of course, solubility data on well-defined minerals. The general results of mineral synthesis at somewhat higher temperatures should also serve as guideposts.

At this point it becomes possible to say something about soil maturity. Mineral equilibria in aqueous systems are notoriously slow in their operation. When one or more solid phases need to disappear before a single true equilibrium can dominate, unstable mixtures can persist over long periods. Evidence of their presence can be used in discussions of soil maturity. Great caution is required, since the operation of Ostwald's rule permits and indeed predicts the appearance of unstable species of higher free energy before the stable ones with the lowest free energy.

Much valuable information can be secured by investigating the mineral distribution in different particle size fractions of the soil, beginning with the fine clay and proceeding through the silts to the sands. Improvements in optical methods supplemented by specific gravity determinations enabled the author to make a start with such work in the 1930's (15), but later developments using X-ray diffraction yield results that are much more certain with respect to mineral identification, as well as more quantitative. From such studies on a number of soils Jackson (11) was led to the formulation of his "weathering sequence" and "weathering index," which we now discuss in some detail.

### Concept of the "Weathering Sequence"

For many years geologists have sought to establish an order among minerals in regard to their ease of weathering. Several such lists are in the literature (Smithson, Barshad, Graham, Pettijohn). The weakness of this earlier approach was that the whole process was regarded as a one-way street. One of the chief advantages of Jackson's concept is that it takes account of the essential reversibility of weathering reactions and their sensitivity to the character of the environment. This has also been stressed by Barshad (2). Because in soil systems recrystallization tends to be slow and imperfect, the most sensitive indicator of current status is the fine fraction of the clay. Thus the mineral composition of this fraction is crucial. The coarse clay and the silt and sand fractions show progressively less of the weathering-in-train. Comparisons of the same kind between horizons of a soil profile and the parent material enable investigators to characterize the whole system through its solid mineral phases. Ideally, therefore, with quantitative information on the various fractions and horizons, we should arrive at a complete measure of the "capacity factors" that have brought the soil from the parent material to its present status. Because in the formulation of reversible reactions all solid phases are given a standard chemical potential of zero, we must look elsewhere for the "intensity factors." These show themselves in the status of the soil solution and have done so since the original parent material began to change. The standard free energy change for any given mineral change,

$RT \ln K$, where $K$ is the equilibrium constant, defines the totality of the intensity factors. These intensity factors, if they have operated over a sufficient period, will show themselves qualitatively in the nature of the new minerals formed. Thus the suite of new minerals formed, as identified in the finest clay, gives a view of the intensity factors, while the quantities determined represent the capacity factors. The present-day intensity factors should not be considered as constant, but as operating through cyclic variations in the solution phase related to the climatic factors. Thus analysis of the equilibrium soil solution now acquires great importance. Of course the intensity and capacity factors of the past may still be revealed in the mineral suite and in its quantitative components. Hence in interpreting mineralogical data through Jackson's weathering sequence, there is considerable flexibility. The very full account by Jackson and Sherman (10) reveals this clearly. Every observation seems to be fitted in, one way or the other.

The weathering sequence proposed for the finest constituents is divided into 13 stages.

1. Gypsum (also other salts).
2. Calcite (also dolomite, aragonite, apatite).
3. Olivine-hornblende (also pyroxenes, diopside, etc.).
4. Biotite (also glauconite, magnesium chloride, antigorite, nontronite).
5. Albite (also anorthite, stilbite, microcline, orthoclase, etc.).
6. Quartz (also cristobalite, etc.).
7. Muscovite (also 10 Å zones of sericite, illite, etc.).
8. Interstratified 2:1 layer silicates and vermiculite (including partially expanded hydrous micas, randomly interstratified 2:1 layer silicates with no basal spacings, and regularly interstratified 2:1 layer silicates).
9. Montmorillonite (also beidellite, saponite, etc.).
10. Kaolinite (also halloysite, etc.).
11. Gibbsite (also boehmite, allophane, etc.).
12. Hematite (also goethite, limonite, etc.).
13. Anatase (also zircon, rutile, ilmenite, leucoxene, corundum, etc.).

Jackson and Sherman comment that chlorites of different composition might occur in different places—the high magnesium members in stage 4 and the more aluminous members in 7 through 9.

When the sequence is examined by consideration of aluminosilicate compositions as in Figure 2, the general course of weathering reactions beginning with feldspar and proceeding through 2:1 lattices to 1:1 lattices and eventually to gibbsite is portrayed. Stages 3 and 4 reflect the ease of decomposition of minerals containing much magnesium and ferrous iron; the solid products of these decompositions appear chiefly in stages 7 through 9.

As Jackson pointed out in his original development of the sequence (11), a given soil colloid often contains a mixture of minerals covering several adjacent stages. The *weathering mean* represents the weighted average. This mean shifts to lower numerical stages as the particle size increases from the fine clay, and in most soils as we go from the surface soil to the parent material (Figure 6). Conditions of sedimentation or of the continued presence of ground water also tend to cause shifts to lower values. Duration of weathering is of course of great importance. When sufficient time has elapsed, the geographic sequences observed as consequences of climatic factors should become very evident in the composition of the mineral soil colloids. Hence the relationships between great soil groups and the mineral composition of the soil colloid should clearly emerge. We now have much quantitative and semiquantitative data

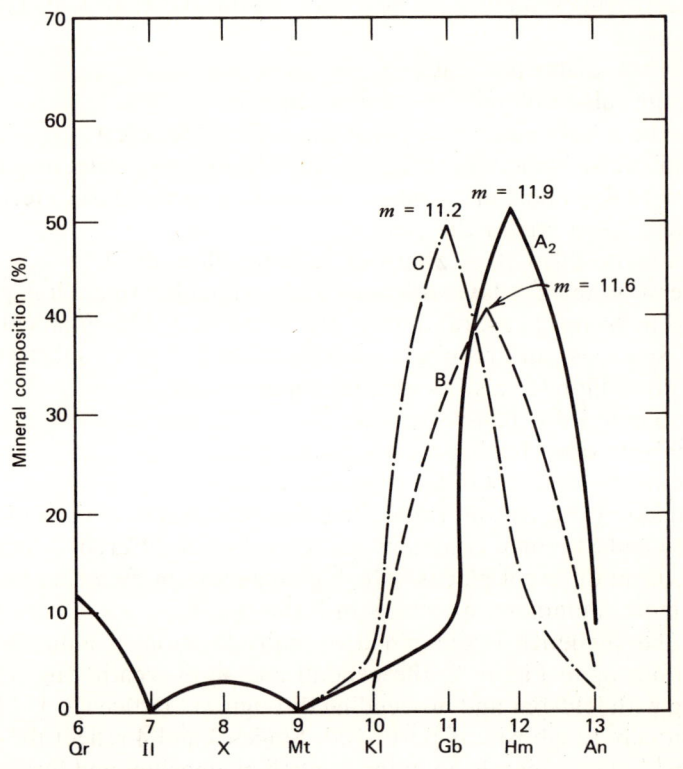

*Fig. 6* Mineral distribution and weathering mean $m$ of three horizons of a ferruginous humic Latosol (Oxisol) from Hawaii. Reproduced from ref. 8 by permission of the American Society of Agronomy.

through which these relationships can be seen. These data are examined later.

Examination of the weathering sequence in the light of modern ideas on mineral reactions reveals the following weaknesses. The detailed separation of stages 5 (feldspars), 6 (quartz), and 7 (muscovite) and of 7 (muscovite), 8 (interstratified minerals), and 9 (montmorillonite) is difficult to justify.

First, the use of stage 7 for primary muscovite as well as secondary sericite and illite can be questioned. The distinct compositional differences between primary and secondary potassium dioctahedral micas are not simply the consequence of hydration and possible substitution of $H_3O^+$ for $K^+$. The secondary micas have distinctly lower Al/(Si + Al) ratios than typical primary muscovites, and they are stable under lower $[K^+]/[H^+]$ ratios in solution. This can be seen in Figure 2.

Second, in the weathering of potash feldspar to give mica or swelling clays or kaolinite, there is loss of silica, and the $Si(OH)_4$ in solution can easily exceed the solubility of quartz. Thus both quartz or chalcedony and one of the above-mentioned products from feldspar can coexist as weathering proceeds. Thus there seems to be little justification for separating a quartz stage from a secondary mica or even a montmorillonite stage. The main sequence to be followed will in any case be determined by both the $[cation^+]/[H^+]$ ratio and the $[Si(OH)_4]$ in the solution phase up to the point at which minerals containing charge balancing cations disappear. At that point we reach the vertical axis of Figure 2, and subsequent weathering consists in movement vertically along the axis by preferential loss of silica.

Thus weathering operates essentially through the absolute and relative losses of balancing cations and silica. If these bear such a relation to one another that starting with feldspar we pass through compositions in the secondary mica–beidellite–montmorillonite range, these are likely to form before kaolinite or halloysite appears. The nature of the cation is very important. Potassium tends to form hydrous micas; sodium or calcium, members of the smectite group. With excess sodium, where conditions favor the formation of alkali soils, the zeolite analcime can appear. This is a genuine weathering product and should find a place near feldspar in the weathering sequence. Mixtures of analcime and smectites have been observed in California soils. The analysis of the soil solution for these cases would define an important boundary line in Figure 5.

Harrison's (7) demonstration of the very sharp change from basic igneous rock to gibbsite under high tropical rainfall has been puzzling soil scientists for many years. In the light of Figure 2 it can now be seen that if losses of silica are comparable to those of bases, compositional changes during extreme weathering may not pass through the regions of swelling clays or micas and may even miss the region of aluminous chlorites. Thus mixtures of allophane and gibbsite or gibbsite alone would soon be reached. By its slope, the arrow drawn from anorthite indicates the initial direction of compositional weathering as calculated from data by Nash and Marshall (19,20). Later stages may very well involve greater relative loss of silica and hence a steeper slope.

Less vigorous removal of silica relative to bases could lead to the formation of quartz and chalcedony together with illitic and beidellitic species, as in the diabase profile in Missouri (Humbert and Marshall, 9).

Whether a given system actually produces intermediate weathering products as indicated by the compositional ranges in Figure 2 will depend on the maintenance in solution for a sufficient time, of the requisite [cation$^+$]/[H$^+$] ratio and [Si(OH)$_4$] activity. Thus the conditions for renewal and replacement of water become critical. Soil physicists have devoted much effort to this aspect. Before taking up soil profiles on a worldwide basis we must review many of their findings (see Chapters 6 and 7).

## REFERENCES

1. Baldar, N. A. and L. D. Whittig, Occurrence and synthesis of soil zeolites, *Soil Sci. Soc. Am. Proc.*, **32**, 235 (1968).
2. Barshad, I., "Soil Development." In F. E. Bear, *Chemistry of the Soil,* American Chemical Society Monograph 126, Reinhold, New York (1955), Chap. 1.
3. Blanck, E., *Einführung in die genetische Bodenlehre,* Vandenhoeck and Ruprecht, Göttingen (1949).
4. Buckingham, E., Contributions to our knowledge of the aeration of soils, U.S. Department of Agriculture Bureau of Soils Bulletin No. 25 (1904).
5. Correns, C. W. and W. von Engelhardt, Neue Untersuchungen über die Verwitterung des Kalifeldspates, *Chem. Erde,* **12**, 20 (1938).
6. Garrels, R. M. and C. L. Christ, *Solutions, Minerals and Equilibria,* Harper & Row, New York (1965).
7. Harrison, J. B., The katamorphism of igneous rocks under humid tropical conditions, Imperial Bureau of Soil Science, Harpenden, England (1934).
8. Helgeson, H. C., T. H. Brown, and R. H. Leeper, *Handbook of Theoretical Activity Diagrams,* Freeman, Cooper, San Francisco (1969).
9. Humbert, R. P. and C. E. Marshall, Mineralogical and chemical studies of soil formation from acid and basic igneous rocks in Missouri, University of Missouri, Agriculture Experimental Station Research Bulletin 359 (1943).
10. Jackson, M. L. and G. D. Sherman, Chemical weathering of minerals in soils, *Advan. Agron.,* **5**, 219 (1953).

11. Jackson, M. L. et al., Weathering sequence of clay size minerals and sediments. I. Fundamental generalizations, *J. Phys. Colloid Chem.*, **52**, 1237 (1948).
12. Kittrick, J. A., Montmorillonite equilibria and the weathering environment, *Soil Sci. Soc. Am. Proc.*, **35**, 815 (1971).
13. Kittrick, J. A., Precipitation of kaolinite at 25°C and 1 atm, *Clays & Clay Miner.*, **18**, 261 (1970).
14. Kittrick, J. A., Soil minerals in the $Al_2O_3$-$SiO_2$-$H_2O$ system and a theory of their formation, *Clays & Clay Miner.*, **17**, 157 (1969).
15. Marshall, C. E., Mineralogical methods for the study of silts and clays, *Z. Krystallogr. (A)*, **90**, 8 (1934).
16. Marshall, C. E. and L. L. McDowell, The surface reactivity of micas, *Soil Sci.*, **99**, 115 (1965).
17. Merrill, G. P., *Rocks, Rock Weathering and Soils*, Macmillan, London (1913).
18. Mohr, E. C. J. and F. A. van Baren, *Tropical Soils*, Wiley-Interscience, New York (1954).
19. Nash, V. E. and C. E. Marshall, The surface reactions of silicate minerals. Part I. The reactions of feldspar surfaces with acidic solutions, University of Missouri, Agricultural Experiment Station Research Bulletin 613 (1956).
20. Nash, V. E. and C. E. Marshall, The surface reactions of silicate minerals. Part II. Reactions of feldspar surfaces with salt solutions, University of Missouri, Agricultural Experiment Station Research Bulletin 614 (1956).
21. Noll, W., Über die Bildungsbedingungen von Kaolin, Montmorillonit, Sericit, Pyrophyllit, und Analcim, *Mineral. Petrograph. Mitt.*, **48**, 210 (1936).
22. Schofield, R. K. and A. W. Taylor, The measurement of soil pH, *Soil Sci. Soc. Am. Proc.*, **19**, 164 (1955).

# 3 Hydrothermal synthesis of minerals in relation to pedology

**GENERAL CONSIDERATIONS**

As pointed out in Volume 1, Chapter 3, congruency in composition between aqueous solution and mineral surface is the exception rather than the rule. For the two- and three-dimensional structures that contain silica, alumina, and alkali or alkaline earth cations, the process of solution is complex. Cations that balance the framework charge pass into solution from accessible surfaces by a process akin to Donnan hydrolysis. Silica also appears, but usually in smaller quantity, by progressive depolymerization of the giant crystalline polymer. Aluminum passes into solution near neutrality in very much smaller amount. Hence the solution phase is much more siliceous than the solid.

In the formation of new solid phases a general rule was propounded by Ostwald: that the least stable will appear first and the most stable last. Hydrothermal syntheses of aluminosilicates afford many examples. One advantage of working at elevated temperatures is that the steps are passed through more quickly, and at sufficiently high temperatures they disappear. Hence the ultimately stable form is favored. In some cases the commonly used device of seeding with the desired stable form fails to produce crystallization. This implies that activation energies for the production of the unstable phases are less than for the stable phase. Hence reactions progress through unstable phases to the stable one, but all may be very slow at atmospheric temperatures. Rise in temperature, however, may provide the activation energy needed for the single step to the stable form. For these reasons as well as through the general reduction in reaction speed at lower temperatures, extrapolations from high temperature hydrothermal synthesis to low temperature conditions are unsure.

A word should be said about the functions of water in hydrothermal syntheses. Geologists have long rated water as a mineralizer or catalyst. It facilitates the breaking of −Si−O− bonds; hence it can accelerate the decomposition of one silicate and the formation of another even when both are anhydrous. Its dissociation into $H^+$ and $OH^-$, which rapidly increases with temperature, is also often a critical factor. Certainly the chemical potential of $OH^-$ ions is of great importance in the synthesis of structures containing OH groups.

An important but rarely considered function of water is its effect on the stability of silicate framework structures. Barrer (1) has shown that some two- and three-dimensional structures containing water are stable only in its presence. Water changes the chemical potential of the framework skeleton, and when it is removed the structure collapses. Some zeolitic structures collapse irreversibly. The montmorillonites and vermiculites afford examples of reversible collapse.

The hydrothermal synthesis of a given mineral can be effected from a variety of starting materials. In practice simple oxides and hydroxides are often used, or else amorphous glasses, colloidal sols, or precipitates. The main problem concerns hydroxides of aluminum and ferric iron, which are extremely insoluble. Silica, even in the form of quartz, is so much more soluble that it presents no real problem. By introducing a soluble aluminum salt at high dilution, Hénin and co-workers (10,11) have been able to synthesize members of the montmorillonite group at temperatures below 100°C. Yet complete success was achieved only in cases in which other, more reactive constituents of the octahedral layer were present (e.g., magnesium, chromium, cobalt, zinc, and iron).

In systematic syntheses, complete coverage of all possible compositions, temperatures, and pressures can rarely be achieved. The situation is much better in the hydroxides and 1:1 lattice clays than in the montmorillonites, vermiculites, chlorites, and micas, because all members of the last group have a great range of compositions. Even in such simple cases as the oxides and hydroxides, gaps in our knowledge remain. A striking illustration of this, which may have considerable relevance in pedology, is the existence at low temperatures of double hydroxides of divalent and trivalent metals known as the Feitknecht compounds (8): aluminum and magnesium readily form such a phase, thereby providing a useful starting point for syntheses of montmorillonites. Yet in coverage of the systems MgO-Al$_2$O$_3$-H$_2$O and MgO-Al$_2$O$_3$-SiO$_2$-H$_2$O (Roy and Roy, 27; Yoder, 34) the double hydroxide was not detected, presumably because the temperatures employed were too high.

The range of compositions, temperatures, and pressures in which a given phase appears is deduced from the results and provides a frame of reference by which natural occurrences of minerals can be evaluated. Frequently the high temperature and pressure data are supplied by experiment, and those under atmospheric conditions come from observations of natural occurrences. The number of variables simultaneously under consideration can be reduced in practice by disregarding external pressure differences in the range 0–300°C; that is, up to about 100 atm, which corresponds to a free energy difference of about 43 cal/mole H$_2$O. In such a system as MgO-Al$_2$O$_3$-SiO$_2$-H$_2$O this still leaves five variables. Usually the three oxides are represented as the apices of the base of a tetrahedron, of which the vertical axis is H$_2$O. Thus the range of compositions for appearance of a given phase is shown as a volume for a given temperature. However in terms of the nonvolatile components it can be projected on the anhydrous base, giving us an area characteristic of a given temperature. The variation of this area with temperature is an important compositional characteristic of the system. Very careful X-ray characterization is needed in defining such areas because there are many instances of limited miscibility of closely related solid phases. The feldspars provide the most familiar example of this, but there are also the dioctahedral and trioctahedral montmorillonites and probably many others among the micas.

A rapid review of composition in relation to the factors that operate in weathering and conversely in synthesis can be obtained by plotting the Al/(Si + Al) ratio on one axis and tetrahedral Al/total Al or the equivalents of alkalis and alkaline earths per aluminum atom on the other (Figure 2). As brought out in Chapter 2, weathering can be thought of as a combination of three processes: (*a*) hydration, which of itself does not affect position on the diagram, (*b*) loss of cations, which moves the composition to the left, and (*c*) loss of silica in solution, which moves the composition upward. From the work of Nash and Marshall (22) on feldspar surfaces, it is clear that under mild conditions loss of cations exceeds loss of silica. Hence there is a dominant movement to the left. However in the case of anorthite release of silica is much easier than for albite or orthoclase; therefore from anorthite the weathering process could conceivably proceed directly to gibbsite without encountering compositional regions of the micas or chlorites. The diagram indicates that this is much less likely to occur with albite or orthoclase. Of course whether a given mineral appears when the weathering stage of other minerals passes through its composition depends on the free energy situation. This may be why the brittle mica margarite is not a common weathering product of anorthite.

When a mixture of minerals weathers together, that which reacts most readily with water largely determines the chemical environment for the species that react less readily. Hence they may be protected for considerable periods in a climatic environment that normally would lead to their removal.

From Figure 2 it can be seen that syntheses in the feldspathoid, feldspar, and zeolite groups demand as compositional variables only the Al/(Si + Al) ratio and the nature of the cation that stoichiometrically balances the aluminum. In the micaceous groups including the 2:1 clays, there is greater possibility of compositional variations—not only that apparent from Figure 2, but also arising from replacements for aluminum (chiefly magnesium and iron) in the octahedral layer. These do not apply appreciably to tetrahedral aluminum as in the feldspars and zeolites.

## THE FELDSPAR GROUP

We are now in a position to consider the compositional variation of a given lattice type as a function of temperature, beginning with the feldspars (Barrer et al., 1–5). Since the ratio of alkali to aluminum is fixed at 1:1 and the solid phase is anhydrous, we do not need the tetrahedral representation. A graph showing temperature in relation to gel composition, silica being the variable, gives a general view of the synthetic range. Within the range other species can and do coexist with feldspar, but where the original gel used for synthesis had a composition close to the theoretical for the feldspar (i.e., $K_2O$, $Al_2O_3$, $6SiO_2$), it was the major product. Figure 7 gives the data for sanidine, albite, and anorthite; the solid circles indicate fair, moderate, or good crystallinity, the open circles poor or very poor, on the basis of X-ray identification. The figures include results for several different reaction times, which explains some apparent but minor inconsistencies in the results. For all three cases the range of $SiO_2$ values at which feldspar appeared was greater at 350–400°C and less at lower and higher temperatures. Whether a considerable range extends to atmospheric temperatures cannot be deduced from these results, but the presence of well-crystallized authigenic feldspars in limestones indicates stability at low temperatures and moderate pressures under slightly alkaline conditions in the presence of salts.

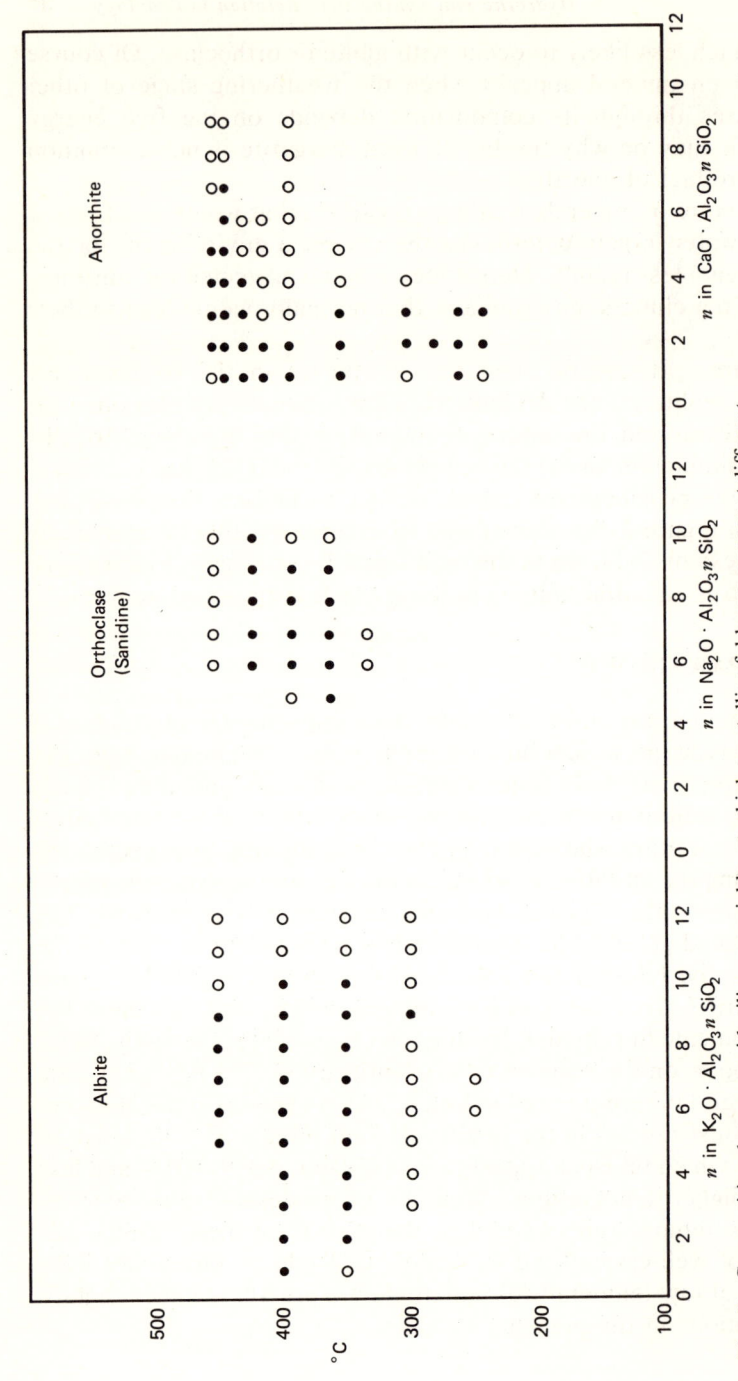

*Fig. 7* Compositional ranges, with silica as variable, at which crystalline feldspars appear at different temperatures. Solid circles indicate fair, moderate, or good crystallinity, open circles poor crystallinity of feldspars.

## THE MICA GROUPS

In the mica groups with their broad range of compositions, synthesis has been achieved only from mixtures whose cation to aluminum ratio falls below unity. Figure 2 indicates the very considerable gap between the feldspars and micas in this respect. Noll (24) synthesized sericite at 300°C using 0.37:1:2 as the ratio of $K_2O$ to $Al_2O_3$ to $SiO_2$, the potassium being supplied as the hydroxide or carbonate. When KCl was used, the product was kaolinite. This sensitiveness to pH is highly characteristic of syntheses of aluminosilicates. The reasons are discussed after other examples have been given.

From field evidence it is clear that sericite, illites, and glauconites are formed at normal atmospheric temperatures. Miss Warshaw (33) has shown that at temperatures between 250 and 500°C a mica phase similar to illite occupies a compositional area on the $Al_2O_3$–$SiO_2$–$K_2O$ triangle similar to that of montmorillonite in the corresponding $Al_2O_3$–$SiO_2$–$Na_2O$ range. However, the sodium mica paragonite is also synthesized between 250 and 500°C for compositions involving 0.3 to 0.4 sodium per atom of aluminum. Under natural conditions paragonite is much less common than sericite or illite. The potash micas can accommodate appreciable quantities of sodium in the lattice. Hence the normal results of weathering where mixtures of potassium and sodium are concerned is the production of potassium-sodium micas with a dominance of potassium, plus sodium-potassium montmorillonites with a dominance of sodium; or, if the cation to aluminum ratio is high, analcime appears. Calcium enters into normal micas in very small amounts. The brittle micas such as margarite probably form only in environments very high in calcium.

## THE CLAY GROUPS

The work of Noll (24) contributed much to our knowledge of the conditions in which members of the clay groups can be synthesized at moderate temperatures. The range of compositions studied was restricted. In the system $Al_2O_3$-$SiO_2$-$H_2O$, ratios of $Al_2O_3$ to $SiO_2$ of 10:1 to 1:10 were used at temperatures from 200 to 500°C. Below 400°C kaolinite was the only clay phase identified, mixed with boehmite at ratios in excess of 2:1 and with silica at lower ratios. Pyrophyllite appeared for appropriate compositions above 400°C.

The relationships of kaolinite, montmorillonite, and the zeolite analcime were strikingly shown by a series carried out at 300°C in which the $Al_2O_3/SiO_2$ ratio was held constant at 1:4 and sodium, potassium, calcium, or magnesium, respectively, was introduced at $R'_2O$ or $R''O/$

$Al_2O_3$ ratios of 0.02 : 1 to 2 : 1. Mixtures of kaolinite and montmorillonite were formed with all four bases at low ratios, the montmorillonite clay increasing with the proportion of base. Finally, when $Na_2O$ and $Al_2O_3$ were at 1:1, analcime was the sole product. Montmorillonite was the sole product between ratios of MgO to $Al_2O_3$ of 0.2:1 to 2:1. In another series at 300°C potassium aluminosilicate compositions were examined, montmorillonite being formed at the ratio $K_2O/Al_2O_3/SiO_2$ of 0.2:1:4 and sericite at 0.37:1:2. Acidic systems of both compositions gave kaolinite.

Strese and Hofmann (31) showed that magnesium-rich members of the montmorillonite group could be synthesized at relatively low temperatures. A considerable literature on syntheses below 100°C has now resulted from the efforts of French investigators, who as already mentioned, have successfully introduced iron, manganese, magnesium, cobalt, nickel, and zinc into the octahedral layer of clays of the montmorillonite group. In most of these cases very dilute solutions were allowed to interact in a medium of controlled pH at temperatures of about 80°C. A detailed investigation of the production of montmorillonite in the presence of excess magnesium from systems of variable pH and aluminum to silicon ratio showed the wide range over which this clay was formed. The presence of magnesium greatly facilitated the reaction and increased the proportion of aluminum fixed by the silica. The presence of salt solutions such as NaCl and KCl favored the production of montmorillonites, and in the case of KCl, the clays produced had unusually high exchange capacities. Under the conditions of these experiments (i.e., presence of excess magnesium at 80°C), reactions near neutrality favored the production of chamosite or antigorite, whereas acidic mixtures gave rise to boehmite.

The most complete systematic series have been carried out by Roy and Osborn for the systems $Al_2O_3$-$SiO_2$-$H_2O$; by Yoder (34), Roy and Roy (27), and Nelson and Roy (23) for the systems MgO-$Al_2O_3$-$SiO_2$-$H_2O$; and by Sand, Roy, and Osborn (29), and Barrer and White (5) for the systems $Na_2O$-$Al_2O_3$-$SiO_2$-$H_2O$. Considerable study has been made of certain ranges of composition in the systems $K_2O$-$Al_2O_3$-$SiO_2$-$H_2O$ (DeVries and Roy, 6) and CaO-$Al_2O_3$-$SiO_2$-$H_2O$ (Barrer and Denny, 4).

An important conclusion regarding the montmorillonite group follows from the summary of data presented by Roy and Tuttle (28) on MgO-$Al_2O_3$-$SiO_2$-$H_2O$ and $Na_2O$-$Al_2O_3$-$SiO_2$-$H_2O$ systems. The range of compositions within which montmorillonites (smectites) appear increases markedly with decrease in temperature from 480 to 200°C. Whether this broadening extends further to atmospheric temperatures is not known. At 200°C montmorillonites (smectites) can be produced whose compositions are only slightly more siliceous than kaolinite.

In studies of the systems $MgO$-$Al_2O_3$-$SiO_2$-$H_2O$ no members of the sepiolite-attapulgite group were synthesized. Mumpton and Roy (21) found that at 200°C montmorillonites are definitely more stable than sepiolites or attapulgites. More recently Siffert and Wey (30) have prepared a sepiolite at room temperature by the interaction of very dilute silicic acid with magnesium chloride under alkaline conditions. Whether it is the stable phase under these conditions has not yet been determined.

Another important question in the montmorillonite group concerns the existence of a complete mixed layer series connecting the dioctahedral aluminum members and the trioctahedral magnesium members. On the basis of analyses of naturally occurring clay minerals, Ross and Hendricks (26) came to the conclusion that two separate series existed, the one mainly dioctahedral and the other mainly trioctahedral. Syntheses by Roy and Roy (27) gave no indication of a separation, although it should be possible to detect the presence of the two separate phases by their different (060) spacings. The matter was discussed by Romo and Roy (25).

A series carried out by the author at 200°C gave clear evidence of the presence of two distinct montmorillonite phases in systems of atomic composition 1.23 Mg, 1.58 Al, 3.6 $SiO_{10}(OH)_2$, and 0.4 Na, 1.23 Mg, 1.58 Al, 3.6 $SiO_{10}(OH)_2$; reaction times were 69 and 53 days, respectively. In each case two distinct 060 bands were observed. A third mixture containing sodium carbonate, having atomic composition 1.0 $Na_2CO_3$, 0.4 Na, 1.23 Mg, 1.58 Al, 3.6 $SiO_{10}(OH)_2$, after 50 days, gave a well-defined mixture of montmorillonite and analcime. The dioctahedral 060 band of montmorillonite had disappeared, since the aluminum had been used up in the formation of analcime. The trioctahedral band of the magnesium montmorillonite remained in its original position. Similar results were obtained with sodium citrate; but sodium acetate, slightly less alkaline, gave no analcime, and both montmorillonites registered as before. Thus a very sensitive pH function shows itself.

The general relationship of analcime to a member of the smectite group, namely, sodium beidellite, is depicted in Figure 5. The two minerals have a common boundary, extending from the triple point albite–analcime–sodium beidellite at a value of log ($[Na^+]/[H^+]$) of 5·5, to about 6·3, where the triple point analcime–sodium beidellite–kaolinite occurs. These values may be uncertain because of the lack of precise thermochemical data for analcime. In any case a trioctahedral magnesium smectite such as a saponite would be expected to exhibit different values.

If it is desired to produce montmorillonites at high alkalinity without the formation of three-dimensional framework structures, strong organic bases can be used (Barrer and Denny, 3). The author found that the compositions 0.4 B, 2.4 Al, 3.6 $SiO_{10}(OH)_2$, and 0.8 B, 2.4 Al, 3.6

SiO$_{10}$(OH)$_2$ (B is tetraethylammonium hydroxide) gave good montmorillonite after 53 days at 200°C. Yet unidentified metastable phases were present after reaction for 5 days, which illustrates the difficulties likely to be encountered at lower temperatures. Even such good mineralizing agents as hydroxyl ions do not preclude the formation of metastable phases—this again emphasizes the importance of activation energy relationships.

A discussion by Romo and Roy (25) raised the question of the true status of mixed layer clays, which are exceedingly common in soils. Theoretically they should be metastable with respect to the single end members. Hence Romo and Roy regarded them as products of partial weathering, not as derived by natural synthesis from molecules in solution. However later work by Warshaw (33) indicated that mixed layer illites could be synthesized. Koizumi and Roy (17) have prepared mixed layer phases under hydrothermal conditions using temperatures over 350°C and compositions that gave good beidellites and saponites at lower temperatures. Thus far mixed layer clays have not been prepared synthetically at very low temperatures.

The phase diagrams for the systems Al$_2$O$_3$-SiO$_2$-H$_2$O have been discussed by Roy and Tuttle (28) and include the kaolin clays as well as the hydrates of alumina. Dickite and nacrite have not been synthesized reproducibly. Halloysite (4H$_2$O) has been produced at room temperatures by precipitation (Karsulin and Stubican, 12) in the presence of seeds of natural halloysite (4H$_2$O). The area of stability of halloysite (4H$_2$O) has only been established by experiments on the natural mineral. As expected, the loss of water from halloysite (4H$_2$O) is suppressed in the presence of steam under pressure, and the zone of stability extends up to 150°C.

Kaolinite has often been produced synthetically, both pure and in mixtures with montmorillonites and with hydrous alumina. Noll's important work indicated the importance of nonalkaline conditions for the formation of kaolinite. Mixtures of kaolin clays and potassium and sodium micas have not yet been produced by synthesis. Such micas require higher log [K$^+$]/[H$^+$] ratios than correspond to members of the montmorillonite group. The conditions can be viewed in the light of Figure 3.

Kittrick (14) has synthesized kaolinite in a very interesting way by making use of the solubility relationships of montmorillonite, kaolinite, and amorphous silica. Under acidic conditions and with small additions of sodium silicate or aluminum chloride in certain cases, it was shown that crystalline kaolinite was formed from montmorillonite. In a reaction period of 2 years at room temperature, large amounts of kaolinite were produced from Belle Fourche montmorillonite under conditions of supersaturation with respect to kaolinite. With undersaturation, no kaolinite was formed. It would be very interesting to carry out similar experiments with beidellitic clays from midwestern soils, since solution studies on the latter indicate that kaolinite may be the ultimately stable phase.

Using the same axes as in Figure 4 it is interesting to compare the stability diagrams for various phases found in soils (Figure 8). For the purpose of this diagram montmorillonite is given the simplified composition $Al_2O_3$-$4SiO_2$-$H_2O$. The range of stability for this composition is very restricted, but as Kittrick showed (13) it becomes much wider for actual montmorillonites with appreciable magnesium in the octahedral layer.

This diagram illustrates also the relative stabilities of kaolinite and halloysite. Clearly halloysite is more soluble than kaolinite. It has a standard free energy of $-898.6$ kcal as compared with $-903.8$ kcal for kaolinite. Thus it might well be expected as an unstable phase in weather-

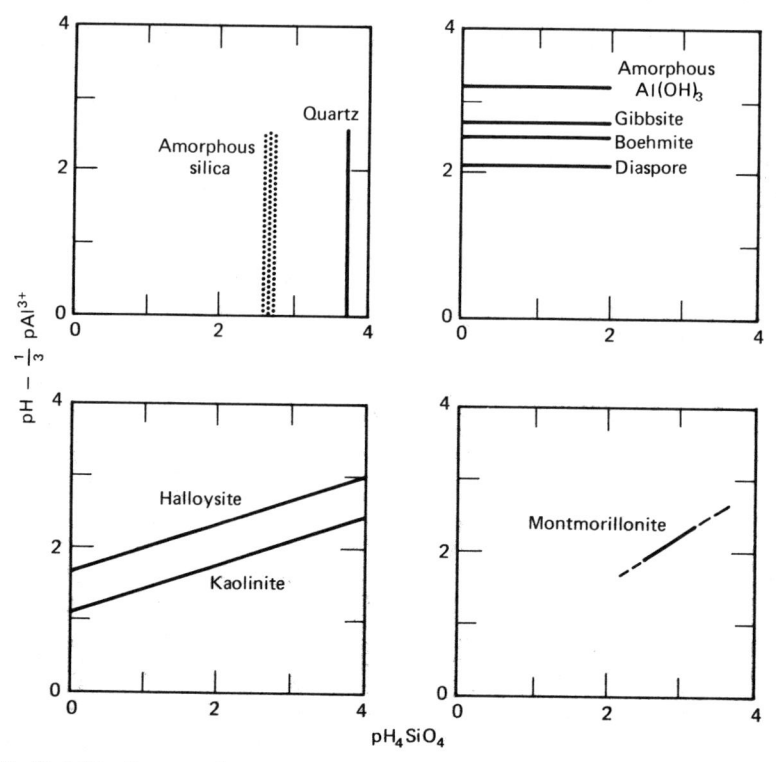

*Fig. 8* Stability diagrams for some minerals in the $Al_2O_3$-$SiO_2$-$H_2O$ system at 25°C and 1 atm. The area above or to the left of the solubility lines represents supersaturation (15). Used by permission, from *Clays and Clay Minerals*, Pergamon Press.

ing processes. Tamura and Jackson (32) found this to be the case in successive transformations from weathering feldspar, passing through allophanes, hydrated halloysite, and metahalloysite to kaolinite. La Iglesia and Galen (18) have shown that the halloysite–kaolinite transformation can be carried out in relatively short periods (< 90 days) at room temperature by the use of the complexing agents oxalic acid and ethylene diamine tetracetic acid (EDTA) applied to the finely ground halloysite.

Eberl and Hower (7) have reexamined conditions for the synthesis of kaolinite. With amorphous starting material and reaction periods up to 100 days, the lowest temperature for recognizable kaolinite was 150°C. Low temperatures of synthesis favored disordered kaolinite. Attempts to synthesize halloysite failed. Natural halloysite was converted to disordered kaolinite in hydrothermal experiments at 152 and 312°C.

There is accumulating evidence that the synthesis of kaolinite occurs at room temperatures in the presence of certain organic compounds. Linares and Huertas (20) synthesized a "pre-kaolin" type of mineral from monomeric silicic acid and aluminum fulvic acid. This kind of reaction may be of great importance in acidic soils. Hem and Lind (9) produced a kaolinite-like microcrystalline product in solutions containing quercetin.

The function of such complexing agents is clearly to make aluminum mobile in molecular form, but at the same time the activity in solution of such common ions as $Al^{3+}$ and $Al^{2+}(OH)$ will be reduced. In terms of the Kittrick diagram (Figure 4), complexing agents will tend to bring conditions from gibbsite stability through kaolinite and montmorillonite toward silica. The equilibria, however, are 3 times more sensitive to pH than to $pAl^{3+}$. It has been shown (19) that both kaolinite and halloysite can be produced from homogeneous solution when the pH is very slowly raised in the range 4–6.5. Thus we are likely to be in a region where the balance between kaolinites and smectites is extremely sensitive both to pH and to the strength of the complexing action with aluminum. As we see later, it is possible that many occurrences of smectites in the leached horizons of Podzols can be explained in these terms.

## REFERENCES

1. Barrer, R. M., Aspects of hydrothermal chemistry. Presented at the VIIth International Ceramic Congress (1960).
2. Barrer, R. M. and J. W. Baynham, The hydrothermal chemistry of the silicates. Part VII. Synthetic potassium aluminosilicates, *J. Chem. Soc.*, 2882 (1956).
3. Barrer, R. M. and P. J. Denny, Hydrothermal chemistry of the silicates. Part IX. Nitrogenous aluminosilicates, *J. Chem. Soc.*, 971 (1961).
4. Barrer, R. M. and P. J. Denny, Hydrothermal chemistry of the silicates. Part X. A partial study of the field $CaO-Al_2O_3-SiO_2-H_2O$, *J. Chem. Soc.*, 983 (1961).

5. Barrer, R. M. and E. A. D. White, The hydrothermal chemistry of silicates. Part II. Synthetic crystalline sodium aluminosilicates, *J. Chem. Soc.*, 1267 (1952).
6. DeVries, R. C. and R. Roy, The influence of ionic substitution on the stability of micas and chlorites, *Econ. Geol.*, **53**, 958 (1958).
7. Eberl, D. and J. Hower, Kaolinite synthesis: The role of the Si/Al and (alkali)/(H$^+$) ratio in hydrothermal systems, *Clays & Clay Miner.*, **23**, 301 (1975).
8. Feitknecht, W., Über die Bildung von Doppelhydroxyden zwischen zwei- und dreiwertigen Metallen, *Helv. Chim. Acta*, **25**, 555 (1942).
9. Hem, J. D. and C. J. Lind, Kaolinite synthesis at 25°C, *Science*, **184**, 1171 (1974).
10. Hénin, S., Synthesis of clay minerals at low temperatures, *Proc. 4th Nat. Conf., Clays & Clay Miner.*, 54 (1956).
11. Hénin, S. and O. Robichet, A study of the synthesis of clay minerals, *Clay Miner. Bull.*, **2**, 110 (1954).
12. Karsulin, M. and V. Stubican, Struktur und Synthese der Halloysite, *Kolloid Z.*, **124**, 169 (1951).
13. Kittrick, J. A., Montmorillonite equilibria and the weathering environment, *Soil Sci. Soc. Am. Proc.*, **35**, 815 (1971).
14. Kittrick, J. A., Precipitation of kaolinite at 25°C and 1 atm, *Clays & Clay Miner.*, **18**, 261 (1970).
15. Kittrick, J. A., Soil minerals in the $Al_2O_3$-$SiO_2$-$H_2O$ system and a theory of their formation, *Clays Clay Miner.*, **17**, 157 (1969).
16. Kittrick, J. A., Soil solution composition and stability of clay minerals, *Soil Sci. Soc. Am. Proc.*, **35**, 450 (1971).
17. Koizumi, M. and R. Roy, Synthetic montmorillonoids with variable exchange capacity, *Am. Mineral.*, **44**, 788–805 (1959).
18. La Iglesia, A. and E. Galen, Halloysite–kaolinite transformation at room temperature, *Clays & Clay Miner.*, **23**, 109 (1975).
19. La Iglesia, A. and J. L. Martin-Vivaldi, Synthesis of kaolinite by homogeneous precipitation at room temperature. I. Use of anionic resins in (OH) form, *Clay Miner.*, **10**, 399 (1975).
20. Linares, J. and F. Huertas, Kaolinite synthesis at room temperature, *Science*, **171**, 896 (1971).
21. Mumpton, F. A. and R. Roy, The influence of ionic substitution on the hydrothermal stability of montmorillonoids, *Proc. 4th Nat. Conf., Clays Clay Miner.*, 337 (1956).
22. Nash, V. E. and C. E. Marshall, The surface reactions of silicate minerals. Part I. The reactions of feldspar surfaces with acidic solutions. Part II. The reactions of feldspar surfaces with salt solutions. University of Missouri Agricultural Experimental Station Research Bulletins 613 and 614 (1956).
23. Nelson, B. W. and R. Roy, Polymorphic 7Å and 14Å phases in the $MgO$-$Al_2O_3$-$SiO_2$-$H_2O$ system, *Proceedings of the Second National Conference on Clays and Clay Minerals* (1953).
24. Noll, W., Über die Bildungsbedingungen von Kaoline, Montmorillonit, Sericit, Pyrophyllit, und Analcim, *Mineral. Petrograph. Mitt.*, **48**, 210 (1936).
25. Romo, L. A. and R. Roy, Essais de synthèse des minéraux argileux dits à "couches mixtes," *Bull. Soc. Fr. Miner. Cristallogr.*, **78**, 433 (1955).

26. Ross, C. S. and S. B. Hendricks, Minerals of the montmorillonite group, U.S. Geological Survey Professional Paper 205-B (1945).
27. Roy, D. M. and R. Roy, Synthesis and stability of minerals in the system MgO-Al$_2$O$_3$-SiO$_2$-H$_2$O, *Am. Mineral.*, **40**, 147 (1955).
28. Roy, R. and O. F. Tuttle, Investigations under hydrothermal conditions, *Physics and Chemistry of the Earth*, Vol. I, p. 138 (1956).
29. Sand, L. B., R. Roy, and E. F. Osborn, Stability relations of some minerals in the system Na$_2$O-Al$_2$O$_3$-SiO$_2$-H$_2$O, *Bull. Geol. Soc. Am.*, **64**, 1409 (1953).
30. Siffert, B. and R. Wey, Synthèse d'une sépiolite à température ordinaire, *Compt. Rend.*, **254**, 1460 (1962).
31. Strese, H. and A. Hofmann, Synthesis of magnesium silicate gels with two dimensional regular structure, *Z. Anorg. Allgem. Chem.*, **247**, 65 (1941).
32. Tamura, T. and M. L. Jackson, Structural and energy relationships in the formation of iron and aluminum oxides, hydroxides and silicates, *Science*, **117**, 381 (1953).
33. Warshaw, C. M., Experimental studies of illite, *Proc. 7th Nat. Conf., Clays Clay Miner.*, 303 (1960).
34. Yoder, H. S., The MgO-Al$_2$O$_3$-SiO$_2$-H$_2$O system and the related metamorphic facies, *Am. J. Sci. Bowen*, 569 (1952).

# 4 The chemical expression of climatic factors

**THERMODYNAMIC FACTORS**

*Changes in the Chemical Potential of Soil Water*

The chemical potential of soil water is affected by pressure or tension, by temperature, by dissolved species, by equilibria between mineral species, and by the surfaces with which it is in contact. Such changes can be formulated in thermodynamic terms through application of the Duhem-Margules equation, as follows:

$$S\,dT - V\,dp + n_{H_2O}d(\mu_{H_2O}) + \Sigma n_{solutes}d(\mu_{solutes})$$
$$+ \Sigma n_{surfaces}d(\mu_{surfaces}) + \cdots + = 0$$

where $S$ is the entropy, $T$ the absolute temperature, $V$ the volume, $p$ the pressure, $n_{H_2O}$ the mole fraction of water, of chemical potential $\mu_{H_2O}$; $\Sigma n_{solutes}d(\mu_{solutes})$ is a composite term made up of the sum of the products of the mole fractions and the changes in chemical potential for individual substrates in true solution, and $n_{surfaces}d(\mu_{surfaces})$ is a similar term for surfaces. We deal throughout with partial molar quantities.

If surface effects are small, and for systems at constant temperature and pressure, we have the simple relationship:

$$n_{H_2O}d(\mu_{H_2O}) = -\Sigma n_{solutes}(d\mu_{solutes})$$

For such cases the mole fractions are easily determined by analysis, and for mixtures of electrolytes the changes in chemical potential are accessible through Debye-Hückel theory.

The complications caused by the presence of interfaces are attributable to several factors. In many cases the value to be used for the mole fraction is indeterminate. Furthermore, the change in chemical potential is a characteristic of a given crystallographic form (group of faces having the same relation to the unit cell). Crystals are usually bounded by faces of more than one form. Thus in applying the Duhem-Margules equation there should be a separate term for each form.

In spite of these complexities, certain extreme cases allowing of simplification exist among soils. Thus in a clay consisting entirely of montmorillonite all lattice units are in contact with water, and the molecular weight can be taken as that corresponding to 24 framework oxygens (see Vol. I, Chap. 3). Hence the mole fraction can be determined in the same way as for a true solution. The surface in contact with the water is the 001 plane, which greatly dominates over the small extent of edge planes. Hence changes in the chemical potential of the clay can be determined. The work of Shainberg (46) illustrates one experimental approach. Mixtures of sodium chloride, sodium montmorillonite, and water were studied. The chemical potential of the water was calculated from measurements of the freezing point depression. That of the sodium chloride was afforded by combining an electrode reversible to sodium with one reversible to chloride. Hence the effect of the clay on the chemical potential of water can be calculated, as well as that of the sodium ions from the sodium clay on the chemical potential of sodium chloride.

The possibilities for fruitful work along these lines are perhaps greater than has been generally realized. Thus finely divided micas could be handled in the same way, provided the total external surface was determined, for instance, by the nitrogen adsorption method. The total area divided by the planar area per lattice unit gives the number of molecular units involved. The same would be possible for kaolins, always assuming that the edge effects are small, or preferably that both areas can be calculated from electron microscope observations.

Coming back to whole soils as assemblages of surfaces of many different kinds, including the air-water interface, let us first examine changes in the chemical potential of soil water.

## CYCLIC CLIMATIC CHANGES AND THE SOIL SOLUTION

Two types of cyclic change in soils are clearly apparent—temperature variations and moisture variations. Each can be characterized by considering periodicities, amplitudes, and their variations.

### Temperature Variations

Temperature variations have been thoroughly explored in a few individual cases as functions of time and depth. Wollny began such investigations in Germany in the period 1877–1883. Examples from Oxford were discussed by Keen (27).

In 1932 Smith (47) gave very complete data for a California site. In the tropics several series have been obtained, including those of Leather (31), Ramdas and Dravid (40) in India, and Braak (7) in Indonesia.

In all cases diurnal variations are superimposed on seasonal variations of the annual cycle. Because of the slow conduction of heat through soils, both the diurnal and seasonal cycles change in amplitude and in their time-phase characteristics, as a function of depth. Heat conductivity changes markedly in relation to soil moisture content. It is relatively low for dry soils, increases rapidly at low moisture contents and more slowly at high moisture contents (Baver et al., 3).

The general consequence is that temperature differences cause movement of water in the vapor phase, from points at high temperature to those at low temperature, unless the corresponding free energy difference for pure water is upset by other factors (different salt contents in different horizons, etc.). Such movement of water can be significant in the overall operation of climatic factors of soil formation and development.

With regard to diurnal variations, attention has been focused on dew, deposited on soil surfaces and on overlying vegetation at night, when the surface layer is cooled by radiation losses. Opinions on the importance of dew as a supplement to rainfall have been reviewed by Mohr and Van Baren (34). They conclude that it originates largely as water vapor from the surface layer and that only a minor part can be regarded as a separate contribution from the atmosphere. No exact measures of the two sources are available. The total dew is usually less than 10% of the annual rainfall.

Seasonal variations, with their slow progression through the soil profile, afford illustrations of several different temperature regimes in cool temperate climates, whereas under tropical conditions, with small temperature variations, a fairly constant pattern emerges. Figures 9 and 10 illustrate temperature differences as they show themselves daily and through the year in relation to depth.

Winter climates that include lengthy periods of frost cause the formation of frozen layers and even ice lenses in the upper parts of the soil. Since the formation of ice is a drying process, and the expansion of the water involved loosens the soil structure, large but temporary improvements in aggregation and aeration are observed after the spring thaw. These factors have long been recognized as favoring the production of a good soil tilth in the spring, particularly on heavy clay and silt loam soils.

When soils are frozen, weathering processes become minimal, more because of the drying than because of the lowered temperature. Thus such periods can be added to those of air dryness at higher temperatures to represent dormant periods for chemical weathering.

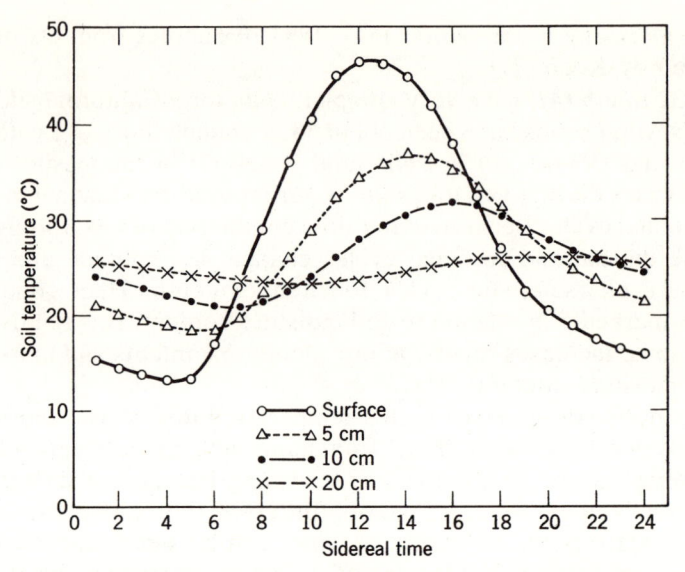

*Fig. 9* Diurnal variations in temperature in a loam soil (51).

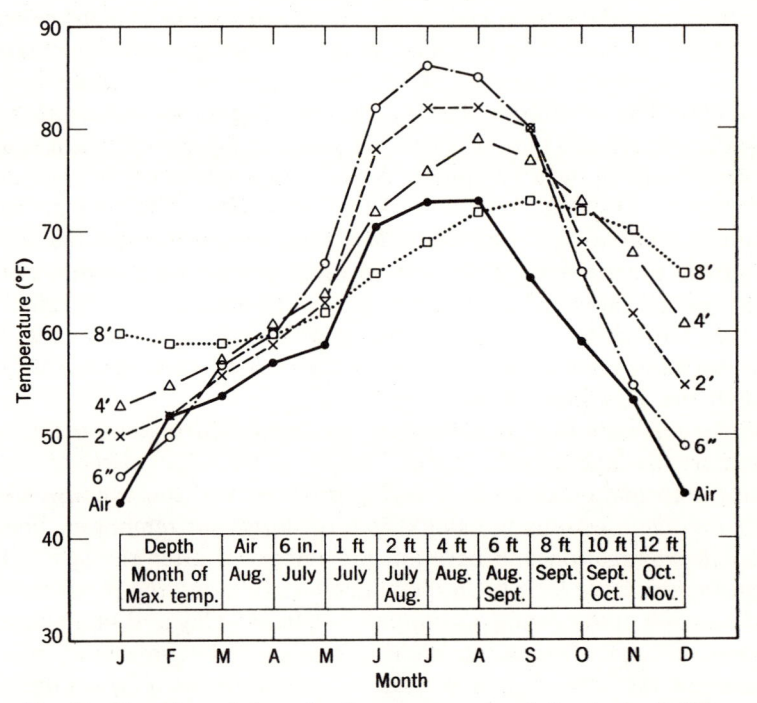

*Fig. 10* Monthly variation of soil temperature in relation to depth (47).

## Moisture Variations

As soon as we seek to define the chemical potential of the soil water, it becomes apparent that its changes with time follow a series of irregular cycles, of varying amplitudes and frequency. We therefore ignore the crude model that depicts the soil as a porous body through which water continually percolates, the amount of percolation being the excess of precipitation over evapotranspiration.

Within the annual cycle of the prevailing climate, certain subcycles can easily be distinguished. First we can separate those intervals in which the soil layer under consideration is dry, either air-dry or frozen. At such times weathering reactions and pedological processes almost cease. The periods of effective moisture content contain small cycles of variation, interrupted by the rainstorms that cause the appearance of drainage water, hence the irreversible loss of soluble and suspended materials. The quantitative relationships of the operative factors vary considerably with depth.

Thus the formation of soil horizons may be considered to be an expression of differential processes related to depth. Upward as well as downward movements of soil water must naturally be taken into account.

It might be supposed that the general chemical effects of moisture variations could be predicted by application of Donnan-type theory as described in Volume I, Chapter 1. The exchange complex, characterized for each cation by a volume concentration of dissociable sites, would then be presumed to remain separate from an aqueous soil solution containing salts and bases but no dispersed colloidal particles. However the processes of equilibration in soils differ from those applied to colloidal systems in Donnan experiments. The latter are arranged to permit observation of ion and salt movements in the liquid phase. In natural soils, equilibration with respect to the chemical potential of water can operate much more rapidly, since the gaseous phase is available in addition to water films. Thus it is scarcely possible to vary independently or to define accurately the "inside" or "outside" volumes. Movement of electrolytes over centimeter distances except with mass flow of water is slow compared with that of water vapor as it moves through the gaseous phase.

## Hysteresis Curves

Since in most soils wetting and drying are not fully reversible, it is important to consider the nature and causes of hysteresis in moisture cycles.

There are two distinct ranges for hysteresis. It was shown by Haines (16) that sand beds either of uniform or nonuniform particle size show a strong capillary hysteresis. This arises from the fundamental fact that in moist systems both thick and thin films can coexist at a given moisture tension. When an initially flooded sand is subjected to increasing moisture tension, the drying process consists first of the progressive emptying of larger, then smaller pores. Then follows the thinning of continuous water films until finally the latter become discontinuous and the appearance eventually changes to that of an air-dry system. The reverse process involves less water for a given pressure deficiency; differences become relatively large in the middle range. When the tension or pressure deficiency is brought back to zero, a certain proportion ($\sim$ 10%) of air still remains trapped in the sand (Figure 11). The free energy differences are only a fraction of a calorie per mole of water.

The second range of hysteresis forms part of the phenomenon of adsorption of gases and vapors by solids. Polar molecules such as water show a wider range of hysteresis than nonpolar molecules such as nitrogen or benzene (Figures 56 and 57, Vol. I). The free energy range in which this type of hysteresis operates lies from the tens to the thousands of calories per mole of water. One of the classic examples of this type of hysteresis is afforded by silicic acid gel.

The effects of hysteresis on water distribution in soils are not as far-reaching as might be supposed. During the year most soils are in the drying part of the cycle much longer than in the wetting part. Thus the drying curve can reasonably be used to express the behavior toward

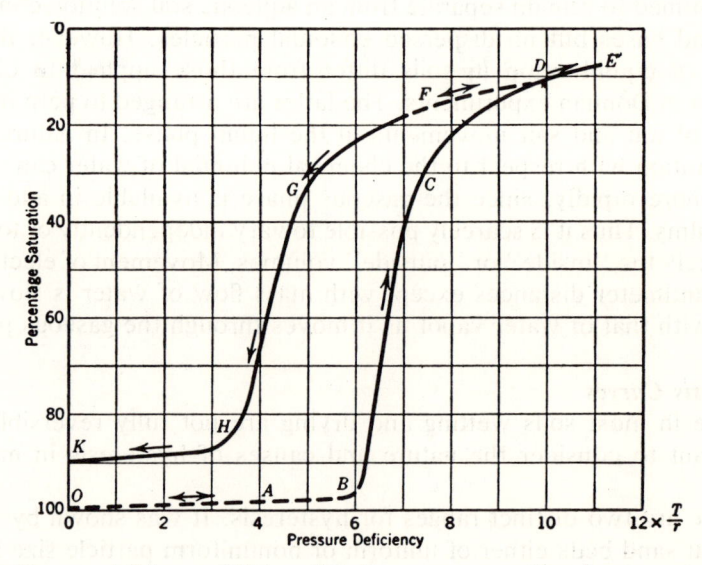

*Fig. 11* Pressure deficiency of Haines (16). Upper curve—wetting; lower curve—drying.

moisture; but of course this would not be the case for waterlogged soils nor for those subjected to high daily rainfall in the tropics.

## MINERAL EQUILIBRIA

Mineral equilibria in soils commonly involve water as a participating molecule. In dealing with equations for these equilibria thus far, we have assumed that the water is in its standard state throughout (i.e., at 25°C and 1 atm pressure). However in soil systems we have cycles of wetting and drying that cover a range of chemical potentials from the standard value to values characterized by measurably low vapor pressures. The formula $(\Delta F)_{H_2O} = 1364 \log p/p_0$ gives the free energy difference in calories per mole of water corresponding to the vapor pressure $p$ as compared with that in the standard state $p_0$. Thus the soil condition of air dryness, which often involves a relative humidity of around 50%, corresponds to a free energy difference of $-1364 \times 0.301 = -410$ cal/mole. This is not a negligible quantity when introduced, for instance, into the mass action equation for the change from orthoclase to muscovite:

$$3KAlSi_3O_8 + 12H_2O + 2H^+ \rightleftarrows KAlSi_3O_6Al_2O_4(OH)_2 + 2K^+ + 6Si(OH)_4$$

from which we now have

$$K_r = \frac{(K^+)^2 \, [Si(OH)_4]^6}{[H^+]^2 \, [H_2O]^{12}}$$

$$\log K_r = 2 \log [K^+] + 6 \log [Si(OH)_4] - 2 \log [H^+] - 12 \log [H_2O]$$

For the example chosen, the value of $\log K_r$ would be decreased by $12 \times \log 0.5$, that is, increased by $12 \times 0.301 = 3.612$. This has the same effect as an increase in $2 \log [K] + 6 \log [Si(OH)_4]$, which of course at constant $[H^+]$ would tend to drive the reaction backward. Thus the change feldspar $\rightarrow$ mica for which the standard free energy change (using $[H_2O] = 1$) is given by $1364 \log K = -1364 \times 4 \cdot 9 \times 2$ (Garrels and Christ, 14, p. 360), or 13,370 cal/mole mica with 24 oxygens, will now carry a standard free energy reduction diminished by $1364 \times 3.612$ or 4930 cal/mole mica, a fall of almost 37%.

Subsequent change of the mica to gibbsite involving 18 moles of water per mole of mica will similarly reduce the standard free energy change for the appropriate reaction by 7395 cal/mole mica.

Thus the thermodynamics of the situation enable one to predict that weathering will be markedly reduced by factors that tend to reduce the vapor pressure of water. The reactions to be affected most will involve the greatest number of water molecules.

There are also certain reactions in which water appears on the other side of the equation.

The dehydration of goethite to form hematite is a very instructive example, set forth in some detail in *Tropical Soils* by Mohr et al. (35). This reaction involves a small free energy change whose value is sensitively affected by the degree of crystallinity of the two solid phases. Considering well-crystallized goethite and hematite, we have:

$$2\text{FeO(OH)} \rightleftarrows \text{Fe}_2\text{O}_3 + \text{H}_2\text{O} + 0 \cdot 2 \text{ kcal}$$

Thus goethite is the stable phase at atmospheric temperature and standard state conditions. The activity of water corresponding to this equilibrium is given by:

$$\log [\text{H}_2\text{O}] = -\left(\frac{0 \cdot 2}{1.364}\right) = -0.17$$
$$[\text{H}_2\text{O}] = 0.68$$

Thus the relative vapor pressure is 68%. Hematite should be the stable phase below this value and goethite above it. In the drying and wetting of soils, this value must frequently be crossed and recrossed (air-dry soils commonly correspond to about 50% relative humidity). The outcome in terms of mineral change will depend on the unknown rates of the forward and reverse reaction. However these results hold only for macro-crystalline hematite and goethite. The effects of particle size variation below 1 $\mu$ have been examined by Langmuir (30). The free energy for the reaction changes sign according to the relative particle sizes of the two minerals. This is clearly shown in Figure 12, where $\Delta G^0$ is plotted against the logarithm of cube edge. These results are calculated from:

$$\Delta G_{298} \text{ (cal/mole)} = (531 \pm 600) - \frac{72 \cdot 3}{x} + \frac{34 \cdot 2}{y}$$

where $x$ and $y$ are the cube edges in microns of goethite and hematite, respectively. In general the particle size of goethite is less than that of hematite. Thus the values of $\Delta G$ may often be negative, lying in the general area between curve II (for equal sizes of crystals) and curve III (goethite $< 1$ $\mu$, hematite $> 1$ $\mu$). The general effect of these considerations and of the sluggishness with which hematite hydrates to goethite is that hematite tends to predominate over goethite in soil systems under ordinary atmospheric conditions.

The effect of temperature changes on weathering equilibria can be inferred from the individual equations and the properties of water. We are mainly concerned with the range from the freezing point of water to about 30°C. In this range the value of $RT$ increases by 11.0%. The vapor

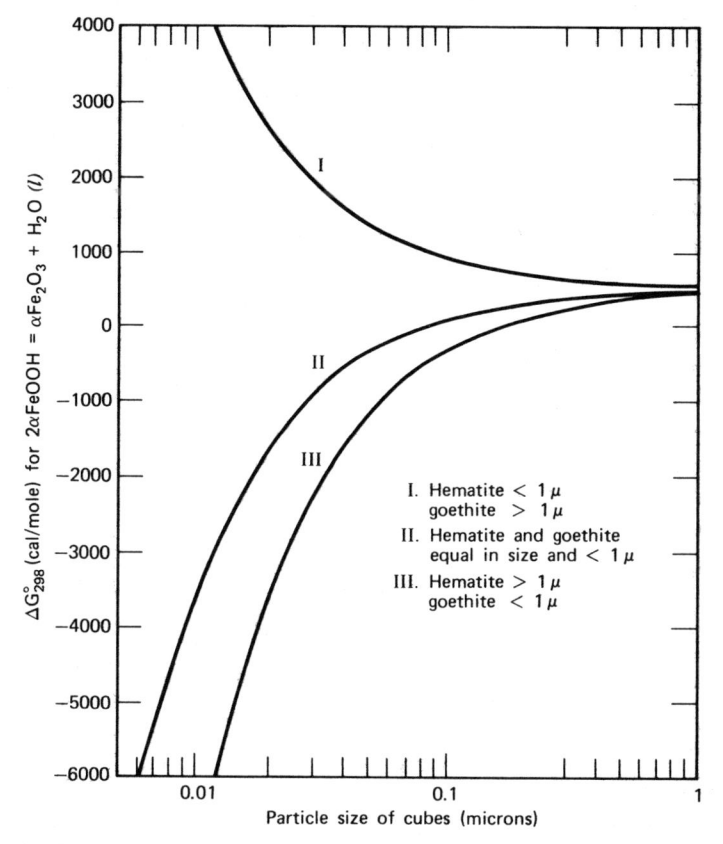

*Fig. 12* Particle size effect on the Gibbs free energy of the goethite dehydration reaction at 25°C. Reproduced from ref. 30 by permission of the *American Journal of Science*.

pressure of water increases from 4.58 to 31.82 mm Hg and the dissociation constant of water rises from 0.114 to 1.469 × 10$^{-14}$. The effects have to be studied in relation to each equation. Helgeson (19) has tabulated the values of log $K$ for certain reactions in the range 25–300° C.

## KINETIC FACTORS

The movement of ionic and molecular species during weathering involves kinetic factors, which fall into two groups: those associated with mass movement of water and those operating through diffusion. To some extent also diffusion can occur in solid minerals, but usually where aqueous pathways are available they dominate the situation. Viewed through kinetic effects, the equations for weathering reactions discussed previously are incomplete, because in every case there is reorganization of aluminum. The individual equation indicates nothing of the pathways by which the aluminum goes from one lattice to another. It must be noted, however, that each of the aluminosilicate phases present will be in a concurrent equilibrium with water, by which aluminum ions of one kind or another exist in the solution phase. The presence of two such aluminosilicate phases in equilibrium reduces the degrees of freedom of this aluminum ion activity in solution. Given complete thermochemical and analytical data, it can be calculated.

Mass movement of water carrying various ionic and molecular species in solution is governed by a group of factors that have been well explored by soil physicists. The operating forces considered are chiefly gravity and capillarity. Movement by convection of liquid water such as might arise through differences in temperature, hence density, is of no practical importance. There is also mass movement of relatively pure water through the vapor phase. Differences in vapor pressure are responsible for such movement. These can be traced back to several causes that operate extensively in soil profiles: (*a*) differences in moisture content, (*b*) differences in salt content, (*c*) differences caused by the interactions of soil surfaces and colloids with water, and (*d*) differences in temperature. All these can be formulated as free energy differences.

## *Mechanisms*

In the consideration of soil water it has long been customary to distinguish between water held by capillarity and that held by adsorptive fields of force. Thermodynamically there need be no distinction, since free energy differences are involved in both cases. Experimentally also there is no sharp line of demarcation. Smooth curves always characterize the relationship between the free energy of the water and the soil moisture. Figure 13 illustrates the situation for sandy, medium texture, and clayey soils. The consequence of the presence of the two mechanisms is that at intermediate moisture contents thick and thin films can exist together, both having the same free energy and vapor pressure. The thicker films are those ascribed to capillarity at points of contact of solid particles. They are characterized by concave water-air interfaces. The thin films are adsorption films on flat or curved surfaces that simply follow the contour of the surface.

The actual thickness of liquid films on solids has been measured for water on flat muscovite mica by Deryagin and Kussakov (12). Table 1 compares the measured thickness under two different tensions with the thickness calculated for water held between parallel plates by capillarity.

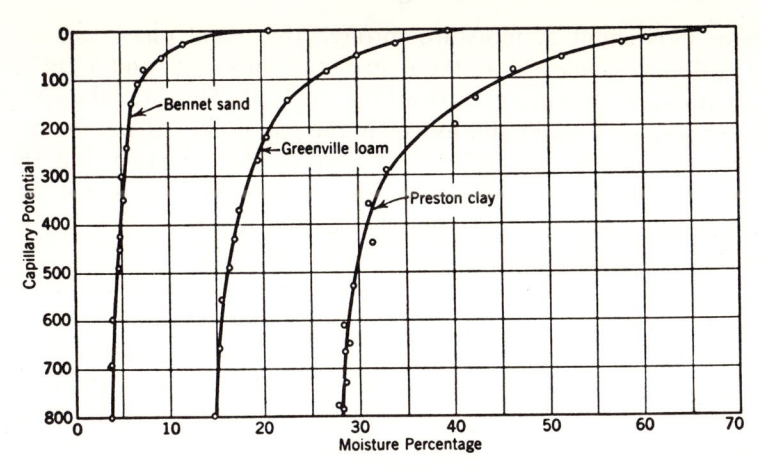

*Fig. 13* Tension-moisture curves (41).

The two sets of figures are 3 to 4 orders of magnitude apart, capillarity corresponding to much thicker layers than adsorption. The free energy differences per mole of water are here very small.

It is, therefore, of some importance to establish where in the scale of pore sizes the capillary model should be abandoned and the adsorptive model adopted. Obviously there is no sharp line of demarcation. The general behavior of soils of medium texture gives practical guidance in the following way. Such soils are characterized by a broad distribution of pore sizes. Consider the experiment in which a limited amount of water is added at a given point to a large mass of air-dry soil. A certain volume of the soil is initially flooded. Water then passes from the flooded portion, progressively wetting more of the dry soil. When a certain stage has been

**TABLE 1** Thickness of Water Films on Flat Mica Plates As Compared with Calculated Thicknesses for Parallel Plates

|  |  |  | Thickness ($\mu$) | |
| --- | --- | --- | --- | --- |
| Tension $P_0$ | pF | $\Delta F$ (cal/mole $H_2O$) | Observed (flat plates) | Calculated[a] (parallel plates) |
| 490 | −0.301 | $2.1 \times 10^{-4}$ | 0.195 | 2940 |
| 13,770 | +1.148 | $5.9 \times 10^{-3}$ | 0.030 | 105 |

[a] Using the formula $P_0 = 2s/d$, where $d$ is the distance apart of the plates, $S$ is the surface tension of water, $d/2$ is the calculated thickness.

reached, movement of moisture through the liquid phase ceases. The moist soil is then uniform in the distribution of water, and a fairly well-defined boundary separates it from the dry soil. This boundary remains fixed, even though water still moves in the vapor phase from the moist to the dry soil. In the dry soil the adsorption films thicken, but the soil indicates by its air-dry appearance that capillary water is not present. Under these conditions the wetted soil has reached "field capacity." What tensions and pore sizes characterize it?

Bodman and Coleman (6) added water at the top of soil columns from which sections could be withdrawn at different depths and different time intervals, and these experiments reveal much about the mechanics of the first part of this process. A sandy loam and a silt loam were studied. Each was characterized by its moisture content–pressure potential curve (Figure 14). [The pressure potential is given in ergs per gram of water, free water at the soil surface being taken as zero. The height of the equivalent column of water is obtained by dividing these values by the gravitational constant 980, assuming the water to have unit density. The scale chosen is such that the values given on the abscissa represent equivalent water column heights within 1000/980. Being tensions, the values are negative. By dividing the values in ergs per gram by the mechanical equivalent $4.19 \times 10^7$ and multiplying by 18, the molecular weight of water, we obtain the free energy difference in calories per mole of water.]

Figures 15 and 16 give the moisture potentials in relation to depth for different time intervals when free water is present at the surface. The moisture potential rapidly assumes negative values that fall on a straight line after a brief time. Finally this curves around and becomes horizontal at the wetting front. The straight line portion is called the transmitting zone. The rate of change of moisture potential with depth is constant, and the rate of infiltration is governed by the linear distance through which the moisture is transmitted. As this distance increases, the rate of infiltration decreases. When water ceases to be added at the surface, the wetting front advances only a limited distance further, then stops. The soil is then approximately at field capacity. The gradient of moisture potential across the boundary between damp and dry soil is very high, as Bodman and Coleman pointed out: from this point onward, redistribution occurs practically only through the vapor phase. Throughout the transmission zone, Bodman and Coleman found that the change of moisture potential with depth was parallel to the linear change of pressure potential with depth and was displaced from it by a moisture potential difference equaling about 27 cm of water tension. This difference was almost the same for the Yolo silt loam as for the Yolo sandy loam. The value is of course lower

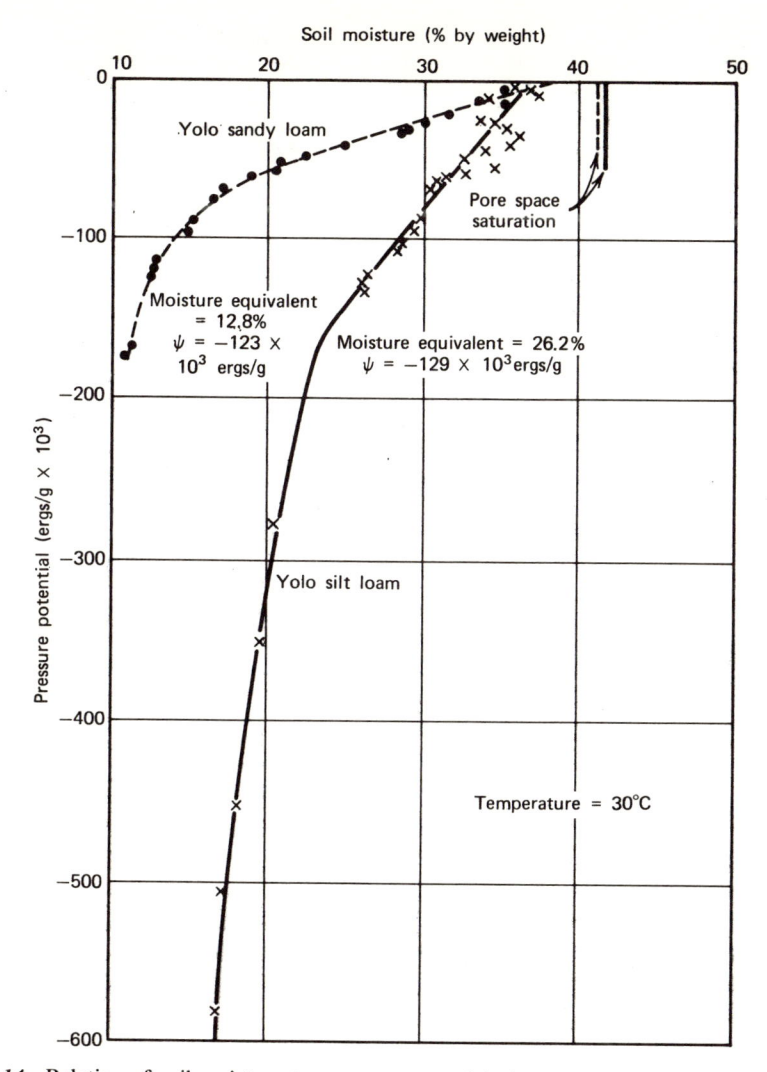

*Fig. 14* Relation of soil moisture to pressure potential, determined on wetting, for Yolo sandy loam and Yolo silt loam. Reproduced from ref. 6 by permission of the Soil Science Society of America.

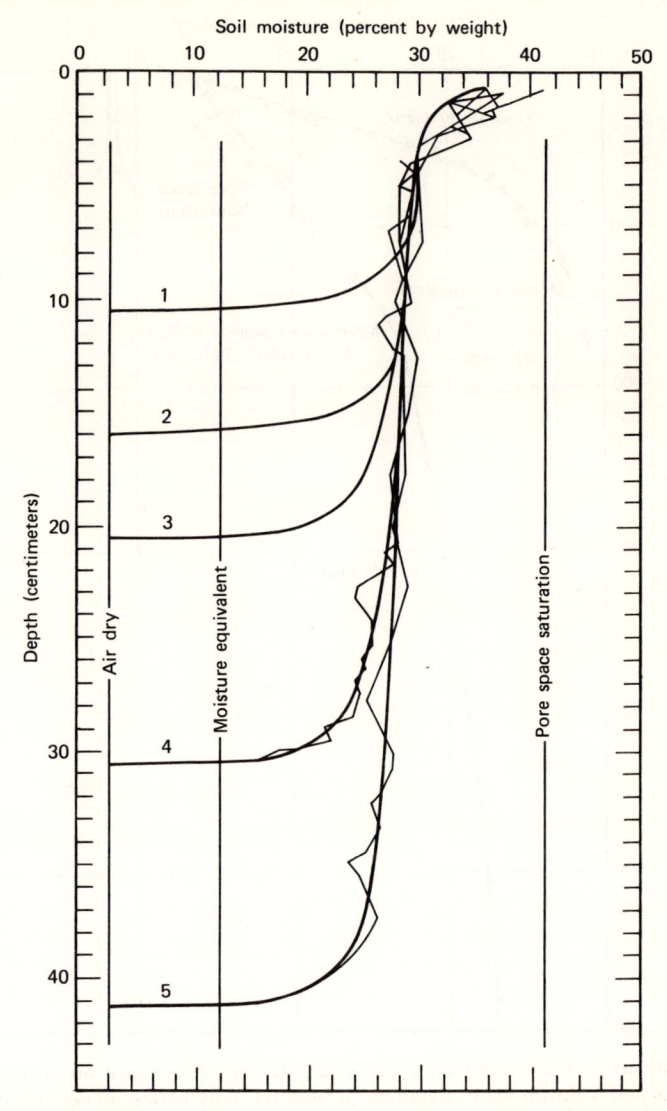

**Fig. 15** Distribution of soil moisture in relation to depth of wetting during infiltration into Yolo sandy loam. Reproduced from ref. 6 by permission of the Soil Science Society of America.

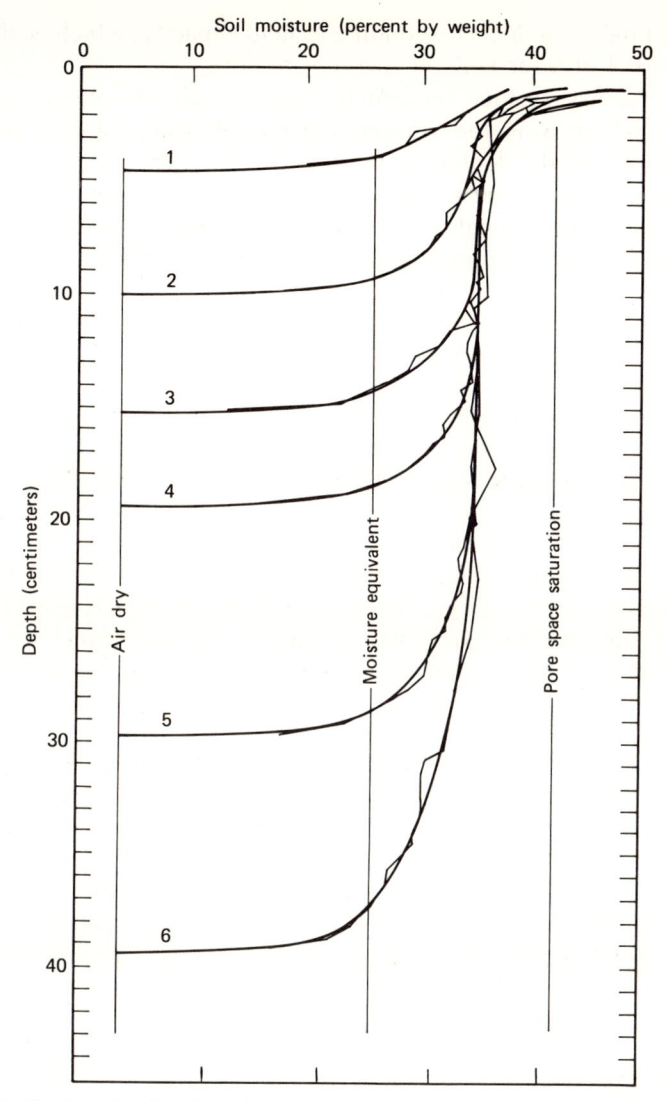

**Fig. 16** Distribution of soil moisture in relation to depth of wetting during infiltration into Yolo silt loam. Reproduced from ref. 6 by permission of the Soil Science Society of America.

than the final tension corresponding to field capacity, which is often taken as about 0.3 atm for soils of medium texture.

The ordinary surface tension formula for circular capillaries is $h = 2S/Dgr$, where $S$ is the surface tension of water against air ($= \sim 72$ dynes/cm), $D$ is the density of the liquid ($= \sim 1.00$), $g$ is the gravitational constant ($= 980$ cm/sec$^2$), and $r$ is the radius in centimeters. Substituting 345 cm for $h$, corresponding to 0.3 atm, the value of $r$ is found to be about $4.3 \times 10^{-4}$ cm or 4.3 $\mu$, or an equivalent diameter of 8.6 $\mu$. Clearly, somewhere around this point we are forced to change from a capillary model to an adsorption model, since by the Poiseuille equation for circular capillaries, movement of liquid water would still be possible although slow (the volume transported is proportional to $1/r^4$).

In both saturated and unsaturated flow this steep function of radius implies that very small pores are relatively ineffective, even though present in larger numbers than large pores. In unsaturated flow, the largest pores are completely inoperative, as is well shown in the Bodman and Coleman work already discussed. Here the 27 cm water tension, which remained constant throughout the transmitting zone, corresponds to the emptying of all pores larger than 110 $\mu$ diameter or 0.11 mm. Hence the range of effective pore sizes was relatively narrow, from 110 to 8.6 $\mu$ as calculated for circular capillaries. Regarding irregular shapes, Donat (13) has shown that soil pores support about twice the tension of circular pores of the same cross-sectional area.

Thus the moisture concerned in mass movement in the liquid phase is, if salt free, almost in its standard state. A tension of 0.3 atm corresponds to a free energy difference of only 0.145 cal/mole $H_2O$. Plants can abstract this water up to the wilting point at about 15 atm tension, which corresponds to a free energy difference of 6.54 cals/mole $H_2O$. This same value would be attained in an aqueous solution of a nonelectrolyte at about 0.67 $M$, the freezing point depression of which is 1.24°C. These various equivalencies are mentioned to emphasize changes in the properties of water upon drying and especially the drastic nature of freezing and air drying.

*Capillary Conductivity*

Pore size distribution shows itself very strongly in capillary conductivity, as is well illustrated in Richards' (41) results for four soils at different tensions (Table 2). Comparing the Superstition sand with Millville silt loam, the former is found to conduct about 4 times as well as the latter at zero tension; but at 60 and 200 cm tension the silt loam conducts better than the sand. Sands have low capillary conductivity at the higher tensions. The figures for the Chino silty clay loam show that upon returning from a high tension to a lower tension, a strong hysteresis appears. This is

**TABLE 2** Capillary Conductivity ($K$) of Four Soils (41)

| Superstition Sand | | Millville Silt Loam | |
|---|---|---|---|
| $H$ (cm of water) | $K$ (cm/hour) | $H$ (cm of water) | $K$ (cm/hour) |
| 0 | 6.59 | 0 | 1.69 |
| 10 | 1.30 | 10 | 0.13 |
| 20 | 0.65 | 20 | 0.12 |
| 40 | 0.37 | 40 | 0.11 |
| 60 | 0.024 | 80 | 0.059 |
| 200 | 0.00012 | 200 | 0.0059 |
| Coachella Loamy Fine Sand | | Chino Silty Clay Loam | |
| 0 | 2.40 | 0 | 0.723 |
| 10 | 0.82 | 10 | 0.060 |
| 25 | 0.24 | 20 | 0.038 |
| 50 | 0.15 | 40 | 0.013 |
| 85 | 0.063 | 80 | 0.0064 |
| 100 | 0.019 | 120 | 0.0030 |
| 200 | 0.00058 | 160 | 0.0014 |
| | | 200 | 0.00115 |
| | | 120 | 0.00116 |
| | | 20 | 0.0049 |

probably a manifestation of slow swelling of the clay, with marked reduction of pore size. It is a common observation in the field that many heavy clays seal themselves almost completely as rewetting proceeds. Where the heavy clay is a subsoil layer this can result in the appearance of "perched water tables" during wet periods, with important consequences as regards soil development.

Several attempts have been made to relate directly the tension or pF-moisture curves for soils to rates of water movement. These are discussed in Baver's book (2,3). Some authors have used a single tension in deriving a porosity factor that exhibited a simple relation to the rate of percolation (Nelson and Baver, 36). The relative ineffectiveness of small pores is well demonstrated in the expression of Smith et al. (48) for the porosity factor, namely, porosity factor = %(pores drained at 10 cm) + %(pores drained between 10 and 40 cm)/4 + %(pores drained between 40 and 100 cm)/10.

More detailed mathematical treatments of flow in relation to pF curves and to pore size distribution are dealt with by Childs (9) and Kirkham and Powers (29).

## Diffusion

The soil water that participates in mass movement is, by diffusion, in constant interchange with thin water films, through which it receives, also by diffusion, the soluble materials to be moved. As we have seen, the stationary water is in capillaries or films less than 110 $\mu$ thick. The diffusion of small molecules or ions moving outward from mineral surfaces thus involves short periods of time, of the order of seconds or minutes; whereas cyclic changes in the whole moisture regime occupy days or weeks. This may be illustrated by potassium chloride, whose diffusion coefficient in water is $1.99 \times 10^{-5}$ cm²/sec at 25°C. The corresponding displacement for 1 min works out at 0.049 cm or 0.49 mm ($\Delta^2 = 2Dt$, where $\Delta$ is the displacement corresponding to the time interval $t$ sec for a molecule of diffusion constant $D$).

A number of investigations of the self-diffusion of deuterated and tritiated water in unsaturated soils have been carried out. A recent study of soils of differing textures by Brees and Graham (8) (Figure 17) indicates that the coefficient of self-diffusion decreases rapidly with increasing clay content up to about 37% and becomes approximately constant at about

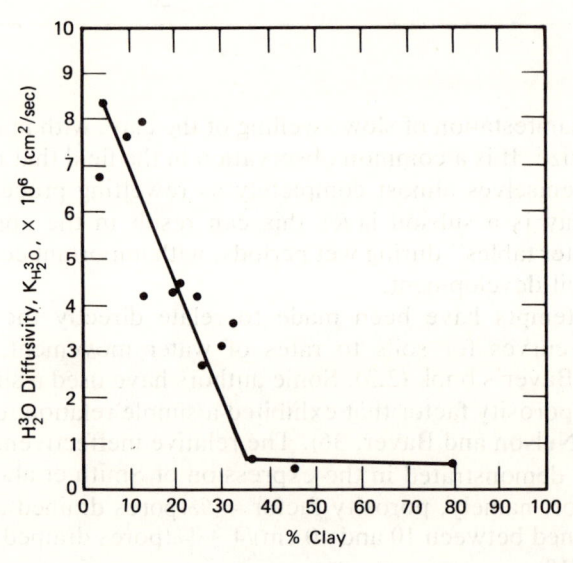

*Fig. 17* Self-diffusion coefficient of tritiated water as a function of clay content. Reproduced from ref. 8 by permission of the Williams & Wilkins Co., Baltimore.

$0.5 \times 10^{-6}$ cm²/sec for higher clay contents. These were swelling clays. Sandy soils gave values near $8 \times 10^{-6}$ cm²/sec. The value for unconfined water is $2.35 \times 10^{-5}$ cm²/sec at 25°C. Thus self-diffusion is reduced considerably in soil systems. This is only partly a matter of geometry; water molecules in close proximity to mineral surfaces are subject to fields of force that reduce this overall mobility. The nature of the mineral surface, including that of the exchange cation, is highly significant.

Thus we can say that for all practical purposes, the whole of the stationary water is always in equilibrium or steady state with the solid surfaces with which it is in contact. Even the capillaries of appreciable size need relatively short periods of time to attain equilibrium with regard to small and medium-sized molecules in solution. Thus small-scale heterogeneities in soils, such as the presence of different mineral grains of sand or silt, cause only very fleeting concentration differences in the soil solution. This is why it is possible to treat processes of downward leaching in many soils as though analogous to the operation of chromatographic columns. For differences to persist over appreciable time spans, there must be active mechanisms to maintain them, such as those of roots or microorganisms. The heterogeneity that is seen in the microscopic study of soil structure is thus largely biological in its ultimate origin, although three chemical mechanisms also contribute. These are (1) differences in solubility or chemical stability by which materials can be dissolved in one place and redeposited in another, (2) differences in state of oxidation leading to chemical depositions, and (3) differences in the electrical double layer, which influence ease of peptization and coagulation.

## SOLUBILITY DIFFERENCES

The formation of concretionary bodies consisting of gypsum or calcite is very common in soils of dry regions. Finely divided material tends to dissolve and to be redeposited on coarser aggregates. For gypsum the solubility of coarse crystals is 0.227% at room temperature, whereas crystals of 0.5 $\mu$ diameter give a value of 0.248%.

In the case of calcium carbonate a second factor probably dominates the situation, namely, the variation in partial pressure of $CO_2$ as maintained at different points in the profile by biological agencies. The solubility of calcite as calcium bicarbonate is a steep function of the $CO_2$ pressure (see Vol. I, p. 93). Thus it will be deposited in regions of low $CO_2$ pressure. The formation of surface efflorescences of calcite is common; it is caused partly by drying and partly by the change from a high $CO_2$ pressure in the soil body to a much lower value in the atmosphere.

Silica represents a much more difficult case. It is ubiquitous in the soil solution, and the stable mineral quartz is also almost universally found in soil profiles. However the silica that is in solution in soil extracts or drainage waters usually exceeds the solubility of quartz (~ 6 ppm), without affording a close approximation to accepted values for amorphous silica (~ 120 ppm). In addition, silicic acid enters into the equations for many mineral reactions, and this might lead us to expect values obtained to correspond to recognizable equilibria. Usually this is far from being the case; for instance, the gibbsite-kaolinite equilibrium demands a fixed activity of silicic acid in solution corresponding to only 1 ppm $SiO_2$. Leachates from tropical soils are much higher in silica than this. Harrison gives values in the range 20–30 ppm in waters percolating through weathered basic and acidic igneous rocks.

The formation of secondary silica in soil profiles may take the form of quartz or of amorphous or cryptocrystalline silica as chalcedony and opal. In the Missouri diabase profile (21) both quartz and chalcedony have been formed in roughly equal amounts. Usually, however, the amorphous forms dominate. A very interesting development (49, 50) has been the recognition of opal phytoliths, formed in grasses and other plant species high in silica, and left in the soil as a residue of plant life.

Geochemists and sedimentary petrographers know that recrystallization of amorphous silica to quartz is very sluggish. The maintenance of highly alkaline conditions favors the production of crystalline quartz, but there are often so many other possibilities for utilization of the silica that this mineral may not appear. Thus in presence of abundant magnesium, minerals of the attapulgite-sepiolite group may be produced. They have been increasingly recognized in deep-sea deposits, and also in certain alkali soils in the Near East.

Differences in solubility according to particle sizes naturally reflect themselves in the standard free energies of formation determined through solubilities. Thus Hem (20) quotes a value of −272.3 kcal/mole for microcrystalline gibbsite of about 0.1 diameter as compared with −273.9 kcal/mole for the coarse crystalline mineral.

## DIFFERENCES IN STATES OF OXIDATION LEADING TO CHEMICAL DEPOSITIONS

### General

The most striking manifestations of heterogeneity in many soils are the regions of high concentration of and cementation by hydrated iron oxides. Frequently other sparingly soluble oxides and hydroxides accompany the iron. Jenne has given a very complete review of these occurrences (22). Such depositions commonly appearing as concretions, and so on, can be treated as electrochemical in origin, since change of valency of the iron or other element is usually involved. In this approach it is often assumed that the ferric ion $Fe^{3+}$ does not move appreciably in true solution in soil profiles. The extremely low solubility product of ferric hydroxide ($\sim 10^{-39}$) and the very limited range of existence of the iron anion $HFeO_2^-$ make it necessary to proceed to high acidities before soluble ferric ions become important in the solution phase (Garrels and Christ, 14). For appreciable mobility of ferric ions, the only mechanism available seems to be through complexing. Microbial action and exudates from living roots provide a wide range of organic molecules with complexing possibilities. But many of these compounds are also strong reducing agents. Hence the ferrous ion $Fe^{2+}$ is readily produced. It can then remain either as such, or as a complex with organic molecules.

The relationships found in soils under reducing conditions have been greatly clarified since Volume I appeared, by the work of Garrels and his collaborators (14) on $E_H$–pH relationships generally, and by that of Ponnamperuma and collaborators (38,39) on soils under waterlogged conditions. What had not been realized in earlier treatments of oxidation-reduction in the soil was the extent to which it is governed by identifiable solid phases. The most important iron phases are ferric hydroxide [$Fe(OH)_3$, unstable]; ferric oxide, hematite [$Fe_2O_3$, stable under dry conditions]; goethite [$FeO \cdot OH$, stable under wet conditions]; ferrosoferric hydroxide [$Fe_3(OH)_8$, unstable]; ferrous hydroxide [$Fe(OH)_2$]; and ferrous carbonate, siderite [$FeCO_3$]. By the use of $E_H$–pH diagrams, first introduced by Pourbaix and extensively developed by Garrels and Christ (14), conditions are very clearly portrayed. In considering soils it is important to include the two iron hydroxide species that are not stable under ordinary atmospheric conditions and are unknown as macrocrystalline minerals: $Fe(OH)_3$ and $Fe_3(OH)_8$ (Ponnamperuma, 39).

The resulting diagram (Figure 18) includes both iron (solid lines) and manganese (dotted lines). Species that appear only at pH values above 10 are omitted. The iron diagram is constructed for conditions in absence of $CO_2$. Ponnamperuma et al. (39) have shown that this represents conditions in most waterlogged soils. Siderite ($FeCO_3$) is not a governing phase under normal conditions. Since the manganese equilibria are much more sensitive to $CO_2$, however, the diagram employed here is that constructed by Garrels and Christ for an initial $CO_2$ pressure of 1 atm. Most features remain similar for lower $CO_2$ pressures. The main difference is that the vertical line separating $Mn^{2+}$ and $MnCO_3$ moves to the right as the $CO_2$ pressure is reduced. From 1 to $10^{-4}$ atm $CO_2$ its pH value goes roughly from 7 to 8.4.

This superimposition of diagrams demonstrates that soils that remain

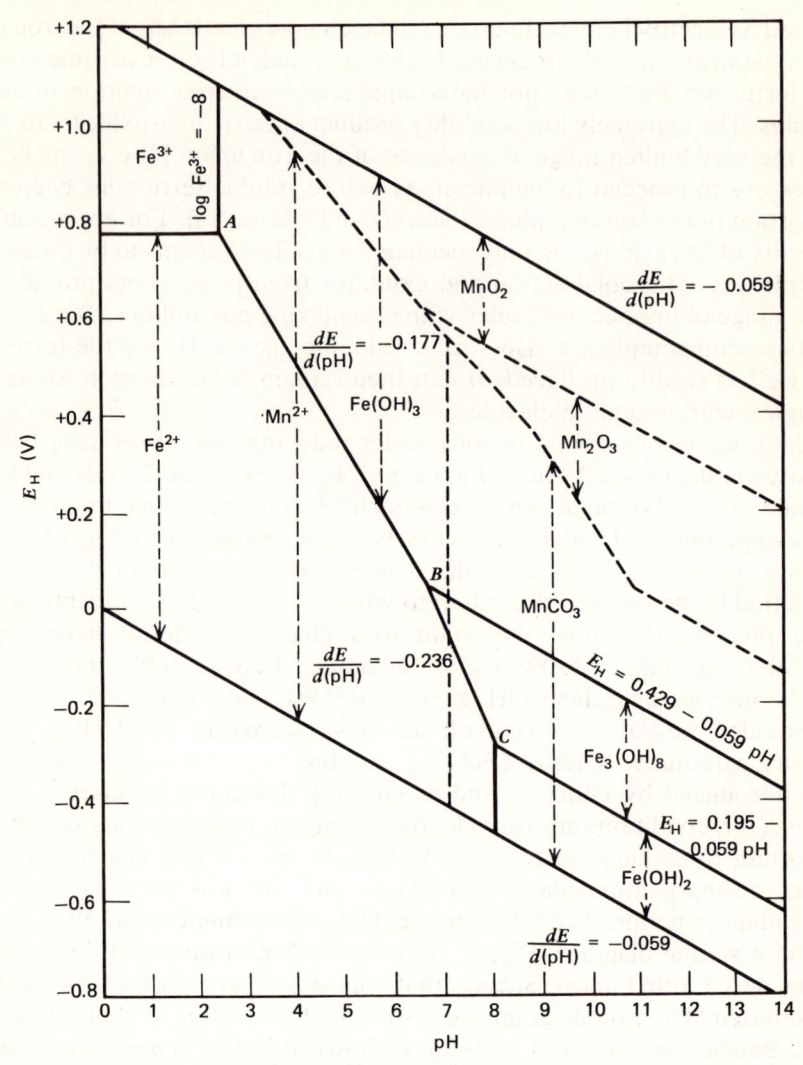

**Fig. 18** Oxidation reduction potential $E_H$ as a function of pH. Solid lines, iron equilibria; heavy dashed lines, manganese equilibria.

neutral or acidic may contain $Fe(OH)_3$ and $MnO_2$. The compounds $Mn_2O_3$ and $Fe_3(OH)_8$ are only stable beyond pH 6·5 under the conditions depicted.

The incidence of reducing conditions leads to sequences that can be predicted from the diagram. Suppose we begin with a soil at pH 5 and an $E_H$ of 0.6 V containing free $Fe(OH)_3$. The first effect of reduction is to produce a small concentration of ferrous ions in solution. The $E_H$ falls vertically until some such line as $AB$ is reached. This line has been drawn to correspond to a ferric ion activity of $10^{-8}$. The triple point $A$ is fixed at $Fe^{2+}/Fe^{3+} = 1$, making $AB$ the boundary between $Fe(OH)_3$ and $Fe^{2+}$ at $10^{-8}$. From $E_H = +0.33$ corresponding to pH 5, the intensification of reducing conditions causes the system to become less acidic as $E_H$ falls, until point $B$ is reached. The new phase $Fe_3(OH)_8$ then appears, and the line changes to a different slope down to the triple point $C$. After this it becomes vertical, since it represents the boundary between $Fe(OH)_2$ and ferrous ions at $10^{-8}$. [It should be noted that a parallel system $A'B'C'$ is sometimes drawn for other ferrous ion activities such as $10^{-6}$. This is probably closer to reality for highly reduced systems, but $10^{-8}$ seems better for $E_H$ values above $-0.3$ V. In the upper part of the diagram, where ordinary oxidized soils are formed, the concentration of the ferric ion in solution is normally considerably less that 10 ppm, which is about $1.8 \times 10^{-7} M$.]

The relationships expressed by these diagrams refer to analytical determinations on true solutions in equilibrium with specified solid phases. Direct determinations of redox potentials in soil-water systems cannot be interpreted without further information. Ponnamperuma et al. (38), point out that in waterlogged soil systems the negative potentials observed directly by inserting electrodes are much lower than those found for true solutions in equilibrium with the soil. They found that the latter were quantitatively in agreement with lines on the diagram for appropriate solid phases; hence their choice of the strictly thermodynamic approach through the removable soil solution. Clearly one factor of significance in the interpretation of measurements made by burying platinum electrodes in soil-water systems is the suspension or Donnan effect. It can amount to several pH units; thus a potential difference of 0.2 V or more can easily result.

This correction of the measured potential for pH, to bring it on to the standard electrochemical scale, therefore differs for a soil-water paste and for the expressed soil solution in equilibrium with it.

### The Production of Concretionary Material

Figure 18 is extremely useful in the consideration of soil granules, concretions, and other features. We begin with the assumption that reducing conditions in soils are in part localized. A reducing center has a certain characteristic oxidation-reduction potential, which, of course, may change with time. It also contains a certain quantity of reducing or reducible substances, which determine the total extent of possible chemical transformations. Thus we have an intensity factor and a quantity factor. Reactions proceed as reducing substances pass outward by diffusion. They involve first, solids of the soil, from which both soluble and very sparingly soluble materials may be produced. Second, soluble reducing substances can react with the soil solution to produce both soluble and sparingly soluble products. In terms of Figure 18, the only metallic ions present in significant quantity in solution at pH values between 3 and 10 will be $Fe^{2+}$ and $Mn^{2+}$. The solid phases to be considered in the same pH range for iron are ferric hydroxide, ferrosoferric hydroxide, ferrous hydroxide, and ferrous carbonate; for manganese they are manganese dioxide, manganese sesquioxide, and manganese carbonate. A further possibility for future investigation would be the existence of a double hydroxide of ferric iron and divalent manganese [Feitknecht-type compound: several examples are known between trivalent and divalent hydroxides; ferrosoferric hydroxide $Fe_3(OH)_8$ is of this type].

Consider a small volume of a normally oxidized soil at the center of which a reducing system with a characteristic $E_H$ begins to function. The solid phases present are assumed to be $MnO_2$ and $Fe(OH)_3$. If the $E_H$ is only slightly below the normal value for the soil under atmospheric conditions, the first reaction is a reduction of $MnO_2$ to give soluble $Mn^{2+}$. Now $Fe^{2+}$ and $Fe(OH)_3$ are automatically involved through the reaction

$$2Fe^{2+} + MnO_2 + 4H_2O \rightleftarrows Mn^{2+} + 2Fe(OH)_3 + 2H^+$$

From the standard free energies tabulated by Garrels and Christ (14), the value of the calculated equilibrium constant is given by log $K_r$ = 5.82, from which we can write:

$$\log [Mn^{2+}] + 2 \log [H^+] - 2 \log [Fe^{2+}] = 5.82$$

Then at pH 6 we have

$$\log [Mn^{2+}] - 2 \log [Fe^{2+}] = 17.82$$

Suppose the $[Mn^{2+}]$ builds up to $10^{-4}$. Then log $[Fe^{2+}]$ = $-10.9$. Thus the ferrous ion concentration will stay at an extremely low value until all the $MnO_2$ has been brought into solution. In actual fact the $[Mn^{2+}]$ does not build up very extensively because of the reaction with the exchange complex to produce exchangeable manganese, which is weakly dissociated.

In the absence of $MnO_2$, or when it has been used up, the $Fe^{2+}$ ion concentration rises and for a given $E_H$, the pH will correspond to a point on AB or a line parallel to it. This condition will continue until some change in $E_H$ occurs.

If $E_H$ falls, eventually a value corresponding to the point B will be attained, and here the new phase $Fe_3(OH)_8$ appears. Depending on the pH we may enter the region of $MnCO_3$ also. The phase $Mn_2O_3$ is possible at higher $E_H$ values corresponding to pH values of 6.5 and higher.

Thus by outward diffusion from a given reducing center, an expanding zone is produced, containing free $Mn^{2+}$ and $Fe^{2+}$ and eventually $Mn_2O_3$, $MnCO_3$, and $Fe_3(OH)_8$ as solid phases. When the reducing substances are exhausted and movement ceases, a gradual return of oxidizing conditions sets in at the outer zone and moves inward. The processes become reversed, and essentially the original condition of the manganese and iron is restored.

However if the original reducing environment contained no solid $Fe(OH)_3$ and $MnO_2$, iron and manganese could be mobilized by reduction from various mineral forms including exchangeable iron and manganese. Eventual oxidation would then change $Fe_3(OH)_8$ and $Mn_2O_3$ into $Fe(OH)_3$ and $MnO_2$. The bulk of the soil still contains no free $Fe(OH)_3$ and $MnO_2$, but there would be small quantities of these phases remaining at the original sites of reduction.

At this point, as attained by either route, consider conditions of widespread reduction in the soil mass: $Fe^{2+}$ and $Mn^{2+}$ are produced and diffuse into the zones containing free $Fe(OH)_3$ and $MnO_2$. They become immobilized as $Fe_3(OH)_8$ and $Mn_2O_3$ until all the $Fe(OH)_3$ and $MnO_2$ have reacted. When conditions of oxidation set in, the total iron and manganese in the zone under consideration are transformed into $Fe(OH)_3$ and $MnO_2$. By repetition of the cyclic process, more and more iron and manganese accumulate in the given zone until it becomes so densely precipitated in the soil fabric that diffusion processes grow less and less effective. Deposition finally occurs on the outer surfaces only. In this way a concretion may form and slowly grow, cycle by cycle.

Figure 18 enables us to draw conclusions regarding the composition of concretionary material. If reducing conditions are mild, corresponding to $E_H$ values above $+0.2$ V, the main deposition will be that of manganese. Highly vigorous conditions, from zero downward, will bring both elements into solution and eventual deposition. In general, of course, $E_H$ values vary with time; therefore the outer parts of concretions particularly may show zonation with respect to the relative amounts of manganese and iron. It will be interesting to put this idea to the test of experiment, especially through extension to other elements such as titanium, copper, cobalt, nickel, and chromium.

The subject of concretion formation may also involve organic matter through possible complexing and precipitating action. In the next section the general topic of organic complexing is taken up. Participation in concretion formation is suggested by the analytical evidence that some concretionary material (but not all) contains a concentration of organic matter as well as of iron, manganese, and so on. Jenne (22) has reviewed the evidence available, which is incomplete and in some cases conflicting. He comes to the general conclusion that the ultimately controlling factors are the solid phase relationships involving metallic oxides and hydroxides as described earlier. He was the first to point out the importance of the relative situation as regards iron and manganese in their $E_H$–pH relationships in soils.

In summary, concretions begin on sites at which $Fe(OH)_3$ or $MnO_2$ is present, either having been present originally or produced through reoxidation of a limited zone of reduction. They then build up gradually by the operation of cycles of reduction and oxidation. In this way diffusion and mass movement of the soil solution bring iron and manganese from a large volume of soil into the highly concentrated zone of the concretion. Mild reducing conditions produce manganese concretions; more vigorous conditions produce iron-manganese concretions. Since conditions vary, in practice zones of differing composition within a given concretion are to be expected.

## THE COMPLEXING OF METALLIC ELEMENTS

The movement of such elements as iron, aluminum, and manganese in soils has been attributed, in part, to the formation of soluble complexes with suitable organic molecules. In 1933 Jones and Willcox formulated podzolic processes in this way (24). More recently the quantitative expression of reaction equilibria for well-defined complexes has become extensive (Sillen and Martell). Soil organic matter has finally been drawn into this orbit. The first application of the modern treatment of complex formation was provided by Coleman, McLung, and Moore (10), who determined formation constants for $Cu^{2+}$–peat complexes. Through improved characterization of the water-soluble fraction, namely, fulvic acid, Schnitzer and Skinner (43,44) have been able to determine stability constants for fulvic acid complexes of $Cu^{2+}$, $Fe^{2+}$, $Zn^{2+}$, $Pb^{2+}$, $Ni^{2+}$, $Mn^{2+}$, $Co^{2+}$, $Ca^{2+}$, and $Mg^{2+}$.

In the formulation of equilibria, attention can fruitfully be focused on the acidic and exchange functions of soil organic matter as represented in the general equation

$$Me^{n+} + OM \rightleftarrows Me^{n+}\,OM + nH^+$$

where $Me^{n+}$ stands for a metal of valence $n$ and OM is acidic organic matter. The overall equilibrium constant $K$ is then given by:

$$\frac{[Me^{n+}\,OM]\,[H^+]^n}{[Me^{n+}]\,[OM]} = K_r$$

The value of $K_r$ can be derived from the individual equations for the complexing of the hydrogen ion and that of the metallic cation. This connection was pointed out by Coleman et al. (10). In this way it can ultimately be linked with the free energies of formation for the hydrogen and metal complexes.

The effect of an increase in pH (reduction in [H$^+$]) can readily be seen, since [OM] can be regarded as relatively constant. Thus the ratio Me$^{n+}$OM/Me$^{n+}$ will increase; more metal will be complexed. If in a soil profile complexing occurs in an acidic environment, the passage to lower horizons that are less acidic will not lead to chemical decomposition of the complex. The latter will pass through unchanged except as it may undergo biological decomposition, react with preexisting solid phases, or be adsorbed as a molecular unit. The latter process is quite probable, since Schnitzer and Kodama (42) have already shown that fulvic acid is held between the lattice layers of montmorillonites. Thus from the theoretical side, part of the groundwork of a chemical theory of Podzol formation is already apparent. Experimental pedology in the form of leaching experiments with complexing agents has already provided considerable evidence, beginning with that of Jones and Willcox (24). They equilibrated successive horizons of Podzols with dilute solutions of oxalic and tartaric acids, the clear extract from one horizon being used for the next down the profile. The concentration of sesquioxides in the solution phase reached its maximum in the uppermost B horizon, then fell sharply in the lower parts. Here active deposition occurred and was explained by the formation of sparingly soluble basic salts with compositions such as FeR$_3 \cdot$ Fe$_2$O$_3$, where R represents the organic unit. Precipitates of this kind deposit spontaneously in solutions of ferric tartrate and ferric oxalate, especially upon exposure to light.

The formation and decomposition of soluble complexes takes on a somewhat different aspect through this work, since the formation of basic salts would clearly be promoted by the presence of solid phase hydroxides in the B horizon.

Kawaguchi and Matsuo (26) emphasized chelation as the governing factor in the movement and deposition of iron. This was illustrated using oxalic acid and soil in varying proportions. Determinations of iron in solution indicated the formation of a soluble complex of formula Fe(C$_2$O$_4$)$_3$, which was rendered less soluble in presence of excess ferric hydroxide. This accounts for the finding of these investigators that the amount chelated was dependent on the ratio between the amount of soluble complexing agent and the solid substrate. Using these concepts, the authors postulated Podzol formation as follows.

1. Beginning with a uniform soil, passage of a chelating agent through the upper part would give a certain concentration of the iron chelate,

which at first would pass through the lower layers and be lost in the drainage. This process would deplete the surface layer in active or easily complexed iron. Then iron mobilized in the upper layer, with the minimum amount of chelating agent, would be immobilized upon contact with the lower layer.

2. Additional iron would be deposited at the upper boundary of the lower layer, which would thus tend to grow towards the surface.

3. This tendency would be increasingly offset by the increasing ratio of chelating agent to iron in the upper layer. The excess chelating agent would dissolve iron from the uppermost part of the developing B horizon and deposit it lower down. This makes for a very sharp boundary between the A and B horizons. The change from B to C would be more gradual, in accordance with field description of Podzols and podzolic soils.

The influence of solid phase hydroxides on the decomposition of iron and aluminum chelates has been discussed more quantitatively by Van Schuylenborgh (45). The reaction can be formulated as a hydrolysis, whose constant $K_h$ is equal to the product $K_{mz}^{-1} \cdot k_{SO}^{-1}$, where $K_{mz}$ is the stability constant of the complex and $K_{SO}$ is the solubility product of the hydroxide. Less acidic conditions favor the hydrolysis of the complex and precipitation of the hydroxide. For a given hydroxide and pH, the actual value of $K_{mz}$ determines whether precipitation will occur.

When the nature of all the solid phases becomes known, it should be possible to use standard free energies in reviewing the processes which dominate the B horizon.

The third area of evidence lies in the composition of roots, leaves, and leaf litter and its effect on soil development, as judged by field observations. Early analysis correlated the low base content of pine needle ash with the formation of highly acidic humus; and the high base content of the ash of leaves of deciduous trees (particularly beech), appears to be connected with the production of more nearly neutral humus. Thus Podzols form under pine forests more readily than under deciduous forests, the latter giving rise to Brown forest soils. Joffe's book (23) gives a comprehensive survey of this kind of evidence, mainly from European sources.

A most remarkable and extreme example of the influence of the leaf canopy is found in New Zealand. Here well-developed Podzols are found under individual trees of the Kauri pine, but the intervening spaces are characterized by soils with no bleached horizon. Not only is the humus formed under the pines extremely acidic, but the drippings from the trees under rainfall were found to contain free organic acids. These easily form complexes with iron, aluminum, and other elements. There is evidence also from other sources that fresh leaves provide much more plentiful sources of free organic acids than humified leaves.

The influence of fermentation in the production of artificial profiles has been very strikingly shown by Bétrémieux and Hénin (4) under both aerobic and anaerobic conditions. Table 3 gives the relative contents of iron and manganese of different layers compared with those of the original soil. Glucose solution was used as the energy source for fermentation. Under normal conditions of drainage, iron and manganese lost from the two upper horizons accumulated chiefly in the 10–14 cm layers; but with a high water table the loss from surface horizons was not balanced by gains in the waterlogged horizons. Thus the formation of a ferruginous B horizon seems to demand oxidizing conditions; reducing conditions maintain iron and manganese in a mobile state.

Bloomfield (5) considers that certain constituents of undecomposed plant materials are the most effective mobilizers of iron in soil profiles. He finds that the activity of tree leaf extracts is correlated with their content of polyphenols, and that the latter are very effective mobilizers under sterile conditions; this in sharp contrast to Bétrémieux and Hénin, who emphasize biological fermentation reactions under more or less anaerobic conditions.

With so much diversity of opinion about the mechanisms for solution and redeposition of iron in Podzols, two urgent needs present themselves. (1) Adequate definition of the chemical environment in successive horizons of natural Podzols: this will give a general overview of the situation. (2) Quantitative measures of the actual contributions of different competing mechanisms under natural and artificial conditions. Here very difficult problems face us. The use of radioactive tracers seems to be called for, but little has so far been done. Hallsworth's review (17) of modern work on Podzols even throws doubt on the essentiality of the bleached A horizon, which is a far cry from earlier concepts of podzolization.

**TABLE 3** Fermentation of Glucose and Content of Iron and Manganese in Two Experimental Profiles Maintained in the Field (4) (Fe and Mn, original soil = 100)

| Soil Condition | Depth (cm) | Fe | Mn |
|---|---|---|---|
| Well drained, in cylinder | 0–3 | 72 | 19 |
| Well drained, in cylinder | 7–10 | 87 | 45 |
| Well drained, below cylinder | 10–12 | 141 | 154 |
| Well drained, below cylinder | 12–14 | 108 | 201 |
| Well drained, below cylinder | 14–20 | 100 | 111 |
| High water table, in cylinder | 0–13 | 72 | 13 |
| High water table, in cylinder | 7–10 | 86 | 27 |
| High water table, below cylinder | 10–14 | 95 | 80 |
| High water table, below cylinder | 14–18 | 102 | 65 |

## COLLOID–CHEMICAL CONSIDERATIONS

The movement of soil constituents in actual profiles has long been attributed in part to the colloid-chemical processes of peptization and coagulation, the two being separated from each other in time and place. In general, a constituent such as soil humic matter was regarded as being peptized in one horizon and coagulated in another. Thus translocation directed by the mass movement of water was assumed to occur. The diffusion of colloidal particles is assumed to be too slow to be of significance.

*Peptization*

Two kinds of peptization have been widely used in pedology in the explanation of processes of soil development. The first operates through the protective action of one colloidal constituent on another. In this process the protected colloidal particles acquire through adsorption coatings that confer on them the colloidal characteristics of the coating material. Under the right circumstances, humus and other organic substances can act as a protective agent for clay particles. The protected clay can then move and subsequently can be coagulated (*a*) by removal of the coating, (*b*) by rendering it sensitive to coagulation on its own account (e.g., by changing the nature of the exchange cation or raising concentration of the external solution), or (*c*) through chemical reactions that radically change the surface properties of the protective layer. Clay can also be peptized by raising its zeta potential (see Vol. I, Chap. 9), either by changing the nature of the exchange ions in the electrical double layer or by reducing the electrolyte content in the solution phase.

Problems of this kind obviously involve all the colloidal and near-colloidal constituents of soils in their mutual interactions. The formulation of pedological processes entails the identification of the dominant processes and the colloidal species associated with them. Processes classified as colloid-chemical operate concurrently with mineral transformations through the solution phase, with complexing reactions, and with multitudinous biological syntheses and degradations. Early work on colloid-chemical aspects of pedology concentrated attention on humus, clay, colloidal hydroxides and oxides, silicic acid, and the soil solution. Different pedological results were explained by different mechanisms. In the 1930's more comprehensive relationships were tried. The general reactions between solid phases as exchangers and the solution phase were expressed by Mattson (32,33) in terms of Donnan equilibria. Reactions between colloidal constituents were formulated through changes in electrokinetic properties; this led Mattson to a group of generalizations known as the theory of isoelectric weathering.

The various colloid-chemical phenomena that have been used to fashion theories of particular pedological processes are briefly discussed below.

### Colloid-Chemical Aspects of Podzolization

During the early years in the rise of colloid chemistry, fundamental advances were rapidly followed by applications to processes presumed to occur in nature. This was plainly seen in the idea of humus as a protective colloid, assisting the removal of constituents from the A horizons of Podzols, which were later deposited in the B horizon where conditions were different. If humus resembled the proteins and colloidal carbohydrates as a protective colloid, such a process would be relatively efficient; that is, a very small quantity of humus would peptize a much larger quantity of mineral matter. This seemed to agree with analytical determinations on Podzols. Most of the humus remained in the $A_0$ or $A_1$ horizons, but some was found in the B horizon, where it might even form a distinguishing layer in the upper part. The mineral matter translocated and deposited was characterized by a lower $SiO_2/Al_2O_3 + Fe_2O_3$ ratio than the clay fraction of the assumed parent material (C horizon) and considerably lower than the residual clay in the A horizon. Clearly the analytical results demanded a selective process, rather than a general peptization of colloidal mineral matter by humus and its subsequent coagulation and deposition.

A colloid-chemical basis for selective processes that operate through mineral constituents, humus, and the soil solution was provided by Mattson (32,33) in extensive studies of mutual coagulation, peptization, and protective action, carried out on sols of colloidal silica, aluminum hydroxide, ferric hydroxide, and humus. The general principle was that described in Volume I, Chapter 9—namely, that systems with a high zeta potential were in suitable condition for peptization and transport, whereas the isoelectric condition was one of coagulation and deposition. It was assumed that ferric and aluminum hydroxides in soils were similar to the precipitates obtained from the reactions of soluble salts of these metals with bases. The compositions of isoelectric mixtures were found to be a function of pH and dependent also on the nature and concentrations of cations and anions in the soil solution. Since the ferric and aluminum hydroxides used were positively charged and differed appreciably in their pH–zeta potential relationships, it was possible to account for differential effects, such as those found in soil profiles. Each characteristic soil pH was compatible with a certain isoelectric composition. Hence peptization and deposition could be considered to be a consequence of pH and salt variation in the soil solution, thus of differing conditions in soil horizons.

Several criticisms of the broad application of these findings to processes of soil formation and development have been brought forward.

1. Regarding the isoelectric condition, the examination of oxide and hydroxide minerals in soils shows that they have different properties from the freshly precipitated hydroxides and carry negative charges over a much broader range than the latter (see Table 30, Vol. I, p. 331).

2. The mineral colloids of most soils are dominantly in the crystalline form and have properties different from those of amorphous mixtures of the same chemical composition.

3. Podzols, which Mattson used as the prime example of the operation of the theory of isoelectric weathering, can also be explained through chelation of iron and aluminum in the A horizons with decomposition of the chelates and deposition in the B horizon. Mattson considered that chelation would lead to passage of iron and aluminum completely through podzolic profiles.

4. Both organic matter and mineral matter can move downward in soils without causing the compositional segregation characteristic of Podzols. Prairie soils and Chernozems are examples. This argument is not conclusive, since the acidic function of humus is a crucial factor in Mattson's analysis, and neutral or cation-saturated humus would not qualify in the same way.

5. Synthesis experiments as well as identifications of the clay minerals in actual soils and weathered rocks all point to the fact that kaolinite is usually formed under more acidic conditions than montmorillonite. But montmorillonite has a higher $SiO_2/AlO_3$ ratio than kaolinite, corresponding, in Mattson's terms, to a more acidic environment and composition.

6. Mattson's treatment of the properties of humus-hydroxide mixtures emphasized the existence of a mutual interaction, manifesting itself over a very broad range in modification of the electrokinetic properties. This is different from the older treatment of protective action of a hydrophile colloid (negative humus) on a positively charged inert particle, as originally demonstrated by Aarnio (1) and more recently by Deb (11). Once a complete coating of humus has been achieved, the electrokinetic properties would be exclusively those of the coating (Vol. I, Chaps. 4, 9). Relatively small amounts of protective colloid are needed for such surface coverage. In contrast, Mattson's complexes require high ratios of humus to mineral matter before the colloid-chemical properties of humus are attained.

In spite of these criticisms and uncertainties, it must be admitted that Mattson's theory of Podzol formation based on isoelectric weathering is capable of explaining qualitatively a large part of the compositional data obtained. How can this be so, when the properties of the amorphous colloidal bodies considered by Mattson are so different from those of the clay minerals identified in soils? In a general way, Mattson's ideas at any given instant in any given soil horizon could be said to apply to that small amorphous proportion of the clay colloids which is freshly produced and in process of change. This implies that podzolization is a more rapid process than recrystallization under the influence of a changed environment characterized by increased acidity and the presence of humus.

As mentioned earlier, the crucial considerations in Mattson's theory are the variations in isoelectric points caused by (*a*) variations in composition and (*b*) variations in pH. A generalized description of the way in which Mattson believes these factors to be responsible for podzolization is given below. Analytical data for particular Podzol profiles are considered in Chapter 9.

The accumulation of organic matter at the soil surface under low temperature and fairly high rainfall implies loss of exchangeable bases and acidic conditions. In the $A_1$ and $A_2$ horizons, below the organic layer, mineral silicates decompose to give $Al(OH)_3$, $Fe(OH)_3$, and silicic acid, which would give a mixed gel having a certain isoelectric point. In presence of acidic humus the negative charge of this mixed gel is increased and its isoelectric point drops to a lower pH value. It can thus peptize and move downward in the profile until it attains a layer characterized by a pH approximating its isoelectric point. Here the mixed gel precipitates, forming the B horizon. Since, however, the organic matter continually undergoes change and decomposition, part of the mineral complex is released and will move to a less acidic environment, which again approximates to its isoelectric point. Here the mineral complex will be precipitated. Thus the B horizon will be characterized by an upper part enriched both in humus and sesquioxides, and a lower part with sesquioxides dominant. The only part of the profile corresponding to the isoelectric point of a highly silicic mineral complex is the $A_1$ or bleached horizon.

This is a very neat colloid-chemical theory, of such apparent generality that its application to nonpodzolic soils should be examined. Mattson saw that it could be applied to laterization in the following way. In absence of acidic organic matter, the weathering of silicate minerals would provide an alkaline medium in which the isoelectric composition would be low in silica, and high in iron and alumina. The possibility of the presence of clays with genuinely acidic properties of their own was not considered. Presumably soils dominated by the kaolin composition would arise under conditions less alkaline than those which produced aluminum and iron hydroxides. In any case the general idea of a movement downward of silicic acid until the isoelectric composition for the appropriate horizon is reached would be applied.

Criticisms of Mattson's ideas were expressed by Kelley (28) on many grounds, ranging from the use of somewhat inconsistent definitions to neglect of the actual properties of colloidal secondary minerals. Mohr and Van Baren (34) have brought forward specific examples of tropical soils with properties quite different from those that would be predicted by the theory of isoelectric weathering. In general, it is difficult to find a clear application of Mattson's ideas to those soils that are dominated by specific clay minerals.

The modern geochemical approach to the interaction of minerals with their aqueous environments indicates that in the long run crystalline products are to be expected, but unstable combinations may have considerable persistence. Modern methods of investigation should give a clear answer regarding the dominant transformations of soils low in organic matter, horizon by horizon. Podzols and other soils high in organic matter will require more sophisticated definitions of the aqueous environment, to include the function of organic constituents. Thus they pose difficult problems.

### Humus in Nonpodzolic Soils

The protective action of humus and its fractions on mineral colloids was early investigated by Aarnio (1) in connection with Podzols, by Odén (37), and more recently by Deb (11). The importance of the cation associated with the humus was particularly emphasized by Gedroiz (15). He applied the great difference in peptizability and solution processes (see Vol. I, Chap. 5) between sodium-saturated humus on the one hand and calcium-saturated humus on the other, to propose a mechanism by which the properties of the black-alkali or Solonetz soils are manifested. The mineral constituents of the soil are strongly electronegative and so also is the humus. Hence the soluble and highly peptized sodium humus moves downward in the profile; but owing to limited leaching under the prevailing dry climate, to the presence of salts, and to the partial irreversibility of drying processes, it does not penetrate far and is believed to be associated with the characteristic columnar structure in the B horizon. The peptization and coagulation properties of magnesium humates lie between those of sodium and calcium humates, and there is considerable evidence of the formation of magnesium Solonetz soils where this cation dominates.

However, soils dominated by exchangeable calcium do show movement of humus to considerable depths as in the Chernozems and Prairie

38. Ponnamperuma, F. N., E. Martinez, and T. Loy, The influence of redox potential and partial pressure of carbon dioxide on the pH values and the suspension effect of flooded soils, *Soil Sci.*, **101,** 421 (1966).
39. Ponnamperuma, F. N., E. M. Tianes, and T. Loy, Redox equilibria in flooded soils. I. The iron hydroxide systems, *Soil Sci.*, **103,** 374 (1967).
40. Ramdas, L. A. and R. K. Dravid, Soil temperatures in relation to other factors controlling the disposal of solar radiation at the earth's surface, *Proc. Nat. Inst. Sci., India*, **2,** 131 (1936).
41. Richards, L. A., Water conducting and retaining properties of soils in relation to irrigation, *Int. Symp. Desert Res. Bull. Res. Council*, Israel (1952).
42. Schnitzer, M. and A. Kodama, Reactions between fulvic acid, a soil humic acid compound, and montmorillonite, *Is. J. Chem.*, **7,** 141 (1969).
43. Schnitzer, M. and S. I. M. Skinner, Stability constants for $Cu^{++}$, $Fe^{++}$ and $Zn^{++}$–fulvic acid complex, *Soil Sci.*, **102,** 361 (1966).
44. Schnitzer, M. and S. I. M. Skinner, Stability constants of Pb, Ni, Mn, Co, Ca, and Mg–fulvic acid complex, *Soil Sci.*, **103,** 247 (1; 67).
45. Schuylenborgh, J. Van, The formation of sesquioxides in soils. In *Experimental Pedology*, E. G. Hallsworth and D. V. Crawford, Eds., Butterworth, London (1965), p. 113.
46. Shainberg, I., Electrochemical properties of $Na^+$ and $Ca^{++}$ montmorillonite suspensions, *Trans. 9th Int. Congr. Soil Sci.*, **I,** 577 (1968).
47. Smith, A., Seasonal subsoil temperature variations, *J. Agr. Res.*, **44,** 421 (1932).
48. Smith, R. M., D. R. Browning, and G. G. Pohlman, Laboratory percolation through undisturbed soil samples in relation to pore size distribution, *Soil Sci.*, **57,** 197 (1944).
49. Smithson, F., Grass opal in British soils, *J. Soil Sci.*, **9,** 148 (1958).
50. Tyurin, I. V., On the biological accumulation of silica in soils, *Sov. Soil Sci.*, **4,** 3 (1937).
51. Yakuwa, R., Über die Bodentemperaturen in den verschiedenen Bodenarten in Hokkaida, *Geophys. Mag. (Tokyo)*, **14,** 1 (1945).

# 5 Quantitative Aspects of Soil Profiles

Soil profiles are open systems; that is, materials can be added or lost. Water is the most important and pervasive of such materials. It can move in and out of the soil both as vapor and as liquid. Carbon dioxide, oxygen, and nitrogen move in the vapor phase and also as dissolved constituents in the water. Many other materials also participate. Oceanic salts are added with the rainwater, together with small quantities of nitrates arising from electrical discharges in the atmosphere. Finely divided solids originating as atmospheric or stratospheric dust settle out under gravity or are carried down in rain. Organic matter is added both as decaying vegetation at the surface and as roots within the soil mass. It can also be lost in the drainage, but usually in small amount. Many soluble constituents move as salts in the soil solution, and from time to time are lost by leaching. Their relationships to the solid phases present are complex, being governed in some cases by ordinary solubility laws, in others by Donnan-type relationships between colloidal materials or surfaces and external solutions, and in still others by depolymerization reactions as applied to silicate framework structures. Two general methods are combined with quantitative analyses to provide the primary data for evaluation: lysimeter studies and resistant mineral studies.

**LYSIMETER STUDIES**

The older method, that of the lysimeter, employs in its simplest form a tank filled with soil, having provision for the collection of drainage water. Several questions are usually answered through lysimeter data. First the quantitative connection between rainfall and drainage can be expressed.

This is exceedingly important. Considerable detail should be sought, and the simple percentage figure is only a beginning. The frequency of drainage in relation to the frequency of rainstorms of different magnitudes provides a first picture of pedological conditions within the lysimeter. Unfortunately, lysimeters have seldom been set up with pedological objectives. They are either viewed as an important adjunct to hydrological investigations, or as providing information of an agricultural-chemical nature, such as the rate of loss of plant nutrients. By making the chemical analysis of the drainage water more complete, however, valuable pedological information can be obtained from most lysimeter installations. Losses can easily be calculated in terms of total elements present or of the most reactive portions, such as exchangeable cations.

The most elaborate installations are those in which an isolated block of soil is suspended on a balance with continuous recording. It is then possible to investigate processes of evapotranspiration through comparisons of cropped and uncropped lysimeters. Crop cover greatly affects the amount and frequency of drainage, in general, considerably reducing them both. In some installations there is no balance; but percolate is collected, measured, and later analyzed.

A second type used mainly by Russian investigators (18) and by Joffe (25) consists of natural soil adjacent to a trench, from which devices for the collection of percolate at various depths can be introduced. This type is especially valuable in comparing chemical compositions of drainage waters at various depths.

The third, or filled type, consists of a watertight box or tank with sand or gravel at the bottom, into which successive layers of soil are introduced. The initial disturbance of the soil profile may seriously affect the actual moisture regime. However it is the only type that allows of soil treatments being incorporated into different layers. For this reason it was chosen by the author and W. J. Upchurch (49) as a means of investigating the role of the exchange complex in relation to the amount and composition of drainage.

*Missouri Results*

The results obtained for the control, for sodium treatment in particular layers, and for calcium treatment in particular layers, are here considered as an example in which the soil solution is viewed in relation to colloid-chemical and mineralogical changes (under the conditions in central Missouri). The details are found in Refs. 34 and 49.

*Soil Characterization*

The soil used (the Mexico silt loam, earlier described as a rolling phase of the Putnam soil, and now classified according to the United States system (7th approximation) as a member of the fine montmorillonitic, mesic family of Aeric Udollic Albaqualfs) is dominated in its properties by a beidellitic clay. The latter contains some potassium, from an illitic component or interlaying of less than 20%. The fine clay ($< 200\ \mu$) contains less mica than the coarse clay. It also contains less kaolinite and no detectable quartz or feldspar; these are present in small amount in the coarse clay. The general properties can be appreciated from Figure 19, which plots total cation exchange capacity against depth. The changes are chiefly a function of the clay content, since organic matter is low, 3.0% near the surface and diminishing rapidly with depth. In the closely related Putnam soil, the clay of the B horizon was earlier shown to contain a greater proportion of the very fine clay than the A or C horizons (49).

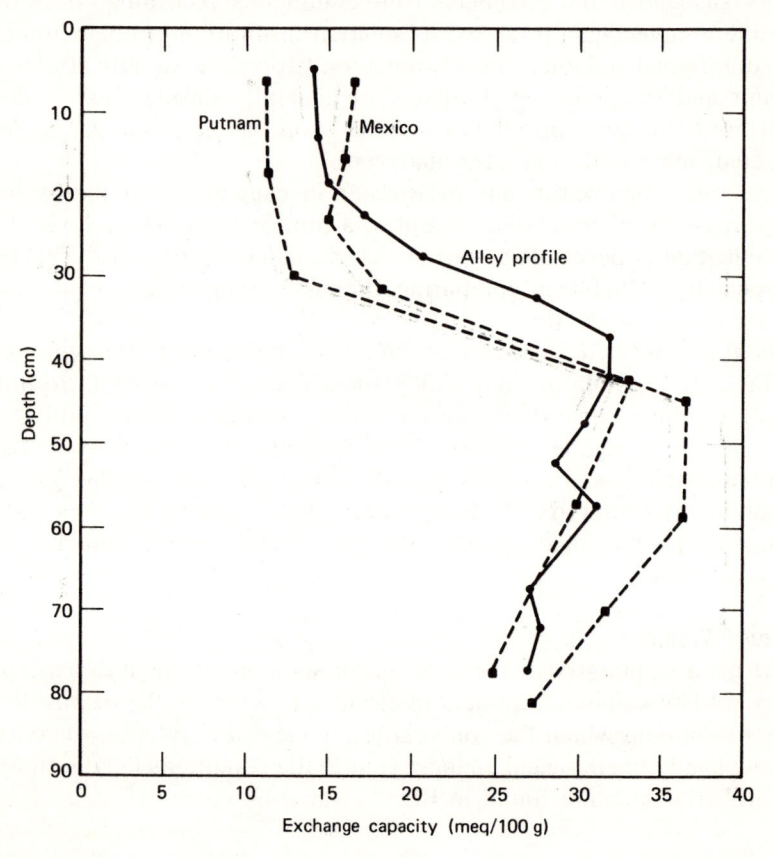

*Fig. 19* Exchange characteristics of undisturbed soil ("alley") between lysimeter plots. The profile is slightly shallower than typical Putnam or Mexico soils in the area.

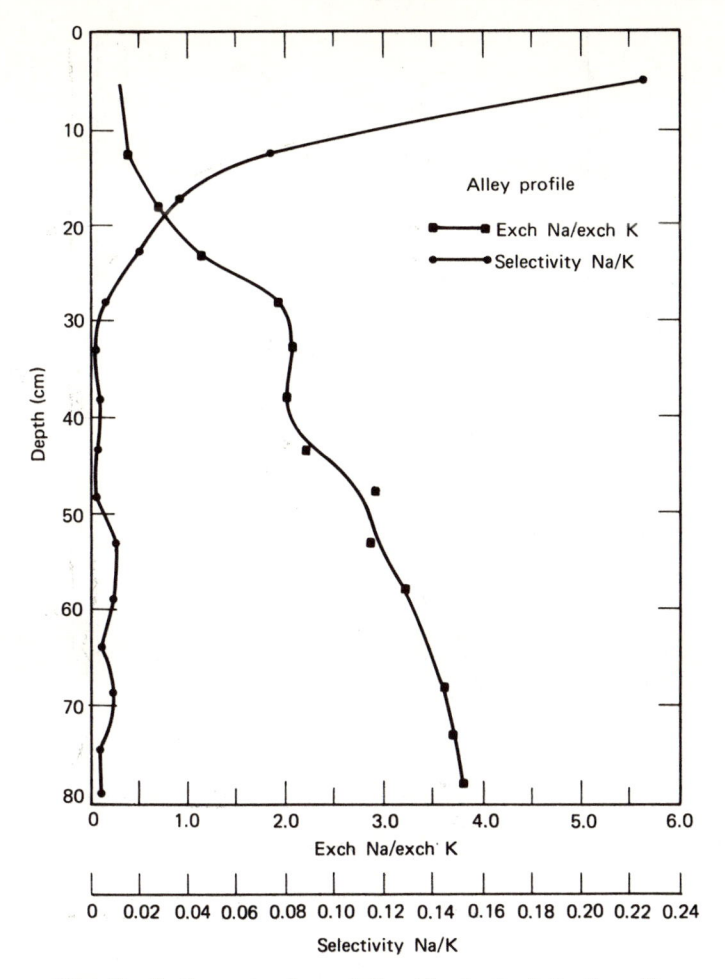

*Fig. 20* Sodium-potassium relationships in the "alley" profile.

Both soils are loessial, and much of the total clay was inherited from the parent loess. There has been clay illuviation and deposition through pedogenic processes.

The exchangeable cations vary absolutely in amount with depth and also show significant variations in their relative proportions. The ratios of exchangeable sodium to exchangeable potassium and exchangeable calcium to exchangeable magnesium are given in Figures 20 and 21, also selectivity numbers for Na/K and Ca/Mg. The Na/K ratios in solution increase strongly with depth and are very much higher in the equilibrated

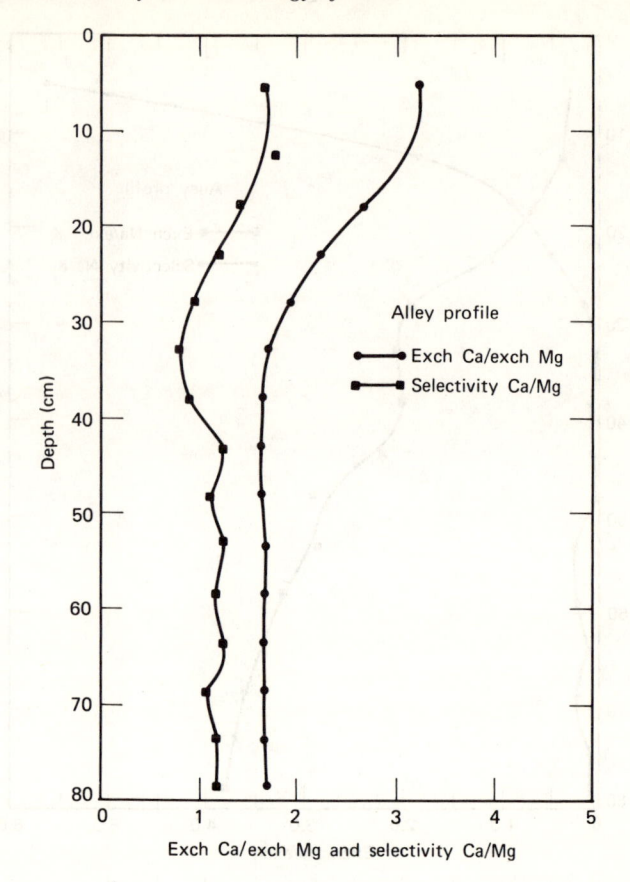

*Fig. 21* Calcium-magnesium relationships in the "alley" profile.

aqueous solutions than for the exchangeable quantities. A very strong potassium fixation tendency is apparent throughout, but especially from 25 cm downward. The changes in the Na/K ratio are expressed as changes in selectivity numbers by dividing the value for the solid phase by that for the solution. Thus in the surface 10 cm the selectivity Na/K is 0.169, or K/Na is 5.92 in favor of potassium as compared with sodium. In the 60–80 cm layers the selectivity K/Na averages 221. Thus if the cationic bonding energy of sodium were treated as relatively unvarying, that of potassium would increase by 1364 log 221/5.92 or 2145 cal/mole between the surface and the lowest layers.

The Ca/Mg ratios decrease somewhat with depth and are a little lower for the solution phase than for the solid. The variations in the solution phase are very slight compared with those for the K/Na ratio. The selectivity numbers for calcium as compared with magnesium are 1.68 at the surface and 1.20 average in the 24–32 in. layer. From about 10 to 16 in. there is a definite zone in which the selectivity Ca/Mg < 1.0 (i.e., magnesium is held more tightly than calcium).

Four factors may have contributed to the changes in K/Na selectivity with depth. (1) There may have been differences in the parent loess. (2) Under the prevailing prairie conditions, roots would remove much more potassium than sodium; the tops would receive this and eventually return it to the upper layers after decomposition. (3) The movement of fine clay from the A to the B horizon would carry down more loosely held sodium than loosely held potassium. (4) Through Donnan hydrolysis the soil solution carries down from the surface more sodium than potassium, because of the lower bonding energy of the former. Factors 2 through 4 are clearly pedogenic, but at present it is not possible to assess them individually.

## The Moisture Regime

Precipitation during the period 1963–1969 is summarized in Table 4. It is distributed throughout the year, with summer rainfall exceeding that in winter. This is more than compensated for by the increased temperature and consequently increased evapotranspiration. The percentage drainage

**TABLE 4** Annual Precipitation (cm) 1963–1969: Lysimeter Area, Columbia, Missouri

|         | January | February | March | April | May  | June |
|---------|---------|----------|-------|-------|------|------|
| Average | 3.9     | 3.1      | 5.8   | 10.3  | 10.6 | 13.1 |
| Lowest  | 0.5     | 0.7      | 1.7   | 4.8   | 5.5  | 4.7  |
| Highest | 9.1     | 5.7      | 9.0   | 15.4  | 15.1 | 25.8 |

|         | July | August | September | October | November | December |
|---------|------|--------|-----------|---------|----------|----------|
| Average | 11.6 | 5.7    | 10.5      | 8.8     | 4.7      | 4.6      |
| Lowest  | 4.3  | 2.5    | 3.3       | 0.8     | 1.3      | 1.2      |
| Highest | 17.8 | 14.2   | 21.3      | 16.6    | 12.9     | 6.4      |

|         | Year  |
|---------|-------|
| Average | 92.1  |
| Lowest  | 61.7  |
| Highest | 140.4 |

measured in the lysimeters varied from 2.70 to 15.5 for the fallow series and from 0 to 4.5 for the series in grass. Thus the number of occasions during the year when drainage actually occurs is very small. Complete throughput of water is to be regarded as a rare event. Most of the time the soil has a moisture content between the wilting point and field capacity, with drying out during the summer, and freezing, which is also a drying process, during the winter. Summer droughts leave cracks extending from the surface to as much as 2 ft into the subsoil. Under continuous frost in winter the soil becomes frozen to a variable depth, normally between 6 and 18 in. After the thaw its structure is temporarily much improved, but it deteriorates considerably during the spring and summer unless stabilized by additions of organic matter.

Under these conditions the soil water generally has ample time to come to equilibrium with the solid phases. Thus ideally the lysimeter drainage should reflect equilibrium conditions in the lowest layer of the soil, assuming that the sand and gravel in which it is collected are completely inert.

## Chemical Composition

Comparisons were made between the composition of water extracted from 1:1 soil-water mixtures under 15 atm pressure with that taken from the lysimeters. The results are presented in Figure 22, using the Garrels and Christ diagram. The compositions of the various samples of water from the control lysimeter are shown as areas enclosed by dotted lines, considering the respective elements, magnesium, sodium; and potassium. The calcium area (not shown) is almost identical to that of magnesium. The compositions of water extracted from the lowest soil samples in the alley and from the plot set up to correspond with the control lysimeter are shown as individual points for the same three elements. Consider first silica. The lysimeter waters average $1.5 \times 10^{-4}$ mole/liter corresponding to $-3.82$ on the horizontal axis of Figure 22. The alley samples (28–32 in.) gave $4.4 \times 10^{-4}$ mole/liter, and the check plot samples $3.0 \times 10^{-4}$ mole/liter. The lysimeter water consistently contains less silica than the extracts. On the vertical axis we have, respectively, log $[K^+]/[H^+]$, log $[Na^+]/[H^+]$, and log $\sqrt{[Mg^{2+}]}/[H^+]$. The lysimeter results are considerably higher than the extracts. Part of this represents a pH difference. The check plot and alley samples gave 7.2 and 7.0, respectively; the lysimeter waters averaged 8.2. However the differences along the vertical axis exceed 1.0 unit, the lysimeter waters being the more concentrated. Thus the lysimeter waters are more concentrated in cations but less concentrated in silica than the extracts. This precludes the possiblity that the one is simply a diluted version of the other. The very large difference in the case of potassium clearly indicates that it is distinctly different from

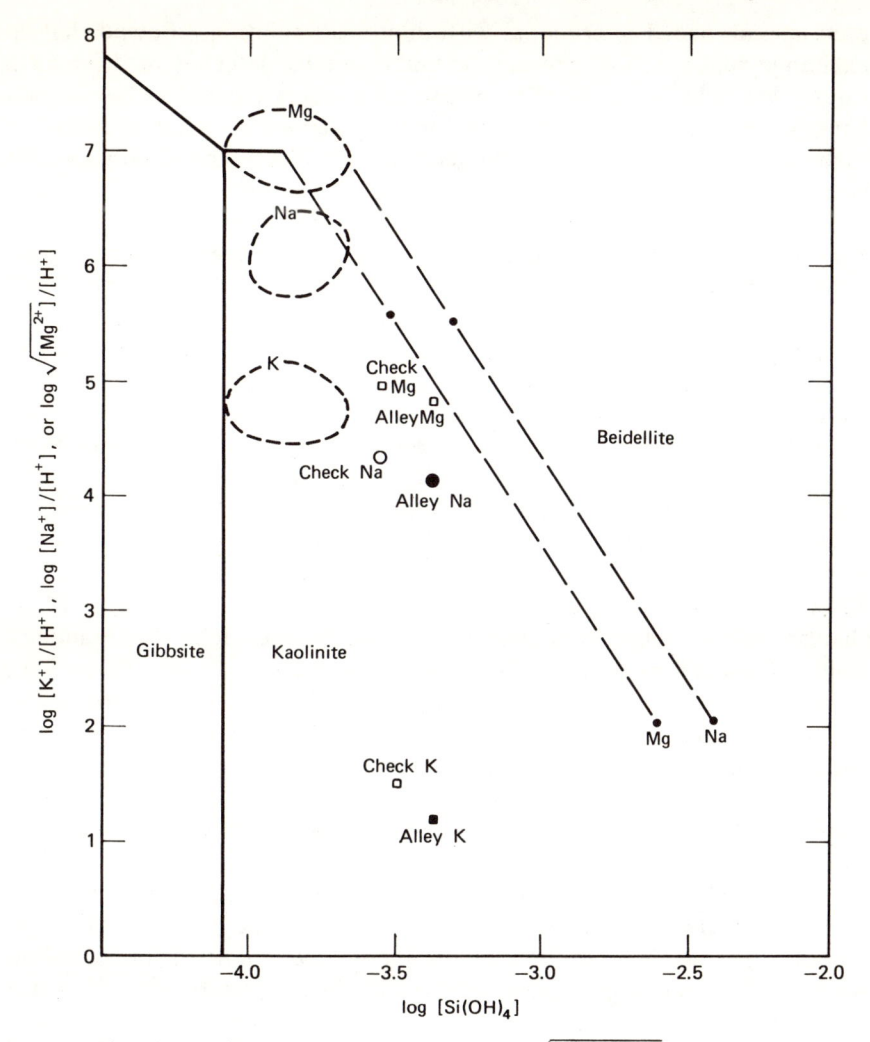

*Fig. 22* Plot of log [K$^+$]/[H$^+$], log [Na$^+$]/[H$^+$], and log $\sqrt{[Mg^{2+}]/[H^+]}$ against log [Si(OH)$_4$], for water samples from the control lysimeter (areas shown) as compared with 1 : 1 soil extracts from the 71–81 cm layers of the check lysimeter and of the "alley" profile. Reproduced from ref. 49 by permission of the Williams & Wilkins Co., Baltimore.

sodium. This in turn suggests that the lysimeter waters are affected by some specific mechanism that has an enhanced effect on potassium. It seems probable that the explanation lies in the exchange properties of the sand and gravel. Since the potassium in solution is so exceedingly low, it would be sensitively affected by even a small exchange function in the gravel. The latter was composed of chert fragments with a brown, weathered appearance. Experiments with dilute salt solutions showed that an exchange function was present. It could not be detected in the quartz sand. Not only does the chert show an exchange function but it may contain, in the unweathered core of the fragments, small amounts of carbonate. This could not be detected through effervescence with acid but is suggested by the higher pH of the lysimeter water as compared with the soil extracts. Thus in designing future experiments with lysimeters, it is important to ensure that the collection zone is completely inert.

Another factor that may operate to differentiate the two types of soil moisture samples is the $CO_2$ pressure, which would generally be different for the pressure extraction; leading to a lower pH and a higher cation content if higher, and, if lower, to the opposite situation, as found previously. Since our soil-water extracts are clearly more reliable than the lysimeter waters themselves, let us examine their chemical compositions in relation to the possible solid phases in the soil.

*Variation with Time.* An exceedingly important question in all silicate-water systems is the attainment of equilibrium. This was examined by making extractions after periods of contact from 18 hours to 36 days. Both 10–18 cm and 51–56 cm layers were used. It was found that changes in the composition from 12 to 36 days were slight. The tendency in all cases was for the log [cation$^+$]/[H] or log [$\sqrt{\text{cation}^{2+}}$]/[H$^+$] to rise with time as did also the log [Si (OH)$_4$] in solution. Thus on the Garrels and Christ diagram a point moves up and to the right. The upward movement indicates that something additional to a Donnan-type hydrolysis has come into operation, since this would give constancy of cation$^+$/H$^+$ or $\sqrt{\text{cation}^{2+}}$/H$^+$ with time. The general dominance of Donnan-type effects can, however, be seen in the ratio of cations to silica in solution; that for the 10–18 cm layer was 4.73 after 36 days and for the 51–56 cm layer was 4.23. These figures are minimal values and in one sense misleading, because the silica is present as a neutral molecule, not as an anion, and no account is taken of the carbonic acid–bicarbonate equilibrium. The latter probably accounts for most of the cations in solution, since the pH is close to 7. It is easy to show how the presence of carbonic acid increases the effects of Donnan hydrolysis by increasing the total cations in solution.

*Variation with pH.* All systems characterized by a simple cation-hydrogen interchange, including Donnan hydrolysis systems, should give straight line functions when for the solution phase p (cation) is plotted against pH, the slope being 1 for monovalent cations, 2 for divalent, and so on. As a rule, however, the weathering of single aluminosilicate minerals does not show such simple relationships. This was strikingly demonstrated for the micas by Marshall and McDowell (33). Soil systems, with their much greater mineralogical complexity, could hardly be expected to behave simply, but a study of the actual relationships might prove enlightening.

Analyses of soil-water extracts were made on soil samples treated with limited amounts of hydrochloric acid or calcium hydroxide. In such experiments the ionic strength is an additional variable. Figures 23, 24, and 25 give the 36 day equilibria for sodium and potassium, calcium and magnesium, and aluminum and silicon, respectively, using soil samples 10–18 and 51–56 cm.

The results indicate complex curves in all cases. Sodium, calcium, and magnesium give sharp maxima for the untreated samples with steep descents on either side corresponding to the acid or alkali treatment. On the acid side this would be expected because of the increased chloride

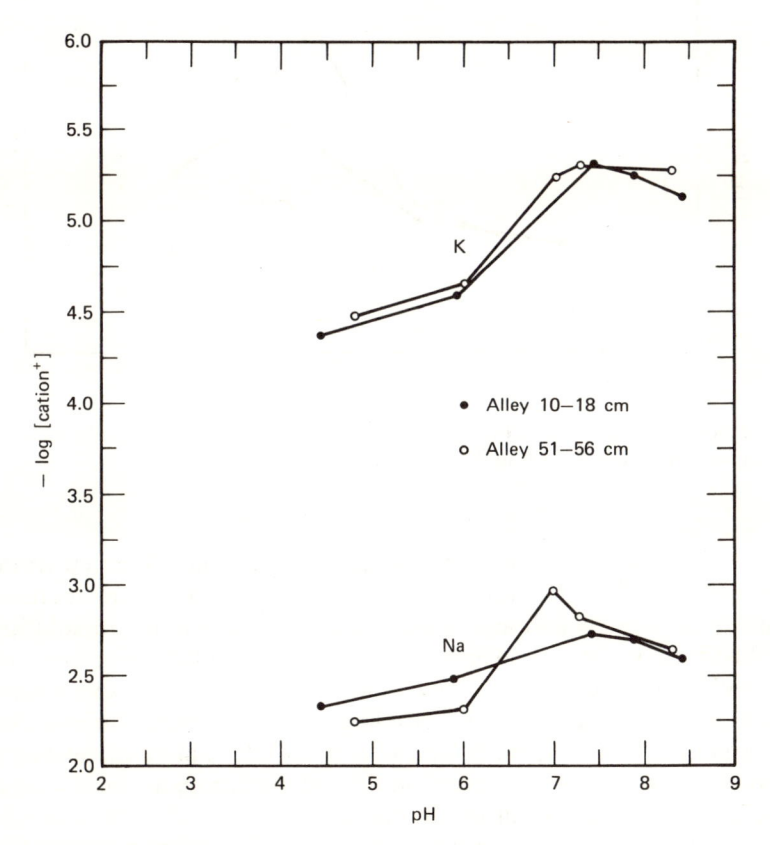

*Fig. 23* pK–pH and pNa-pH relationships for two depths of the "alley" profile: equilibration with HCl, water, or Ca(OH)$_2$ for 36 days.

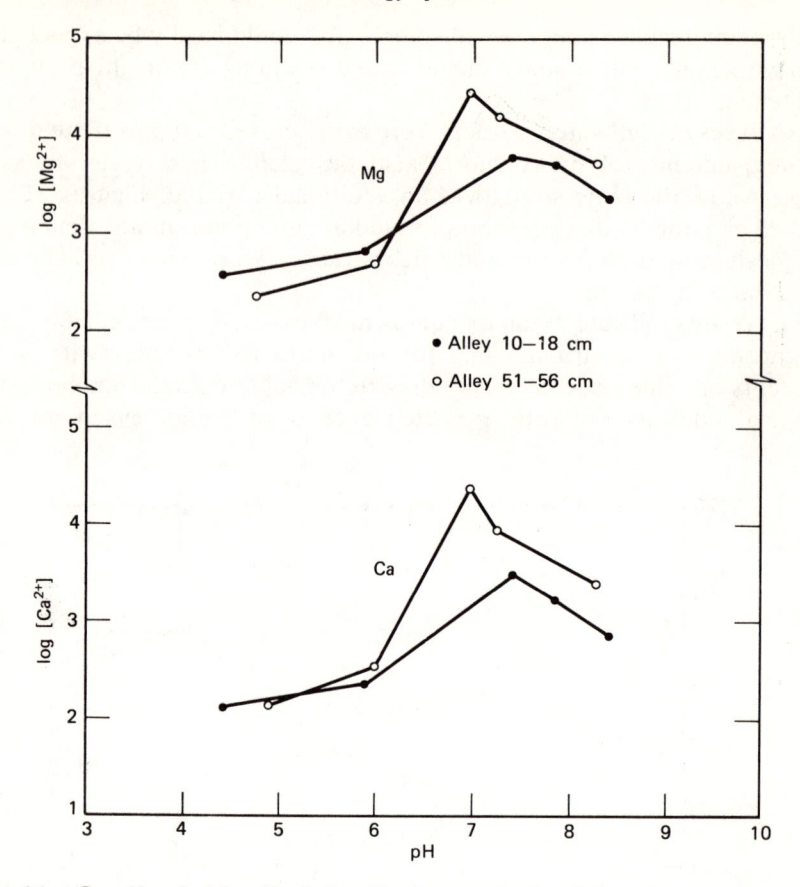

**Fig. 24** pCa–pH and pMg–pH relationships for two depths of the "alley" profile: equilibration with HCl, water, or Ca(OH)$_2$ for 36 days.

ions in solution, which must be balanced by cations liberated from the exchange complex. On the alkaline side there will be an increase in bicarbonate ions, which also must correspond to cations in solution.

Potassium in the 51–56 cm sample behaves differently; the curves approach constancy of pK. The behavior of the 10–18 cm sample is similar to that of sodium. The bonding energy for potassium is clearly very great in the 51–56 cm layer and is evidently rapidly increasing with addition of base, so that even additional bicarbonate anions are not sufficient to produce a fall in the curve.

The silica curves show, on the acid side, a very slight increase of slope with pH; on the alkaline side the slopes increase and then decrease. There

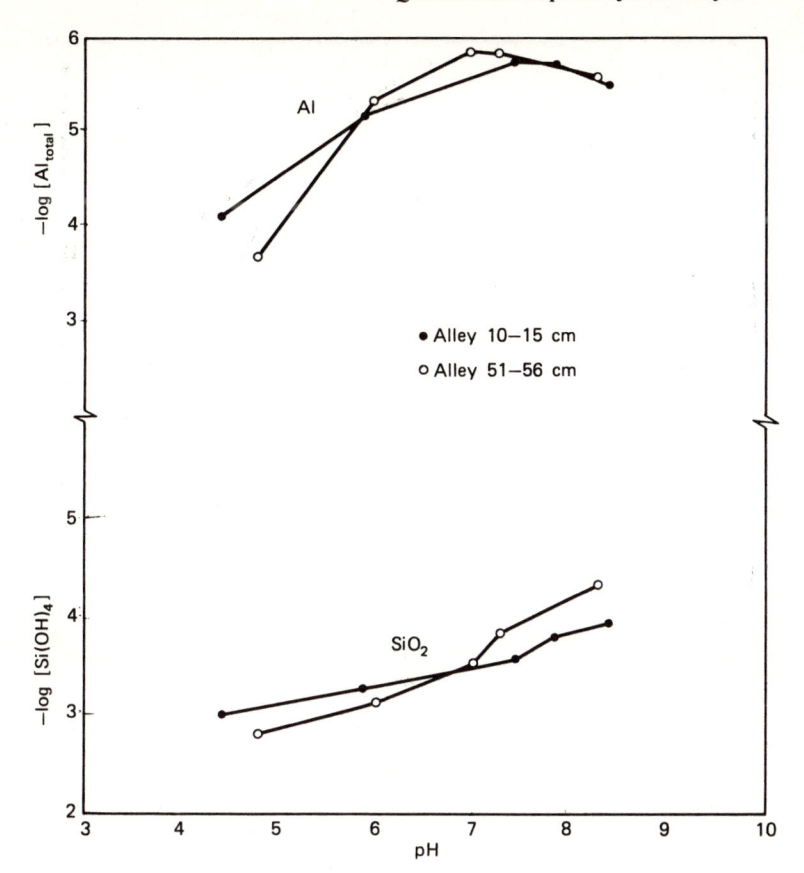

*Fig. 25* Soluble aluminum and silicon in relation to pH for two depths of the "alley" profile: equilibration with HCl, water or Ca(OH)$_2$ for 36 days.

is no region in which the silica in solution is independent of pH, as is found for amorphous silica.

The aluminum curves are exceedingly revealing. They show a flatter maximum than the sodium or calcium curves, and on the alkaline side there is a close approximation to the same constant slope for both samples. This slope characterizes the situation for different reaction periods and for all the depth samples in the alley profile. Because of this constancy, it has been compared with the theoretical slopes for aluminum ions of different compositions. These relationships are clearly visible in Figure 26, which indicates that from pH 6.9 to 8.4 an ion characterized by one negative charge per six aluminum atoms dominates the situation. In

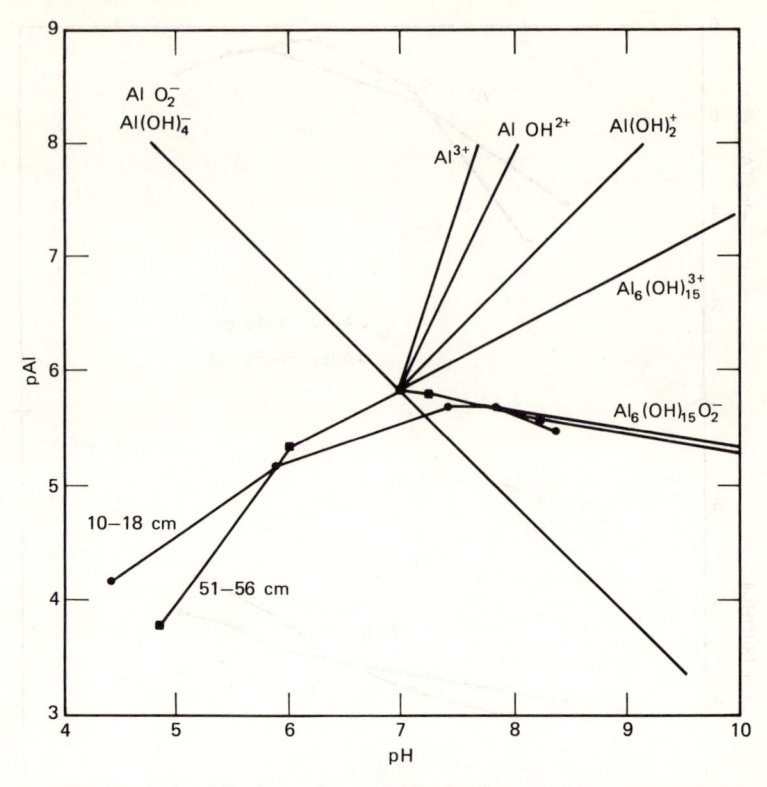

**Fig. 26** pAl–pH relationships for various soluble aluminum species compared with results for two depths of the "alley" profile. Note correspondence with $Al_6(OH)_{15}O_2^-$ over pH range 7–8.5.

general, with nonlinear curves the slope at a given point characterizes the ratio of charge to atoms of aluminum, which may involve more than one ionic species. It seems likely that the complex found in this soil over the limited pH range studied is tetrahedral with regard to the aluminum atom carrying the negative charge and octahedral in respect of the other five aluminum atoms. With increasing alkalinity it seems probable that the proportion of tetrahedral atoms would increase, until finally the tetrahedral ion $AlO^-_2$ or some hydrated form of this dominates the solution.

*Variation Within the Soil Profile.* In Figures 27 and 28 the Garrels and Christ diagrams are used to portray the variations in log [cation$^+$]/[H$^+$], log [$\sqrt{\text{cation}^{2+}}$]/[H$^+$], and log [Si(OH)$_4$] with depth, the depth being indicated by successive numerals corresponding usually to 5 cm soil samples (however, no. 1 was 0–10 cm; no. 2, 10–18 cm; no. 3, 18–20 cm). The potassium values with their great vertical spread are distinctive. The

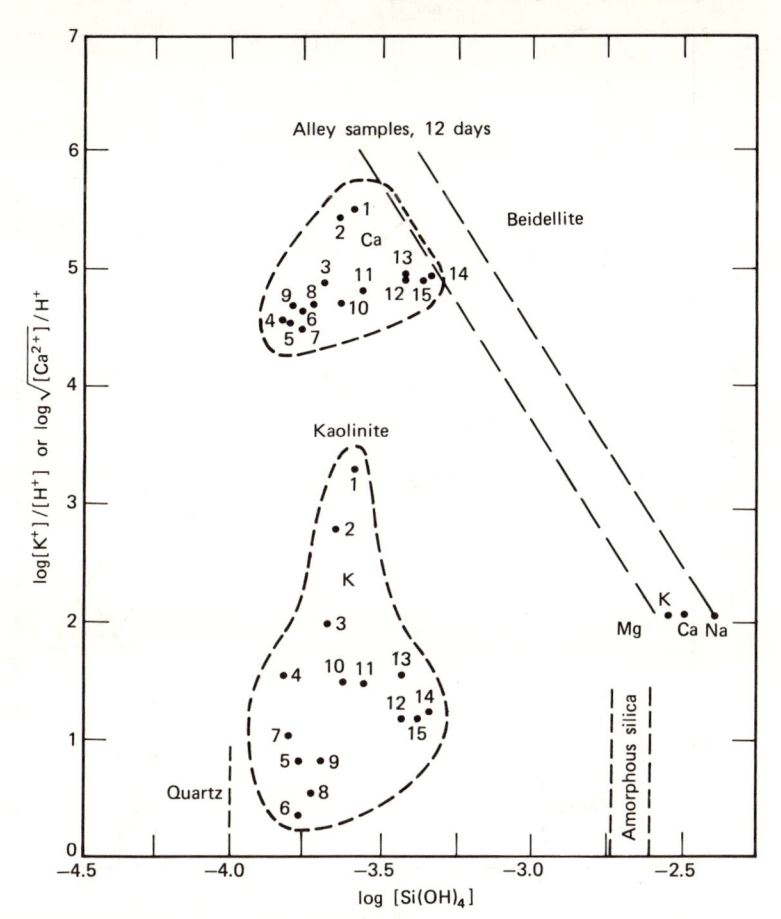

*Fig. 27* Garrels and Christ diagram for calcium and potassium using "alley" samples equilibrated for 12 days: 1 = 0–10 cm, 2 = 10–18 cm, 3 = 18–20 cm, 4 = 20–25 cm, 5 = 25–30 cm, 6 = 30–36 cm, 7 = 36–41 cm, 8 = 41–46 cm, 9 = 46–51 cm, 10 = 51–56 cm, 11 = 56–61 cm, 12 = 61–66 cm, 13 = 66–71 cm, 14 = 71–76 cm, 15 = 76–81 cm.

bonding energy of potassium relative to hydrogen near the surface is moderate. It increases regularly with depth to about 36 cm, then decreases and levels off at intermediate values for 51–81 cm samples. Close examination reveals that over a narrower range of bonding, the calcium and magnesium both follow this pattern. Sodium is quite different. Its bonding progressively increases with depth down to about 25 cm, then slowly decreases. The silica in solution decreases with depth to 35 cm, then increases to the highest values around 70 cm.

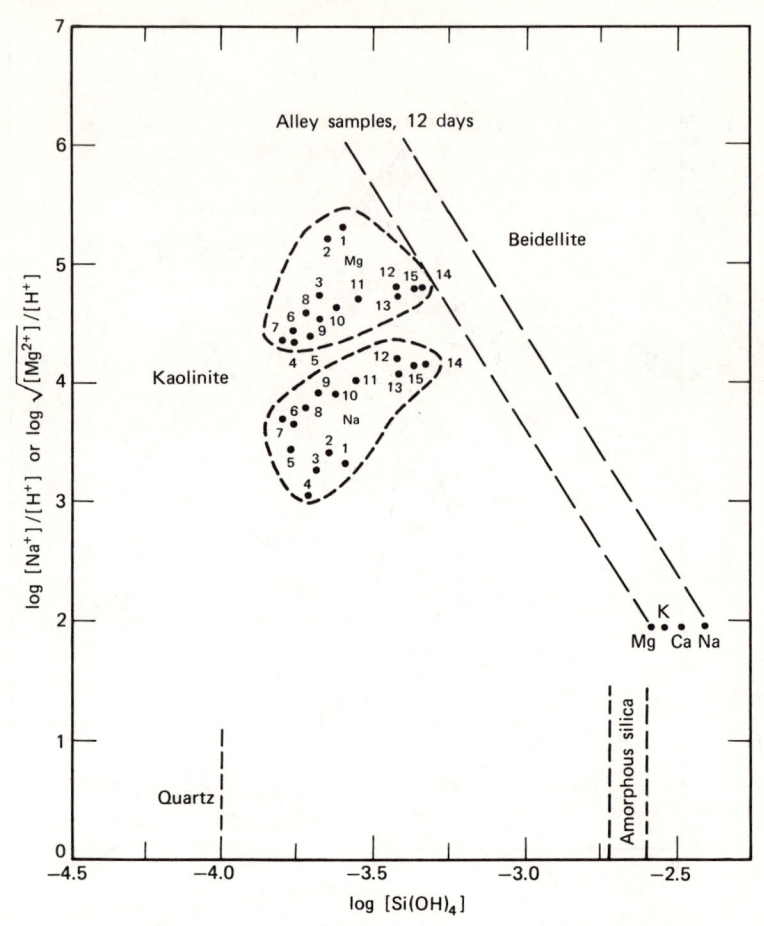

108    *Physical Chemistry and Mineralogy of Soils*

*Fig. 28*  Garrels and Christ diagram for magnesium and sodium using "alley" samples equilibrated for 12 days (numbers as Figure 27).

The similarity in relationship with depth for potassium, calcium, and magnesium suggests that a common pedological factor has operated, one that is different for sodium and for silica. The general concept of a range in bonding energies for each cation can be used in conjunction with the annual vegetative cycle of prairie soils to explain this. The plant species take up abundant potassium and smaller amounts of calcium and magnesium from the root zone and transport most of it to the tops. When these die, part of their content of these cations is quickly dissolved by rainfall and part remains to be incorporated in humus, which releases it later upon decomposition. Thus there is loss of the more loosely bound cations of the root zone and their accumulation in the upper horizons. The deepest layers are little affected. Potassium shows these effects very strongly: since the loosely held potassium is small in amount to begin with, losses from the subsoil fall on a steep part of the bonding energy curve. The same is true to a smaller extent for calcium and magnesium. Sodium is not appreciably taken up by the prairie vegetation, but since it is the most loosely held of the four cations, to some extent it is lost by leaching from the upper horizons. The sodium content of the lower horizons is maintained, or even slightly increased. Silica, like potassium, calcium, and magnesium, is appreciably taken up by the grasses and is liberated when they decompose, but largely in the form of sparingly soluble opal. Loss of silica from the root zone probably makes no difference to the release of that remaining; thus the total range of silica variations is much less than that for the cations. Little can be said about soluble aluminum in its relationship to this profile, since the total content in solution is much less than that of silica. The ratio of silica to aluminum in solution shows some variation in accord with profile characteristics. Changes occur at 18, 36, and 61 cm, the average values for Si/Al in solution being 0–18 cm, 159; 18–36 cm, 250; 36–61 cm, 199; and 61–81 cm, 359.

*Effects of Additions.* Soil samples were taken every 2 in. in the control plot, the plot with sodium addition in the 10–30.5 cm layer and the plot with calcium addition in the 10–30.5 layer. Figure 29 shows the exchangeable sodium and the extractable sodium as functions of depth for the control, the sodium plot, and the calcium plot. Clearly exchangeable sodium moves both downward and upward from the original 10–30.5 cm layer. Its effect on both exchangeable and extractable sodium is evident even at 76–81 cm depth. In the subsoil it has reacted very fully with the heavy clay so that only a moderate rise in the extractable sodium is found, but this continues right to the base of the profile.

Calcium additions provided a complete contrast. Exchangeable and extractable calcium were increased only in the layer of application, therefore movement in the profile is minimal, certainly less than 5 cm. The amounts of sodium and calcium used (6 meq/100 g soil) were insufficient to provide an excess of bicarbonate or carbonate.

Data were obtained also from plots treated with potassium and with magnesium at two different depths. Figure 30 gives the potassium relationships. Clearly potassium has very limited effects outside the zones of application. Its mobility is only slightly greater than those of calcium or magnesium, in accord with the high bonding energy and fixation in this soil.

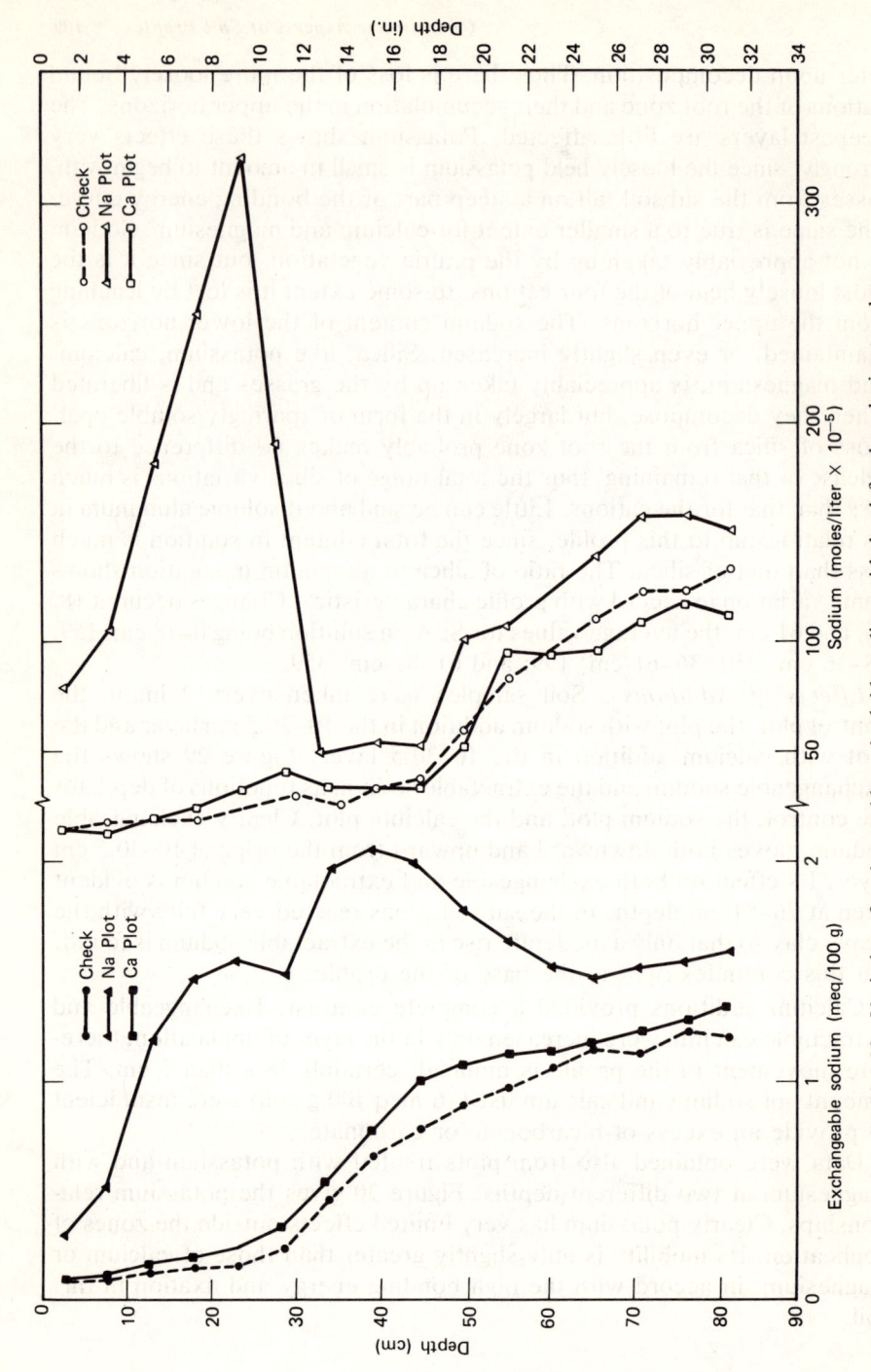

*Fig. 29* Variation of exchangeable and soluble sodium with depth for three lysimeter plots.

## Discussion

Do criteria based on the composition of the extractable soil solution bear sensitively on questions of soil formation and development? Let us now examine critically the data just presented.

The Mexico profile was formed in loess that probably was laid down under three slightly different sets of conditions, as suggested by mechanical analysis figures within the silt fraction. These indicate small changes in particle size distribution at 30.5–35.5 cm and at 61–66 cm. Accompanying such differences might be differences in original clay, especially in terms of the exchangeable cations. These can all be classed as depositional differences on which pedological changes would be superimposed.

Depositional differences imply that there is a constant list or suite of original minerals whose proportions might vary according to the incidence of depositional factors. In itself, such variation should not change the composition of the equilibrium soil solution, since for each reaction that can be formulated, the solid phases are given unit activity. Thus theoretically, as long as all are present, no change should occur. Within the clay group, however, subtle variations are possible. The particle size, the magnitude and distribution of the charge on the framework, and the proportions of exchangeable cations that balance this charge can all vary in different parent materials and, with regard to the exchangeable cations, can sensitively respond to soil-forming factors. Such variations are strongly reflected in Donnan-type equilibria and also in the electrokinetic properties that greatly affect movement of solid phase constituents.

Donnan equilibria are usually formulated as though attention need be focused only on completely dissociated ions of a given valence; but actually, since different ions are dissociated to different degrees, this finds expression in the value of $A$, that is, the volume distribution of effective charge, or, in terms of Gouy theory, in the charge per unit surface. However in studying actual soils, since salt contents are seasonally and spatially variable, ionic ratios are more revealing than absolute concentrations. This brings us then to pH–p (cation) relationships, to ratios of exchangeable ions, and to selectivity numbers between pairs of ions.

Thus in the Garrels and Christ diagrams for the Mexico soil, the qualitative similarity but quantitative difference between potassium on the one hand and calcium and magnesium on the other, in their relation to depth, seem to be pedogenically related to the nutrient cycle in a prairie soil. This also explains the difference between the depth function of sodium and of these elements. Because of the extreme variation in soluble potassium with depth, the K/Na selectivity number is also a sensitive reflector of profile differences. Since calcium and magnesium are so similar, the Ca/Mg selectivity varies only over a very narrow range.

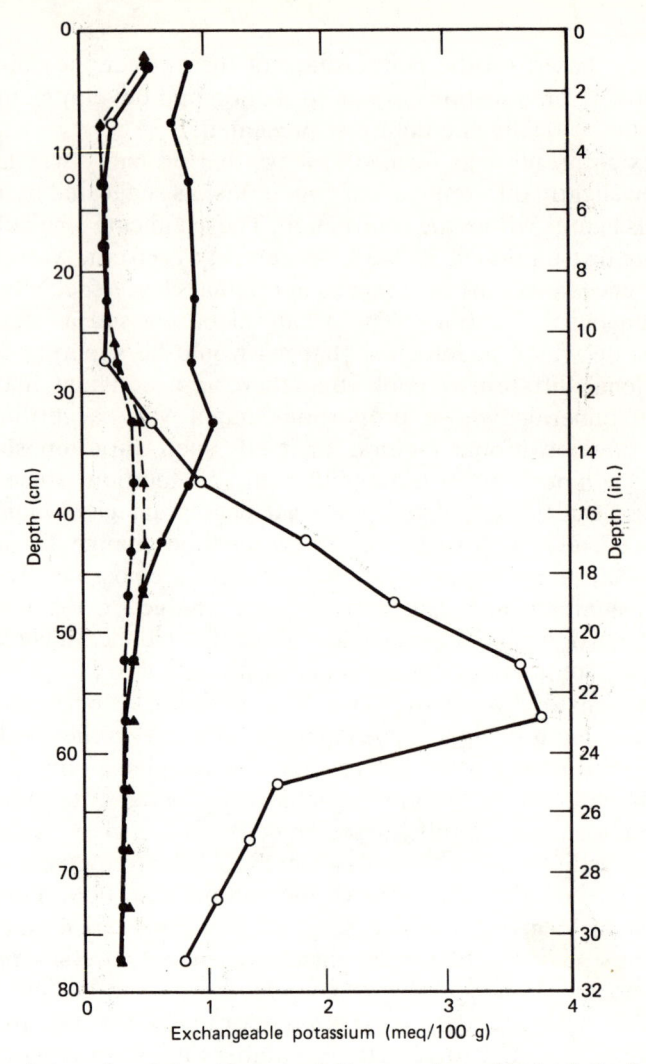

*Fig. 30* Variation of exchangeable and soluble potassium with depth for four lysimeter plots.

Silica poses difficult problems, thus far unsolved. Since both quartz and amorphous silica (opal or chert) are present, as well as a variety of silicate minerals, many equilibria can affect the silicon level in solution. Serious obstacles to understanding are presented by the general sluggishness of the amorphous silica $\rightleftarrows$ H$_4$SiO$_4$ in solution $\rightleftarrows$ quartz reactions and the absence of knowledge on the kinetics of other reactions involving silica.

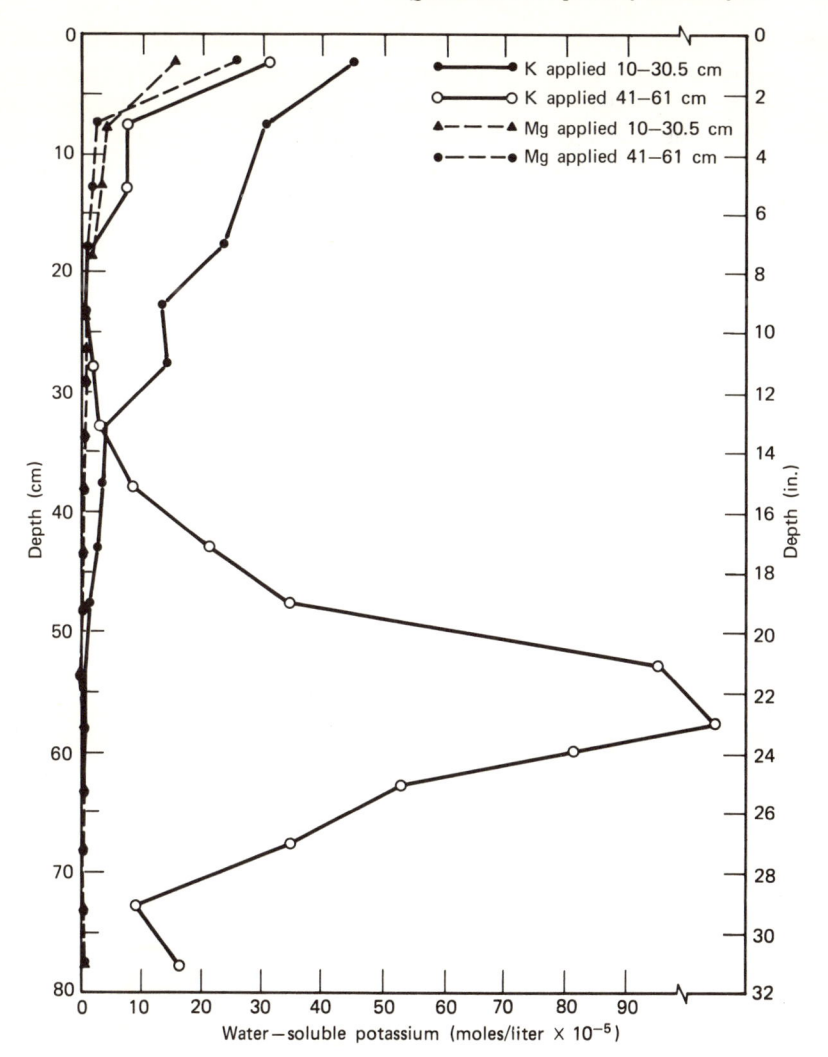

*Fig. 30 (Continued).*

As pointed our earlier, this profile gives some evidence of the cycle of silica through vegetation, but it is suggestive rather than conclusive.

In the genetic study of diverse soil profiles it is clear that the following lines of investigation need to be combined: (1) study of the physical characteristics of soil profiles, beginning with the field description, then study of the structure as shown in thin section, proceeding to quantitative measures of the behavior toward water, including electrokinetic properties; (2) characterization through particle size distribution; (3) mineralogical characterization of both clay and nonclay; (4) quantitative measures of the nature and distribution of organic matter; (5) exchangeable cations and anions and measures of their relative bonding; (6) use of the extractable soil solution, involving ionic ratios, selectivity numbers, and Garrels and Christ diagrams, as outlined above for the Mexico soil. We are very far from possessing such a comprehensive overview of any soil. Chapters 7 through 10 review the present position.

## INDEX MINERAL STUDIES

### *General*

Soil formation and development processes manifestly involve accessions to and losses from profiles in their entirety; but even more striking are differential movements of various constituents within profiles. These movements are chiefly responsible for the observed physical and chemical horizonation. Quantitative measures have been sought from early times.

The general geographic fact, emphasized in the Russian work, that under given climatic and vegetative conditions similar profiles develop on different parent materials, seems to imply that observed horizonation is purely pedogenic. For strictly quantitative comparisons, however, something more is needed. First it is necessary to show that there was no antecedent horizonation, that is, no vertical variation in parent material. Second, this having been demonstrated, constituents not affected by pedogenic processes must be chosen as immobile indicators in the calculations both of differential movements and of integral effects.

The choice of the immobile indicator has varied with different investigators. Thus Kossovich, clearly having podzolic conditions in mind, suggested that total silica could be regarded as immobile. Later, discussing the many conditions under which colloidal and soluble silica are mobile, Rode chose crystalline quartz, separated by strong acid treatments, as a more uniformly valid immobile indicator. Investigators of tropical soils, impressed by the mobile character of silica and the corrosion of quartz grains, leaned toward aluminum, ferric iron, or titanium as the immobile constituent (Harrison, 21; Merrill, 37). These divergent choices, and the obvious uncertainties in the arguments used to support them, led for awhile to stagnation in this approach. Finally critical examination and improvement of the methods used by sedimentary petrographers brought about the successful use of index minerals (Marshall, 30; Haseman and Marshall, 20, 32; Marshall and Jeffries, 31; Barshed, 2). Köster (28) has made a critical comparison of different methods and confirms the validity of the index mineral procedure.

The first step involves the use of the mineral suite present as a means of examining the relationship of the soil horizons to the assumed parent material. For practical results the particle size fractions used should contain the greatest variety of minerals, particularly the resistant minerals of the heavy fraction (S.G. > 2.70). Experience has shown that the resistant heavy minerals are found as single crystal grains in greatest abundance in fine sand and coarse silt fractions (particle size range 0.2–0.01 mm). Although the content is usually below 5%, a great variety of mineral species of diagnostic value are found here. Some are highly resistant to physical and chemical weathering; others less so. Thus the heavy mineral suite itself is in part indicative of the weathering history of the sample as well as of its origin. It is frequently worthwhile to subdivide the fine sand into two or three grades and determine the heavy minerals in each. This information provides a second criterion of uniformity as described below.

Microscopic identification and the counting of individual grains is necessary in applying these methods. Statistical limitations on accuracy should therefore receive close attention. Brewer (6) has given details of thorough statistical treatment. A rapid survey of the situation in any given case is obtained by use of the Poisson relationship as follows (31). The mean error or standard deviation is the square root of the number of grains of a given species counted. Probability tables can be used (Fisher) to determine by what factor the standard deviation must be multiplied to give any specific probability. Provided more than 20 grains of the given species have been counted, it appears that the mean error multiplied by 2 brings us to the 5% point. This means that the odds are 19 to 1 that an error of twice the mean error would not be exceeded. Thus if 100 grains of tourmaline were seen in a series of counts, the mean error or standard deviation is ±10 grains. Multiplying this by 2 leads to the following result: the odds are 19 to 1 that the tourmaline will lie between 80 and 120 grains.

Thus large numbers of observations are required for real precision, and with microscopic methods some compromise usually must be adopted. In favorable cases (e.g., zircon) much more precise chemical determinations can be carried out on the separated heavy mineral fractions.

The following rules slightly amplified below were formulated by Marshall and Jeffries (31) based on experience to 1945. No changes have seemed necessary since then.

1. Characteristic of differences in geological origin are qualitative and quantitative differences in the suite of the most resistant heavy minerals. Figure 31 illustrates such differences for heavy minerals of three soils (fraction 0.1–0.02 mm) from sandy parent materials: (*a*) present-day Missouri River sand, (*b*) the Tilsit silt loam from La Motte sandstone of Cambrian Age with later addition of loess, (*c*) the Boone fine sandy loam from St. Peters sandstone.

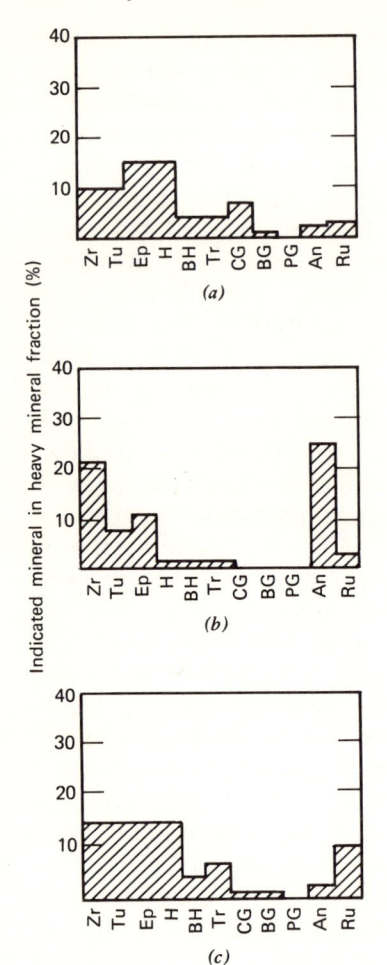

*Fig. 31* Individual minerals in the heavy minerals of sand fractions from different geological formations in Missouri. (*a*) Missouri River sand, 0.1–0.02 mm fraction. (*b*) Tilsit silt loam, 0.1–0.02 mm fraction. (*c*) Boone fine sandy loam, 0.1–0.02 mm fraction. Mineral key: Zr, zircon; Tu, tourmaline; Ep, epidote; H, hornblende; BH, basaltic hornblende; Tr, tremolite; CG, colorless garnet; BG, brown garnet; PG, pink garnet; An, anatase; Ru, rutile.

2. Characteristic of differences in deposition are qualitative similarity of the suite of resistant heavy minerals, and quantitative differences in mineral ratios. Quantitative differences are also to be expected in the particle size distribution of any given resistant mineral. The differences in resistant mineral ratios caused by differences in deposition (Figures 32*b*, *d*, and *e* are similar in character to those found in different particle size ranges for a single soil horizon (Figures 32*a*, *b*, and *c*). This would be expected.

3. Characteristic of differences solely in degree of weathering are qualitative and quantitative similarity in the suite of resistant heavy minerals,

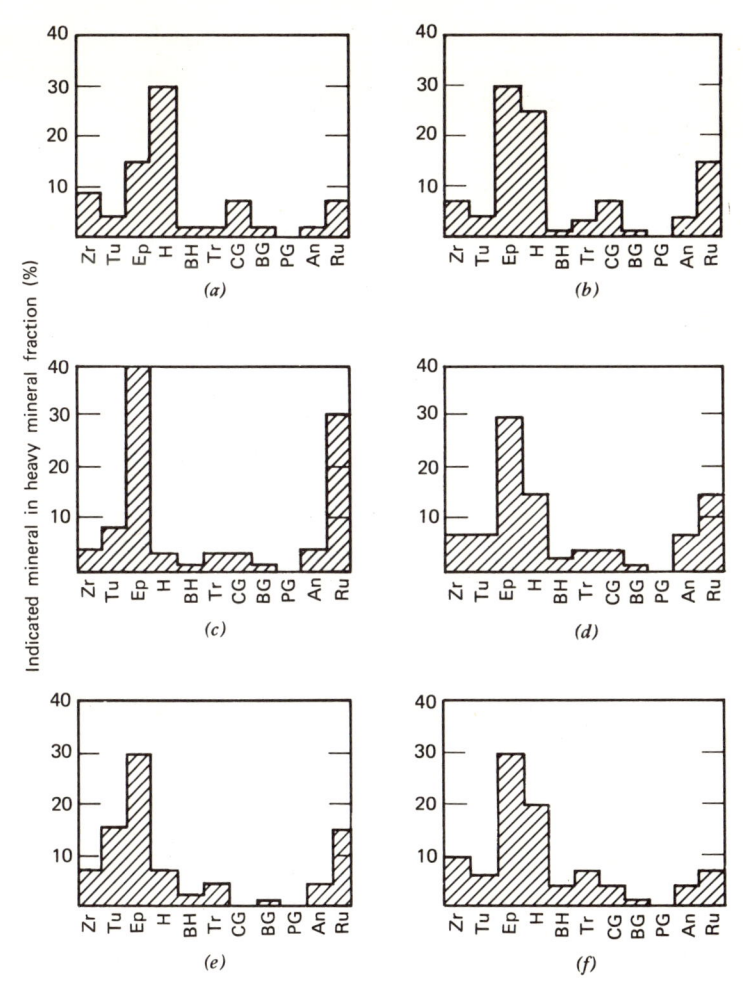

*Fig. 32* Individual minerals in the heavy minerals of sand fractions of the Grundy silt loam soil of Missouri. (*a*) Grundy silt loam, 147–154 cm horizon, 0.125–0.046 mm fraction. (*b*) Grundy silt loam, 147–154 cm horizon, 0.046–0.025 mm fraction. (*c*) Grundy silt loam, 147–154 cm horizon, 0.025–0.011 mm fraction. (*d*) Grundy silt loam, 30–41 cm horizon, 0.046–0.025 mm fraction. (*e*) Grundy silt loam, 274–290 cm horizon, 0.046–0.025 mm fraction. (*f*) Marshall silt loam, 0.1–0.02 mm fraction. (*a*), (*b*), (*c*) Comparison of different particle sizes for the same soil horizon. (*b*), (*d*), (*e*) Comparison of different horizons using the same sand fraction. (*b*) and (*d*) are loess, (*e*) is glacial till, (*f*) is loess. Mineral key: Zr, zircon; Tu, tourmaline; Ep, epidote; H, hornblende, BH basaltic hornblende; Tr, tremolite; CG, colorless garnet; BG, brown garnet; PG, pink garnet; An, anatase; Ru, rutile.

and qualitative similarity but quantitative differences in the more easily weathered minerals. In some cases a mineral may weather out of a finer fraction before disappearing from a coarser.

The lighter fractions (S.G. < 2.7) also provide good indications of weathering differences because they usually contain mixtures of different feldspars and micas as well as quartz.

Although detailed mineralogical data are not yet available on a worldwide basis; it is clear that under common pedological conditions weathering processes continue at different rates in different horizons, leading first to changes in mineral ratios and in particle size distribution of weatherable minerals, and eventually to complete disappearance of certain species.

Microscopic observations made by Haseman and Marshall (20) indicated that in the Grundy soil derived from relatively uniform loess down to 175 cm, weathering of the amphiboles was more advanced in the 0–56 cm horizons than below this. In the much older Ashe soil from granite (Humbert and Marshall, 22), hornblende of particle size 0.05–0.02 mm had disappeared from the 0–10 cm and 10–25 cm horizons, that of particle size 0.25–0.05 mm had disappeared from the 0–10 cm horizon, and that of 0.5–0.25 mm was still present in the surface horizon, although much reduced in amount as compared with the weathered granite at 100 cm depth. Apatite is similarly a good indicator of weathering differences in different horizons.

Mixed parent materials can often be recognized by heavy mineral determinations, and in some cases even their proportion may be estimated. Marshall and Haseman concluded from the epidote content that the surface horizons of the Tilsit soil from the La Motte sandstone contained about 16% of loess down to 43 cm.

## Choice of Index Minerals

Both sedimentary petrologists and soil scientists have contributed to the choice of index minerals. The use of zircon as the most resistant heavy mineral is accepted with certain limitations. The variety of this mineral that has been strongly affected by radioactive transformations (metamict zircon) is known as hyacinth. It weathers much more readily than grains with the ideal composition and characteristics (D. Carroll, 11). However the ability of normal zircon to survive more than one geological cycle indicates its general suitability as an index mineral for the much briefer pedological changes. Under extreme tropical conditions involving basic igneous rocks, Druif (12) observed some corrosion of zircon grains. Raeside (44) considers that inclusions in zircon grains render them much more susceptible to weathering.

The second mineral usually considered is tourmaline. It also survives more than one geological cycle—rather surprisingly in view of its structure and chemical composition (Vol. I, Chap. 2). Thus the ratio of zircon to tourmaline is now widely used as a criterion of uniformity.

Anatase and rutile, the two common forms of $TiO_2$, are also highly resistant. However it is known that secondary anatase as well as leucoxene can arise through weathering reactions. Rutile is a common inclusion in quartz grains, thus can be liberated through cracking or weathering.

The garnets are orthosilicate minerals that cover a broad range of chemical compositions. They are common constituents of secondary rocks. However Raeside (44) considers that certain members disappear fairly readily under acidic conditions of weathering. Mickelson (39) came earlier to the same conclusion in a comparison of three soil profiles from secondary materials in Ohio. Dryden and Dryden (13) concluded that in sandstones, garnets disappeared more rapidly than some amphiboles. They place sillimanite and monazite after zircon and tourmaline with regard to resistance to weathering, with kyanite and hornblende following in the sequence, and all the foregoing preceding the garnets. Humbert and Marshall (2) found in soil profiles from both a granite and a diabase that the garnets (in very small amount) seemed to be roughly constant in proportion, whereas zircon and tourmaline increased on proceeding up to the soil surface. This would indicate that garnets are slowly disappearing by weathering.

Minerals of the epidote-zoisite group frequently occur in small quantity in sediments and soils. Since they are calcium aluminosilicates, their considerable resistance to weathering is rather surprising. Epidote is formed among decomposition products of plagioclase feldspars (saussuritization). In the granite and diabase profiles studied by Humbert and Marshall the increase in quantity going up from the rock to the soil surface roughly paralleled those of zircon and tourmaline. The Marion soil profile from loess (Haseman and Marshall, 20) contained relatively abundant epidote. There was only a slight decrease in the ratio of epidote to zircon toward the soil surface. Thus weathering of epidote under recent climatic regimes in Missouri is only slight.

The amphibole group is generally considered to weather markedly under acidic conditions. Common green hornblende of 0.05–0.02 mm size showed a distinct decrease toward the soil surface for Missouri granite and diabase profiles (Humbert and Marshall, 23). A brown mineral of this group, basaltic hornblende, was found by Haseman and Marshall to be much more resistant to acidic attack than the common green hornblende.

Although opaque grains are common in the heavy mineral fractions of soils, few species can be used reliably as index minerals. Those containing iron are subject to oxidation-reduction transformations and possibly also to solution by organic complex formation. Nevertheless well-defined single crystal grains of magnetite ($Fe_3O_4$) and ilmenite ($FeTiO_3$) have considerable persistence in sediments. Magnetite was used by Yassoglou and Whiteside (53) in a detailed mineralogical study of fragipan soils in Michigan. The ratio of magnetite to garnet was found useful as an indicator of uniformity.

The only mineral of the light fraction used as an immobile indicator is quartz. When relatively short periods of pedogenesis are involved and the conditions are not alkaline, it seems to be a safe choice. A question arises, however, with respect to whether quartz of a particular size fraction is as suitable as the total quartz. Quartz grains of the larger sizes often show internal cracks; hence physical weathering through changes in temperature and moisture is a possibility, and it might proceed at different rates in different horizons. Missouri granite (22) provided a well-authenticated example of this. Thus in some cases total quartz is preferable to quartz of a particular fraction.

Total quartz is often a major constituent of soils, and quantitative determination by X-ray powder techniques offers promise. Igneous rocks, especially basic igneous rocks, give rise to alkaline ground water relatively rich in silica. The deposition of authigenic quartz then becomes a possibility. In the Missouri diabase soil there was very marked accumulation of quartz in all the fractions from 0.5 to 0.02 mm. Even larger quantities of chalcedony accompanied it. Thus quartz is not a universally valid immobile indicator.

Köster (28) compared several methods proposed for obtaining quantitative information on weathering. The particular example was an Indian Laterite derived from granite. He concluded that the index mineral method as applied using quartz grains was valid. Corrections were applied for the surface corrosion of quartz in certain horizons. The index mineral method gave results in agreement with quantitative petrographic determinations. None of the other methods (constant titanium, Barth's constant standard cell, constant volume) showed good agreement. He therefore concluded that the index mineral method provides the only reliable basis for quantitative calculations.

Immobile indicators other than single minerals have also been used, but confidence in the results depends on other lines of evidence being adduced. For instance, the whole of a certain sand fraction might not be objectionable under the right circumstances. Haseman and Marshall (20) discovered that in the Grundy profile, changes in the fine sand quantitatively paralleled those of the zircon (determined as zirconium on the separated heavy minerals). Thus under these relatively mild weathering conditions the fine sand could have been used as an immobile indicator. Such use implies that easily weatherable minerals form a negligible proportion.

Barshad (2) has used quartz plus albite (total sand fractions), the justification being that different mechanical fractions gave the same relationships in the quantitative evaluation of the profile. This indeed is one indicator of initial uniformity—namely, that the particle size distribution of a resistant mineral should remain the same throughout. Haseman and Marshall examined zircon distribution in the Grundy profile. Table 5 compares the two criteria of uniformity for the Marion and the Tilsit.

**TABLE 5**  Zircon-Tourmaline Ratios and Zircon Distribution

| Soil | Horizon (cm) | Zircon/ Tourmaline | Zircon in total heavy fraction (%) 0.125– 0.046 mm | 0.046– 0.025 mm | 0.025– 0.011 mm |
|---|---|---|---|---|---|
| Marion | 10– 20 | 2.49 | 18 | 53 | 29 |
|  | 38– 49 | 1.34 | 26 | 39 | 35 |
|  | 90–100 | 2.00 | 17 | 48 | 35 |
| Grundy | 0– 30 |  | 33 | 35 | 32 |
|  | 30– 43 |  | 35 | 36 | 29 |
|  | 43– 56 |  | 33 | 37 | 30 |
|  | 56– 81 |  | 26 | 34 | 40 |
|  | 81–102 |  | 24 | 38 | 38 |
|  | 102–142 |  | 22 | 38 | 40 |
|  | 142–163 |  | 22 | 46 | 32 |
|  | 163–173 |  | 21 | 44 | 36 |
|  | 173–193 |  | 24 | 41 | 35 |
|  | 274–290 |  | 45 | 37 | 18 |
| Tilsit | 0– 20 | 2.1 | 5 | 83 | 12 |
|  | 20– 43 | 2.2 | 3 | 81 | 16 |
|  | 43– 73 | 2.2 | 5 | 85 | 10 |
|  | 73– 89 | 2.8 | 17 | 76 | 7 |
|  | Weathered sandstone | 3.2 | 25 | 74 | 1 |
|  | Fresh sandstone | 3.3 | 31 | 67 | 2 |

Here both criteria show evidence of depositional differences. The zircon distribution for the Grundy profile is also given. Barshad (2) believed that these last data were sufficiently variable to warrant the assumption of three parent materials down to 160 cm, whereas Haseman and Marshall, conscious of the effects of slight variations in the operation of the elutriator, considered that essential uniformity was demonstrated. Brewer (6) came to the same conclusion by examining the curves obtained by plotting total zircon and total heavy minerals against depth in the profile. Both curves became almost horizontal around 60 in. depth, with no sudden changes in the upper horizons. Thus reasonable uniformity of parent material was demonstrated. Brewer also examined other data from his own work and that of Barshad, thereby throwing some doubt on the uniformity of parent material in two cases (Sheridan and Adelanto soil profiles from California). Such critical studies are important in drawing attention to the danger of shortcuts in the painstaking work of demonstrating uniformity.

The author agrees with Brewer (6) that profiles should be studied to considerable depths and that calculations can even be based on underlying consolidated rocks. Estimates of changes may be quantitatively very different, depending on the horizon chosen as reference. In the Grundy profile, the 63–69 in. layer selected as parent material gave very different results from the 32–42 in. layer. Where pedogenic changes should be separated from rock weathering remains a matter of opinion. In shallow, immature soils they are clearly intermingled. In very deep profiles such as the Laterites studied by Harrison (21) and by Köster (28), a spatial separation is evident.

## *Method of Calculation*

Beginning with Streng in 1858, geologists have used the assumption that certain constituents can be regarded as immobile and have calculated mass changes caused by weathering. Obviously any completely immobile element could be chosen as an indicator in the same way as a mineral; but as soon as the chemical reactions involved in weathering are considered, few elements qualify. Under certain restricted circumstances it may be possible to adduce convincing evidence in favor of using a particular element. For instance, if it could be shown that zirconium occurs *only* as the mineral zircon and that grains of the latter show no weathering, this element as determined chemically could be used. However, zirconium can be found in much more abundant and reactive minerals than zircon—indeed in both light sand fractions and clay fractions from soils (52). A similar case obtains with the boron in tourmaline; chemical methods applied to whole soil horizons need careful justification.

If the percentages of the immobile indicator in the parent rock and the weathered product are respectively $R_p$ and $R_a$, then $W_a = R_a/R_p$, where $W_a$ is the mass of the original rock that gave rise to 1 g of the present-day weathered product. If the percentages of another element $E$ in parent rock and weathered product are respectively $E_p$ and $E_a$, then $W_a E_p$ represents the content of $E$ in original rock corresponding to 100 g of weathered product and $E_a$ is that in the product. Thus the loss by weathering per 100 g of original rock is $(W_a E_p - E_a) 1/W_a$, or $E_p - E_a (R_p/R_a)$. Such calculations have frequently been applied to the weathering of primary rocks and the loss of carbonates from limestones. Many examples based on immobile alumina were discussed in Merrill's classic *Rocks, Rock Weathering, and Soils* (37). Extremely valuable comparisons using $Al_2O_3$, $Fe_2O_3$, and $TiO_2$ as immobile constituents were included in Harrison's study of primary rock weathering under tropical conditions (21). Soil scientists have used these relationships at various times. Nikiforoff and Drosdoff (41) defined $R_p/R_a$ as the parent material quotient $Q$.

Soil profiles having horizons of definable thickness lend themselves readily to a more complete treatment in which a vertical prism of the soil, of horizontal surface area 1 cm$^2$, provides the unit for calculation (32). Then the depth of a given horizon measured in centimeters multiplied by the bulk density (volume weight) gives the total mass of the horizon, and the sum of these values is the total present-day mass of the soil unit. Thus by the inclusion of depth and bulk density measurements, calculations can be expressed both on a weight and a volume basis. Marshall and Haseman expressed this as follows (32):

Let the experimentally determined resistant mineral indicators be expressed as fractions or percentages by weight $R_a, R_b, \ldots, R_p$ of the present-day layers whose total weights are $W_a, W_b$, etc., per square centimeter surface. Then the weights of the original layers $W_a, W_b$, etc., which gave rise to the present-day layers, are given by $W_a = W_a(R_a/R_p)$, $W_b = W_b(R_b/R_p)$, etc. Thus we reconstruct an original profile whose total weight is $W_a + W_b + $ etc. We are therefore in a position to assess the net gain or loss of the profile as a whole as well as that of each separate layer [Figures 33 and 34]. If we care to make the further assumption that the volume weight was originally uniform and equal to that now found for the assumed parent material, we can calculate the original thickness of the profile and of its layers. Except for diagrammatic purposes, however, it is probably best to consider only the weight changes.

The foregoing calculation applies to the whole soil dried at 110°C or calculated on any other suitable basis. The same procedure can be applied to individual parts of the soil, to fractions obtained in mechanical analysis, and to elements determined chemically.

Brewer (6) has set out the detailed equations for such calculations.

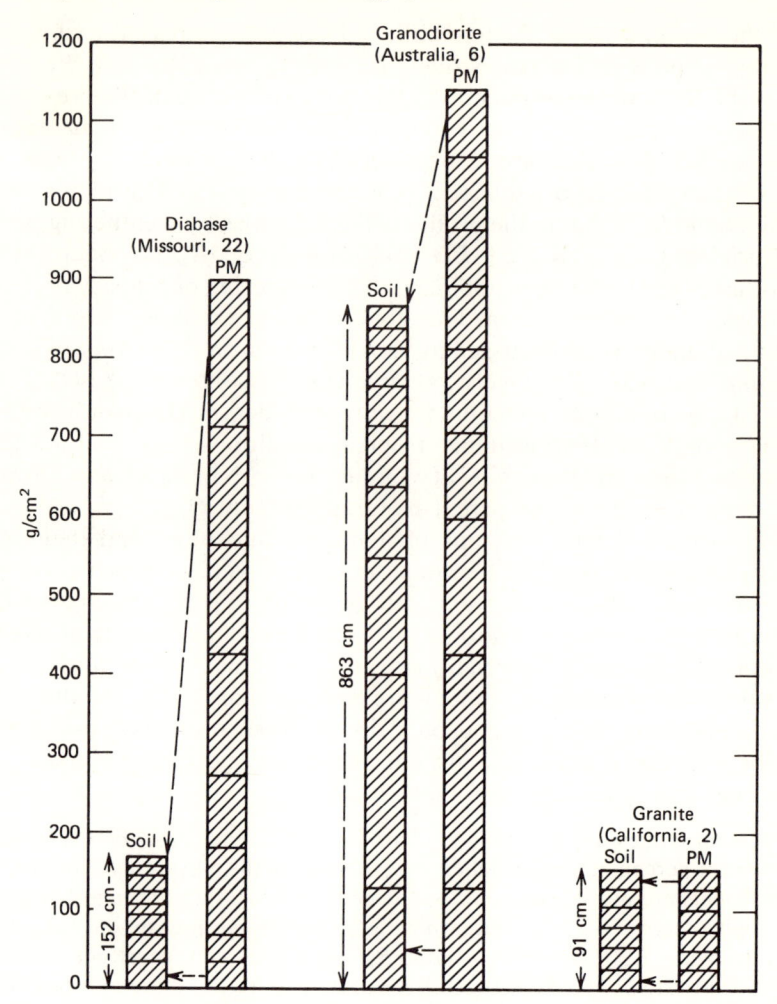

*Fig. 33* Comparison of soil with parent material (PM) calculated by the index mineral method, for three igneous rock profiles.

## Soil Formation and Soil Development

With such data at hand, the investigator is in a position to express soil formation and soil development in quantitative terms. Thus soil formation can be regarded as made up of three additive components: soil formation = losses from profile + transformations within profile + gains of profile.

The net losses that pass into the drainage water consist almost entirely of mineral elements, and they can therefore be assessed by use of the index method with appropriate analyses.

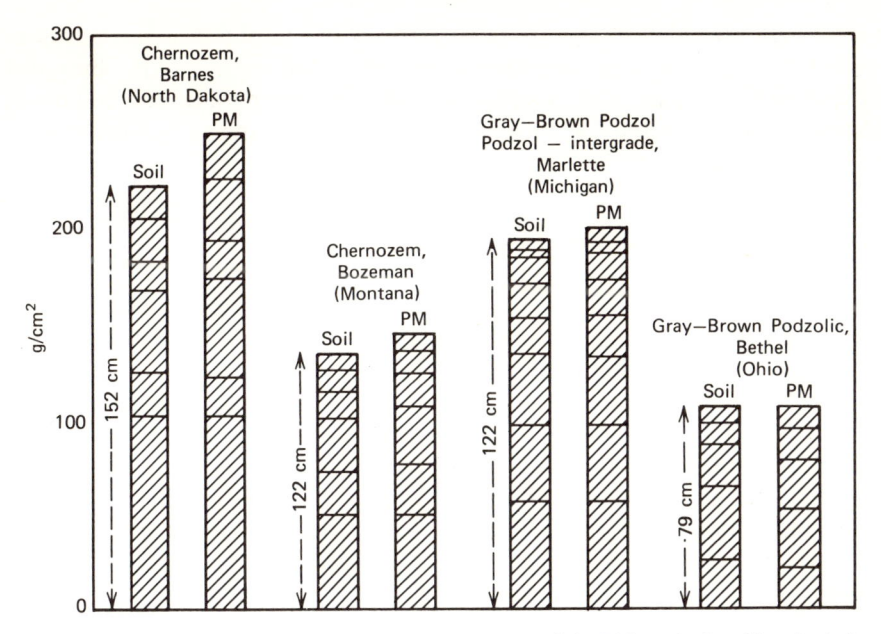

*Fig. 34* Comparison of soils with assumed parent materials (PM), calculated by the index mineral method.

We are rapidly approaching the time when transformations among the minerals can be assessed. Quantitative X-ray methods applied to the sand, silt, and clay fractions are available, and although less accurate than chemical determinations, they are steadily improving. Organic matter transformations are part of relatively rapid cyclic changes. Since the organic matter now present is only a small fraction of the total amount that has been transformed from $CO_2$ back into $CO_2$ during pedological time, we cannot measure these total transformations as we do those of the minerals.

Possible gains come chiefly from the atmosphere and contribute both to the organic matter and, through hydration and oxidation, to the mineral matter. Soil itself may be added by wind action as in the formation of loess. Cyclic salts from the oceans are carried inland in rain. In alkali soils, mineral elements are gained through the ground water.

Thus numerous processes of soil formation operate concurrently, and many of these can now be assessed.

Soil development is the part of soil formation that leads to differentiation into horizons. It arises because the characteristic losses, transformations, and gains may vary from horizon to horizon, both absolutely and relatively. We shall therefore have a series of measures of soil development corresponding to those of soil formation but applied to the horizons in detail. Obviously very complete chemical and mineralogical data on all horizons are required for a proper interpretation, especially with respect to cause and effect.

### *The Constant Volume Hypothesis*

If it be assumed that weathering causes no change in volume, calculations can be based on bulk density determinations combined with the appropriate chemical analyses and particle size distributions. Wild (52) has used this assumption in studies of soils from granitic rocks. Oertel (42) combined particle size distributions with spectrographic determinations of copper, vanadium, and zinc and found a close parallelism between clay contents and these trace elements, horizon by horizon. The use of gallium as an immobile indicator led to the conclusion that a constant volume relationship between parent rock and weathered product prevailed below the upper B horizons of six red-brown earths from South Australia. Brewer (6) gives a detailed discussion of the limitations of these attempts to base calculations on the assumption of constant volume. It seems clear that it cannot be used generally and that its usefulness in limited cases will depend on the validity being established by index studies. Where weathering produces swelling clays, volume changes are to be expected. Indeed, whenever most of the original skeleton disappears by weathering, it would be rare for the original volume to be maintained.

Brewer (6) has shown that by using the larger sand grains as indicators of the skeleton, volume changes can be estimated. Haseman and Marshall (20) calculated apparent volume changes horizon by horizon for the Grundy profile; the A horizons had shrunk and the B horizons had swollen, compared with the assumed parent material.

As already mentioned, Köster (28) included the constant volume hypothesis among those tested for an Indian Laterite. He concluded that it was inapplicable to this case but might find use where the fabric of the original rock could still be recognized by petrographic methods.

### *Barshad's Method: A Critique*

The first chapter of the first edition of F. E. Bear's *Chemistry of the Soil* (3) consists of an article by I. Barshad entitled "Soil Development." This very valuable review contains much original material, one important component of which is a new method, based on chemical analyses and clay percentages, for calculating clay formation from nonclay of the parent material. At the Sixth National Conference on Clays and Clay Minerals held in 1957, Barshad (1) applied the new methods to soils from various parts of the world and considered the results in the light of the major factors of soil formation. The second edition of Bear's book (1964) contains a revised and amplified Chapter 1, "Chemistry of Soil Development," in which the new method is retained and its applications are discussed. The mathematical development is the same as in the first account, but in additional paragraphs the author discusses certain limitations, both in the treatment of the equations and in the application to profiles containing clays of different composition. In attempting to understand these limitations, doubts arose in the present author's mind over the term $k$ in Barshad's equations. An alternative formulation was therefore tried, and it became clear that the equations are not of general validity; but perhaps the method can be applied in modified form to certain special cases. Nevertheless much may be learned from the examination of this kind of treatment.

Barshad begins by calculating the composition of the nonclay for each horizon element by element, given analyses of the whole soil, and of the clay, and the percentage clay. The next step is the formulation of an equation relating the amount of a given element in 100 g of nonclay of the parent material to the amounts of this element in the clay and nonclay of a given horizon. Using Barshad's symbols we have:

$$(a_1 \times 100) - (a' \times f) = a_2 (100 - f) k$$

where $a_1$ = elemental oxide of the nonclay fraction of the parent material (%)

$a'$ = elemental oxide of the clay formed in a given horizon (%)

$a_2$ = elemental oxide of the nonclay fraction in a given horizon (%)

$f$ = amount of clay formed in the given horizon from 100 g of nonclay parent material

$k$ = a proportionality factor affecting the percentage of the elemental oxide in the nonclay due to gains and losses other than those involved in the formation of clay itself

This is essentially intended as a material balance equation for the given element. The novelty consists in eliminating the solution loss term, which would normally be additive, and replacing it by the multiplier $k$, which is defined in an inclusive way. It is then implied that through this device $k$ becomes the same for different elements. In consequence Barshad sets up equations for two different elements and solves them simultaneously for $f$ and $k$. These equations represent hyperbolas.

128    *Physical Chemistry and Mineralogy of Soils*

Consider now the exact material balance for a given element in terms of its *additive* components. In general, we shall have for any elemental oxide $a$: grams of $a$ in 100 g of nonclay of parent material = grams of $a$ in clay formed + grams of $a$ in residual nonclay + grams of $a$ lost or gained from the given horizon.

Using Barshad's notation we now have

$$a_1 = \frac{fa'}{100} + \left[\frac{a_2}{100}(100 - f - L)\right] - l_a \qquad (1)$$

where $L$ = total losses (or gains) from 100 g of the nonclay of the parent material not included in $f$.
$l_a$ = losses (or gains) of the given element $a$ not accounted for in the clay or the residual nonclay

Writing Barshad's equation in the same form, we have

$$a_1 = \frac{fa'}{100} + \frac{a_2}{100}(100 - f)\,k_a \qquad (2)$$

We can now evaluate $k_a$ by equating the last terms:

$$k_a(100 - f)\frac{a_2}{100} = \frac{a_2}{100}(100 - f - L) - l_a \qquad (3)$$

$$\text{hence } k_a = 1 - \left(\frac{1}{100 - f}\right)\left[L + \frac{l_a}{a_2/100}\right]$$

For a second element $b$ we should find similarly

$$k_b = 1 - \left(\frac{1}{100 - f}\right)\left[L + \frac{l_b}{b_2/100}\right] \qquad (4)$$

It is clear that $k_a$ cannot equal $k_b$ except in the three following special circumstances.

1. $\dfrac{l_a}{a_2} = \dfrac{l_b}{b_2}$. This would be fortuitous and highly improbable.

2. $l_a/(a_2/100)$ and $l_b/(b_2/100)$ are both small compared with $L$. This would be possible if neither element were lost appreciably to the solution phase and removed from the given horizon by leaching. In that case we have two linear equations like (1), but omitting $l_a$ and $l_b$, which can be solved simultaneously for $f$ and $L$. However it is not easy to find indisputable evidence to justify the two choices.

3. The terms in the square brackets in the equations for $k_a$ and $k_b$ are both close to zero. Then $k$ approximates to 1 in each case. Since changes

in mass normally accompany the formation of clay, this case has no usable significance.

Case 2 clearly provides the only practical application for Barshad's ingenious use of analyses. Its limitations are serious. Thus losses of silica in solution are ubiquitous in soil profiles as judged by lysimeter results. Similarly calcium, magnesium, sodium, and potassium are lost by various mechanisms and appear in drainage water. This leaves only aluminum, iron, and titanium as possible choices of the major elements normally determined by analysis. Aluminum and iron clearly cannot be used in upper parts of soil profiles where podzolization is suspected. Wherever reducing conditions might prevail over part of the annual cycle, iron would again be excluded. Thus the restrictions are severe. Some of the cases evaluated by Barshad may come into the admissible category, but this cannot be judged from the published results because the choice of elements for the soils investigated is not given. It is necessary to go back to the original analytical data and to consider each soil and each horizon in the light of the foregoing considerations.

The author has examined data for the B and C horizons of the Cecil and Chester soils quoted by Barshad, using both Al-Fe and Fe-Ti percentages from the original analyses. The results were so discordant, with some values of $f$ out of the range of possibilities, that further attempts were abandoned. One reason for the difficulties was that in solving two simultaneous equations for $f$ and $L$ or $f$ and $k$, the final value frequently depended on a small difference between two relatively large numbers. Thus errors in the analyses become greatly magnified. As far as $f$ is concerned, Barshad's procedure for the solution relates to the same equation as the modified procedure, namely

$$f = \frac{100(a_1 - (a_2/b_2)b_1)}{(a' - (a_2/b_2)b')}$$

This expression arises in Barshad's procedure by dividing one equation by the other, to eliminate $k$, and in the modified procedure by multiplication and subtraction to eliminate $L$ assuming that $l_a$ and $l_b$ are very small. Hence the uncertainties in $f$ with a given pair of elements in no way depend on which procedure is used. The modified method might prove practicable if highly accurate results were available for pairs of demonstrably immobile elements, possibly Ti-Zr or Zr-Hf, but no such analyses on soils and clay fractions are known to the author. It would first be necessary to show by the mineral index method that such elements are indeed stationary.

The fundamental relationship of the modified Barshad method to the index mineral or index element method should now be elucidated. In the latter, a single immobile element provides the basis for calculation. In Barshad's method the nonclay of the parent material clearly is not treated as unchanging, since clay is formed from it. Any element in the parent nonclay may become a constituent of the clay. Provided such an element is not removed from or added to the given horizon, we can indeed set up a material balance equation with two unknowns, $f$ and $L$.

When a single index element or index mineral is used, calculation yields first the original mass of the parent material that gave rise to unit mass of the existing horizon. The simple extension to individual elements then enables one to calculate losses (or gains) element by element without consideration of individual minerals or individual particle size fractions of the soil. However if, for instance, the clay fraction has also been determined on the assumed parent material and on the given horizon, the clay that has resulted from pedological action (formation plus possible translocation) can be calculated. These items when added together for the whole profile give the total clay formed. However it is not possible with these data alone, to separate clay formation from clay translocation as these processes have operated to produce the individual horizons.

In applying and extending the index mineral method, Barshad makes use of the additional proposition that clay is formed from nonclay, just as he does in developing the method based on analyses. The percentage of clay is usually known horizon by horizon, permitting the total change in clay of the profile to be related to the change in nonclay of the parent material. The ratio of total clay formed to total loss of nonclay is then used to calculate the clay formed in each horizon. This number is compared with the existing clay figures to calculate gains or losses, either on the basis of 100 g of nonclay of the parent material, or on a horizon basis. The use of the foregoing ratio is equivalent to the assumption that the kind of clay formed is uniform throughout the profile. This is not always true, but it is probably a reasonable assumption for the Sheridan profile (a Prairie soil from granite) worked out in detail by Barshad.

Barshad's analytical method as modified above would give a figure for clay formation from the nonclay of the parent material, horizon by horizon. If feldspar gives rise to kaolinite in one horizon and beidellite in another, the derived values of $f$ and $L$ will differ. As knowledge of such values accumulates, some characterization in terms of them will become possible. Barshad concluded that if the clay in different horizons differed greatly in composition, it would be best to base calculations of clay formation on the horizons that showed least change. The underlying idea is that initial clay formation is a uniform process that may be modified by later pedological differences. This may be true for slight to moderate degrees of development, but in general we should expect different mean chemical environments to prevail in different horizons, with the consequence that weathering processes proceed differently—certainly in intensity, and sometimes also in kind.

The exclusive assumption that nonclay gives rise to clay implies that no mechanisms operate by which clay minerals can crystallize, aggregate, or decompose to form silt or sand size material as determined by mechanical analyses. Many possibilities for such changes come to mind and should be considered in individual cases. Thus formation of chalcedony or even quartz crystals is known to occur in many soils, and the silicon concerned may have arisen from original clay, as well as from primary minerals such as feldspar.

## Clay Formation in Relation to Factors Affecting Soil Profiles

Barshad has given a valuable review of data on clay formation derived from a variety of quantitative studies, some by the index method and others by his analytical method including the "clay from nonclay" postulate (1). There are therefore some quantitative uncertainties, but the general conclusions are well supported in each case.

*Depth function.* In all the cases examined, ranging from Podzols to Tropical Red Earths, the maximum clay formation occurred either in the surface or the immediate subsurface horizon. It then decreased with increasing depth.

*Temperature.* Clay formation from granitic materials increased with rising temperature. In relation to Great Soil Groups the order was Podzols < Gray-Brown Podzolic < Brown Earths < Pariries < Red and Yellow Podzolic Soils and Latosols < Laterites.

*Precipitation.* Examples from South Africa of soils on granite and on gneiss showed very strongly the increase in clay content with increased rainfall.

*Drainage.* Soils with poor drainage exhibit greater clay formation than those with good drainage. This is illustrated with two Ohio soils, Miami (well drained) and Bethel (poorly drained), and with two subtropical black clays from South Africa.

*Vegetation.* Grass vegetation was found to favor clay formation as compared with forest. The comparison chosen was of a Gray Brown Podzolic soil (Miami, forest) and a Chernozem (Hastings, prairie). Although the former was formed under the higher rainfall, it showed less clay formation than the Chernozem.

*Parent material.* Examples of red earths from South Africa showed a very marked decrease of clay formation from basalts through dolerites to granites. In general the more basic rocks weather most rapidly, but texture is also an important factor as shown by various comparisons from the Piedmont region of the United States.

*The time factor.* A table compiled to cover the clay formation discussed in the foregoing examples showed that even the highest rates are quite low in terms of grams of clay formed per 100 g of parent material per year: 2 mg was the highest value given, corresponding to 50 g of clay in 25,000 years. Barshad points out that mineral constituents taken up by plants and later released by decomposition fall in a similar range of amounts.

### Soil Profiles Evaluated by Index Methods

The number of comprehensive evaluations of soil profiles remains small. This is partly because it takes so long to eliminate those which show evidence of geological or depositional differences, partly because of the great expenditure of time and effort devoted to a single well-chosen example, and partly because only a very few investigators have pursued the index method.

Complete data for any one soil profile are voluminous, especially if chemical analyses and quantitative mineralogical determinations are available in addition to mechanical analyses. In most cases the change in clay content as compared with the assumed parent material has been calculated horizon by horizon and for the profile as a whole. Semiquantitative data on the nature of the clay commonly form part of the investigation.

### Changes in Total Mass

Figures 33 and 34 present examples of changes in total mass. Four main factors would be expected to operate simultaneously on profiles as a whole: namely, losses or gains in solution, changes in hydration with formation of clay from nonclay, transformations among clays, and accession of organic matter. Horizon by horizon, we also must consider differential changes such as movements of clay and organic matter. The studies thus far reported are so limited on a worldwide basis that comparisons of different great soil groups are completely inadequate.

Three examples of soils from igneous rocks are given: a soil from diabase in Missouri (Humbert and Marshall, 22) of considerable depth and old geologically; a podzolic soil from granodiorite in New South Wales (Brewer, 6); of great depth and geological age; and a relatively young Prairie soil (Sheridan) from granite in California (Barshad, 2). As might be expected, the diabase (calculated on the basis of zircon + tourmaline + epidote of the 0.05–0.02 mm fractions as the index) shows very great percentage losses, even though there is authigenic formation of quartz and chalcedony throughout the profile. The granodiorite (calculated on zircon as index) shows somewhat smaller losses, as might be expected, since it contains quartz originally. The granite soil (quartz + albite as the index) gave much smaller changes, in which the losses in the upper parts of the profile were almost balanced by gains below 46 cm.

For comparison with the Sheridan we have the Grundy Prairie soil from loess (zircon as index), which also shows losses in the upper horizons, but rather considerable gains in the lower, down to 140 cm (Haseman and Marshall, 20). The latter figures are probably too high, caused either by inaccuracies in the mechanical analyses for fractions separated by the elutriator or by slight depositional differences, as maintained by Barshad. The net gain for the whole profile is thus suspect.

Two Chernozem soils (with quartz as index) are compared; the Barnes from glacial till in North Dakota (Redmond and Whiteside, 45) and the Bozeman from loess in Montana (Bourne and Whiteside, 5).

Two Gray-Brown Podzolic soils from glacial till have been studied, the Marlette from Michigan, with constant quartz as index (Cann and Whiteside, 10), and the Bethel from Ohio with tourmaline as index (Mickelson, 39).

Franzmeier, Whiteside, and Mortland have given data on several Podzols from Michigan (16).

## Changes in Nonclay Fractions

As previously pointed out, Barshad has used net losses of nonclay (multiplied by a determined average factor) as a measure of clay formation. Such a figure, horizon by horizon, is then combined with the actual clay, calculated also per 100 g of nonclay of the parent material, to give a figure for gain or loss by translocation. The uncertainties in this procedure have been mentioned. The changes in nonclay per 100 g of parent material have been calculated horizon by horizon for the profiles under consideration. In general the losses are greatest either for the surface or for a layer immediately below this. If the parent materials contained carbonates, the losses should be corrected to a noncarbonate basis for comparison with other profiles. This is of some importance because of the widespread assumption that mineral changes in profiles are close to zero in presence of free carbonates. Very few figures by the index method are available through which this assumption can be tested.

The relative changes in individual minerals of silt and sand fractions have been determined in certain cases studied by the index method, and also in some cases in which a complete evaluation by the index method was not attempted. In this way valuable information on relative rates of weathering of different common soil minerals can be accumulated. At present we have only a very small beginning.

Matelski and Turk (35) made counts and measured the dimensions of heavy mineral grains in the sand fractions of seven Podzols from Michigan, all formed in sandy materials. The A, B, and C horizons were compared in each case, but no proof of depositional uniformity was reported. The opaque minerals were determined as a group, and figures for the following individual minerals and varieties were obtained: for all seven soils, dark green hornblende, gray-green hornblende, epidote, brown garnet, pink garnet, colorless garnet; for four soils, tremolite, zircon, muscovite, and tourmaline, which were present in very small amount. The general conclusion was that weathering of hornblende was greater in the B horizon than in the A. In the particle size distributions studied, the lowest diameter for hornblende was about 0.037 mm, but only 4 out of 42 examples showed particles smaller than 0.054 mm. The number of epidote grains counted and measured was much smaller than for hornblende, and little evidence for the weathering of this mineral is apparent. No grains of this mineral smaller than 0.054 mm diameter were found.

Yassoglou and Whiteside (53) studied certain podzolic soils from Michigan in which fragipan horizons are present. Two of the three profiles chosen showed good uniformity of parent materials as judged by constancy of the ratio of magnetite to garnet. Weathering of the minerals in the fine sand fraction was minimal; even the feldspars and hornblende were almost constant in proportion throughout the profiles. The properties of the fragipan horizons were the consequence of pedogenic factors. Loss of carbonates and clay resulted in compaction of the residual skeleton, many of the pores between sand grains being partially filled by silt, with deposition chiefly of nonexpanding clay to form interparticle bridges.

The Chernozem soil from loess studied by Bourne and Whiteside (5) also showed minimal changes in the sand and silt fractions. The quartz and feldspar proportions remained almost constant, but the most weathered horizons gave evidence of reductions in heavy minerals and in undetermined constituents that probably included micas.

Losses of plagioclases were noted by Cann and Whiteside (10) in the case of a Podzol–Gray Brown Podzolic intergrade soil (Marlette). It was estimated that slightly more than 50% had been lost from the subsurface layers between 6 and 23 cm, with much smaller losses below. The surface horizon showed only a 19% loss. The orthoclase group, like quartz, appeared to remain unweathered.

Much more extensive quantitative changes in nonclay fractions are found in older soils from igneous rocks. The relatively shallow granitic soil studied by Humbert and Marshall (Ashe stony loam, 21) exhibited less striking reduction in feldspar percentages than the diabase soil from the same area. Figures 35 and 36 show that in neither case is there complete loss of feldspar. The remarkable feature of the diabase profile is the accumulation of authigenic quartz and chalcedony in the sand and silt fractions.

The granodiorite in New South Wales studied by Brewer (6) contained feldspar and hornblende in all horizons, but biotite had completely disappeared from the surface and only traces of it were found down to 50 cm.

This persistence may be connected with two factors, one of course

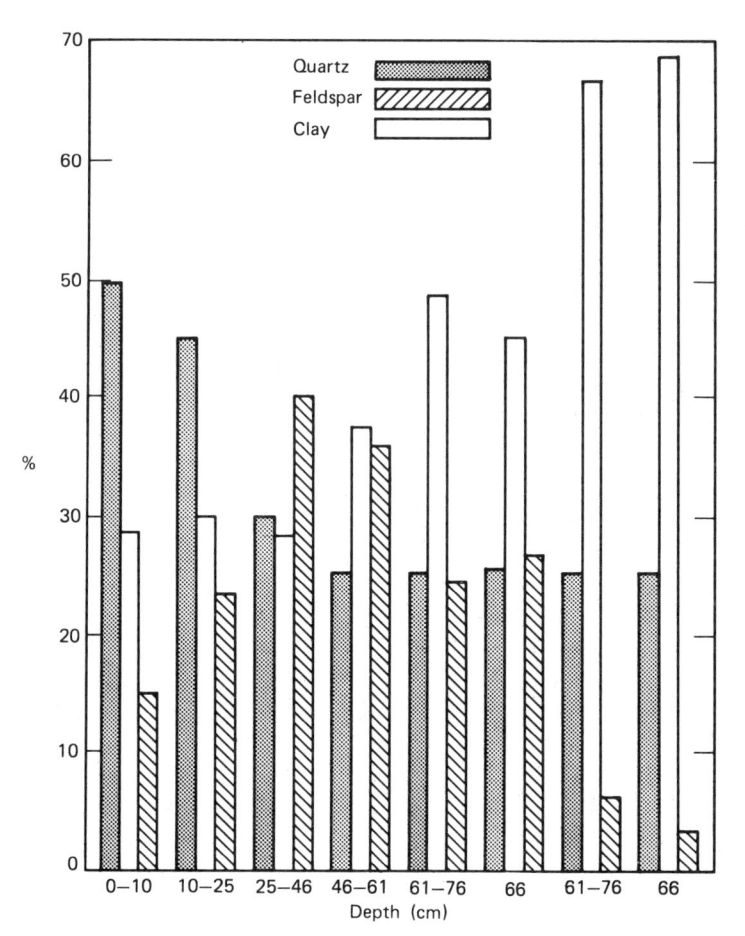

*Fig. 35* Distribution of quartz and feldspar in the sands and coarse silt and the clay in the Ashe stony loam, a residual granite soil in Missouri.

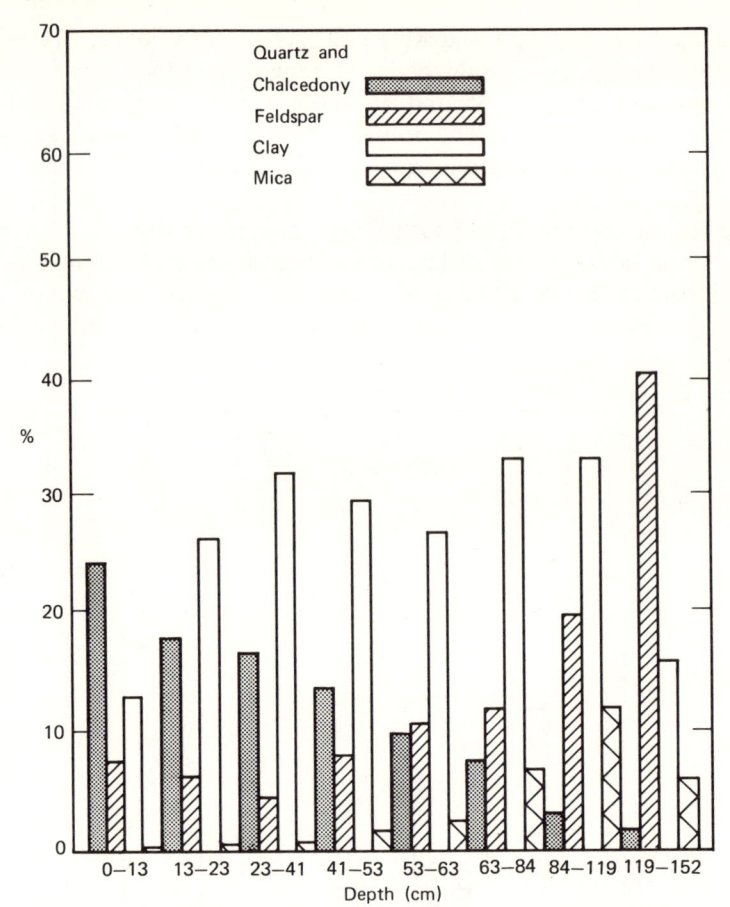

*Fig. 36* Distribution of quartz and chalcedony, feldspar, and mica in the sands and coarse silt, and the clay in a residual diabase profile in Missouri.

being relatively mild conditions of present-day weathering with only moderate rainfall. The other may arise from the manner in which igneous rocks break down. Instead of progressing uniformly downward, the most active sites of weathering follow cracks and fissures that gradually widen until stones and boulders separate areas of more complete breakdown. The cyclic renewals of soil water affect the stones and boulders much less than the soil material between them; thus a given horizon may simultaneously contain material at several different stages of breakdown. This situation may be contrasted with what Harrison (21) found in the weathering of igneous rocks under extremely high rainfall in the tropics, namely, a complete change to the final lateritic products within a few millimeters of the rock surface.

## Changes in Clay Content

*General.* Figures 37 and 38 show what has been achieved by index methods in determining soil clay. There are three cases for loess and two for calcareous till. The net change in clay content per 100 g of parent material has been determined and is represented along the abscissa. The ordinate represents depth (cm) and is read downward. The change for the profile as a whole, expressed also per 100 g of parent material, is indicated by a horizontal arrow. The clay content of the assumed parent material is also shown. Generalizations from such sparse data would be dangerous. The greatest net gain in clay per 100 g of parent material is shown by the Sheridan loam from California developed on granite (about 14 g), followed by the Australian soil from granodiorite (about 7.5 g). The two soils from

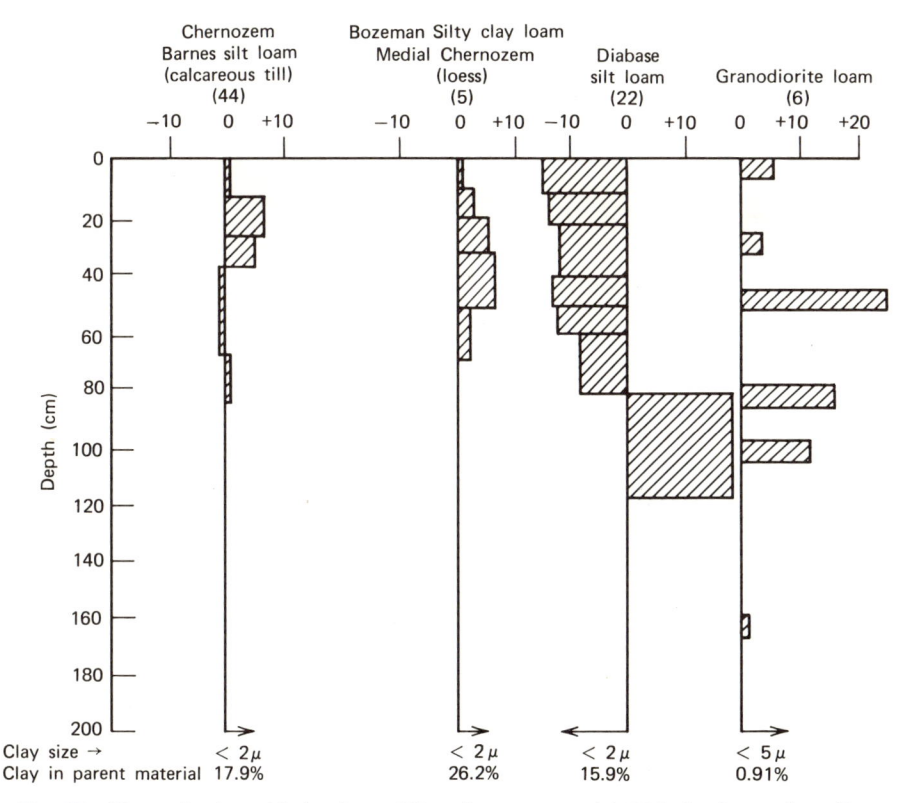

*Fig. 37* Change in clay with depth per 100 g of parent material (PM), for four soil profiles.

138    *Physical Chemistry and Mineralogy of Soils*

loess gave almost equal values (4–5 g), although one was a Chernozem (Bozeman) and the other a Prairie soil (Grundy). The Chernozem from calcareous till (Barnes) was closely similar to these, whereas the Podzol intergrade soil (Marlette loam) gave a very slight loss of clay. The soil formed from diabase in Missouri was very different, showing a large net loss (−11 g). The figures were calculated from the data of Humbert and Marshall (22) assuming constant zircon + tourmaline + epidote of the 0.05–0.02 mm fraction.

The net changes in clay represented in the diagrams include both formation and translocation. As Barshad points out, the calculated variations in the nonclay fractions throw considerable light on clay formation. But an important consideration is the demarcation between clay and nonclay, especially in relation to translocation. It is clear from Haseman and Marshall's results that silt of the fractions 2–5 $\mu$ and 5–10 $\mu$, like the clay, is lost from the A horizons and accumulates in the B. The quantita-

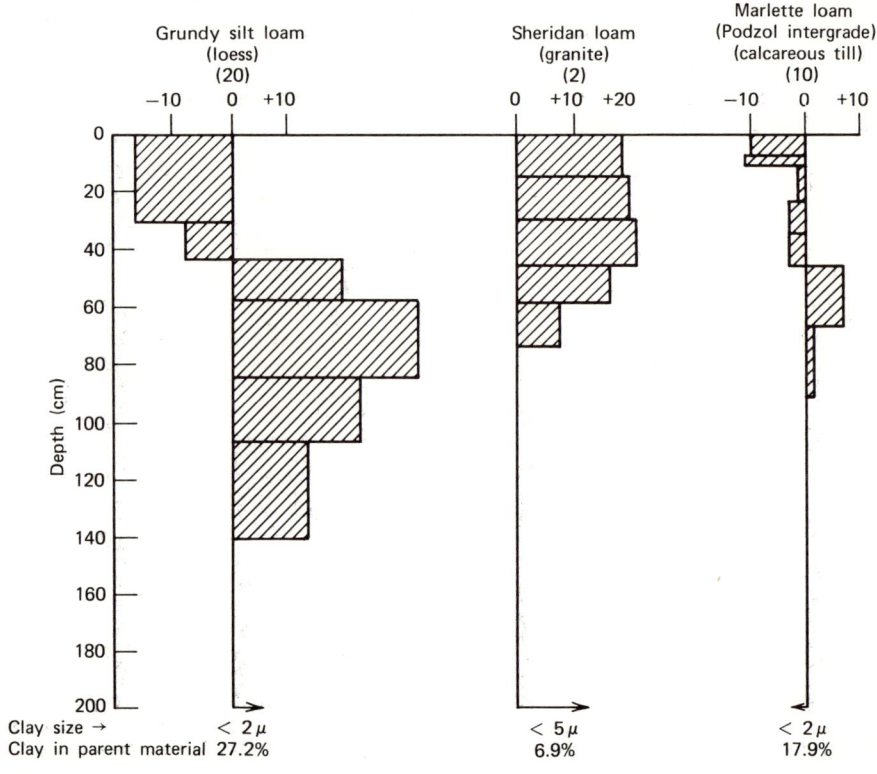

*Fig. 38*  Change in clay with depth per 100 g of parent material (PM) for three soil profiles.

tive aspects of the formation of clay from nonclay depend considerably on the size chosen as the upper limit for the clay. Hence a clear-cut calculation enabling one to separate clay formation from translocation becomes impossible, horizon by horizon, although Barshad's procedure in relating net loss of nonclay to net gain of clay for the whole profile is useful. It is clear that for accurate evaluation we need both precise mechanical analyses and complete quantitative mineralogical determinations on all fractions. It then becomes possible to follow in detail the relative rates of weathering of different minerals. Barshad's discussion of the Sheridan profile shows how the particle size distribution of the hornblende and mica changed with depth as weathering proceeded. Brewer's study of a soil from granodiorite gives a more detailed picture. Quantitative figures were obtained for quartz, feldspar, hornblende, and biotite, in relation to particle size distribution and to depth in the profile. Humbert and Marshall had earlier obtained quantitative figures for the major minerals in several particle size fractions of a soil from diabase and from granite in Missouri.

Shifts in the particle size distribution of a given mineral with increasing weathering may take several forms. The simple removal of successive layers would cause a shift to smaller sizes culminating in a sudden disappearance for units comprising only a few unit cells. This process has never been demonstrated. Normally weathering proceeds in cleavage cracks as well as externally, and the final disappearance occurs at much larger sizes than might be expected. Feldspars, pyroxenes, and amphiboles are rarely detected by X-ray methods in fine clay ($< 200\ \mu$). Quartz grains of sand size frequently contain cracks and break down physically at an early stage of weathering to silt, as in the granite soil studied by Humbert and Marshall. But even when large proportions of quartz are found in the sand and silt fractions, the amounts in the coarse clay are much less and generally become inappreciable in the fine clay. Nor does quartz necessarily dominate the latter in cases of its authigenic formation, as for example in the Missouri soil from diabase. Yet in the case of a Podzol from New Zealand the clay fraction was dominated by a mixture of quartz and cristobalite.

A major physicochemical factor here is probably the general increase in solubility with diminishing particle size. Iler (23) has shown by calculation that the solubility of quartz in water follows a steep curve as particle size diminishes. Taking the normal solubility as 0.0006% for large particles, it rises to 0.0007% for 100 m$\mu$, to 0.0028% for 10 m$\mu$, and to 0.012% for 5 m$\mu$. Amorphous silica probably follows a much less steep relationship but with larger solubilities. In the case of aluminosilicate minerals the complications in the solution process are such that no theoretical estimates of the change in solubility with decreasing particle size have been made. Feldspars have been recorded in the coarse clay of soils from glacial deposits but very rarely are found in the fractions smaller than 200 m$\mu$. Calcium carbonate is normally removed before fractionation, but it has been recorded in coarse clay.

In considering soils as assemblages of minerals with different particle size distributions, Jackson (24) was led to a demonstration of one aspect of his concept of the "weathering sequence" and the "weathering mean." The other complementary aspect came through the study of depth functions in relation to mineralogy. We deal with this whole concept in some detail later.

*Clay Movement.* The uncertainties in Barshad's procedure for estimating both clay formation and clay movement have focused attention on a procedure that distinguishes illuviated clay from sedentary clay. Buol, Brewer, Fitzpatrick, and others have shown that in many cases quantitative determinations can be carried out by microscopic methods in thin sections of soil horizons. They depend partly on the preferred orientation of illuviated clay when deposited on the walls of soil pores: this makes their recognition under polarized light relatively easy. Brewer gives details of the optical procedures (6). Clays of the 2:1 structure are presumably more mobile, more easily forming identifiable skins or cutans than kaolinitic clays or amorphous materials. This would be expected from their electrokinetic properties as described in Volume I (Chap. 9). However orientation in the illuvial deposition of clays may be of varying degrees of perfection and greatly affected by other mobile constituents, especially humic matter. Pallmann considers silicic acid to be a protective colloid in such movement (43). Thus quantitative determinations by the study of thin sections vary greatly in accuracy. There must always be a certain measure of uncertainty in interpretation, because clay constituents may conceivably move in true solution.

The inclusion of thin section studies adds much to the evidence provided by physical and chemical methods and is very directly indicative of mechanisms, even when quantitative measures are difficult to obtain. Thus Bourne and Whiteside (5) found oriented clay films in the B horizons of the Bozeman Chernozem formed from loess. They could not be recognized, however, in another Chernozem (Barnes) from North Dakota (Redmond and Whiteside, 45). Coatings of a nonoriented character were suspected.

Frei and Cline (17) found clear evidence of clay movement and deposition in a Gray-Brown Podzolic–Brown Podzolic soil sequence in New York (Ontario series).

Soils from much more arid regions have also been investigated. Nettleton, Flach, and Brashar (40) found argillic B horizons without oriented clay skins in a number of desert soils from California, Arizona, and Oregon. They concluded that swelling and shrinkage of montmorillonitic clays tended to reduce clay orientation.

Clay skins in the B horizon of the Cecil soil (Khalifa and Buol, 26) proved to contain more fine clay (< 0.2 $\mu$), chiefly kaolinite, with less gibbsite than the soil as a whole. Their thickness was from 20 to 300 $\mu$.

The differential character of clay movement from the A to the B horizons has been suspected for many years through studies of particle size distribution. The B horizons usually contain a higher proportion of the finest particles. Claypan soils containing montmorillonite and illitic minerals exhibit very large differences (Table 6). The question remains,

**TABLE 6** Clay Content and Clay Size Distribution in a Putnam Silt Loam (Planosol)

| Horizon | Depth (cm) | % < 2$\mu$ Clay | Coarse/ Fine | Coarse (%) 2$\mu$– 0.5 $\mu$ | Coarse (%) 0.5– 0.1 $\mu$ | Fine (%) 0.1– 0.02 $\mu$ | Fine (%) <0.02 $\mu$ |
|---|---|---|---|---|---|---|---|
| A | 0– 7.6 | 20.5 | 1.15 | 38 | 16 | 19 | 28 |
|   | 7.6–15.2 | 20.1 | 1.13 | 39 | 14 | 15 | 32 |
|   | 15.2–25.4 | 21.7 | 1.13 | 36 | 17 | 21 | 26 |
|   | 25.4–35.6 | 26.3 | 1.04 | 33 | 18 | 22 | 27 |
| B | 35.6–45.7 | 42.2 | 0.74 | 23 | 19 | 17 | 40 |
|   | 45.7–55.9 | 57.7 | 0.54 | 18 | 17 | 21 | 43 |
|   | 55.9–66.0 | 51.3 | 0.62 | 21 | 17 | 17 | 44 |
| C | 66.0–76.2 | 41.4 | 0.72 | 27 | 15 | 18 | 40 |
|   | 76.2–86.3 | 37.0 | 0.87 | 26 | 20 | 16 | 37 |
|   | 86.3–96.5 | 36.3 | 0.79 | 28 | 16 | 18 | 38 |

however, whether the clay was moved as such or was transported in solution and resynthesized in place. The analysis of drainage and lysimeter waters indicates a much greater $SiO_2/Al_2O_3$ ratio than that found in the fine clays; thus there does seem to be a strong case for movement as clay and not as components in true solution. In addition, syntheses by weathering of nonclay minerals must be considered, and this may be the dominant factor in some cases.

Two distinct types of movement are readily distinguished in extreme cases but are superimposed in many actual situations. The type of differential movement mentioned earlier can be idealized through the model of a filter bed with a static distribution of pore sizes. But many soils shrink sufficiently in dry periods that deep cracks are formed. Heavy rain then washes surface material into the cracks. Indeed in typical self-mulching soils dry earth may fall into the cracks, as exemplified by the Houston clay. This is a highly montmorillonitic, calcareous soil from shales, found extensively in eastern Texas.

Experimental confirmation of the filter-bed type of clay movement was obtained by Bodman and Harradine for six soils from California (4). These were packed into columns and upon completion of leaching they were sectioned. Mechanical analyses were carried out on each segment. The results relating to less than 1 $\mu$ clay are represented in Figure 39,

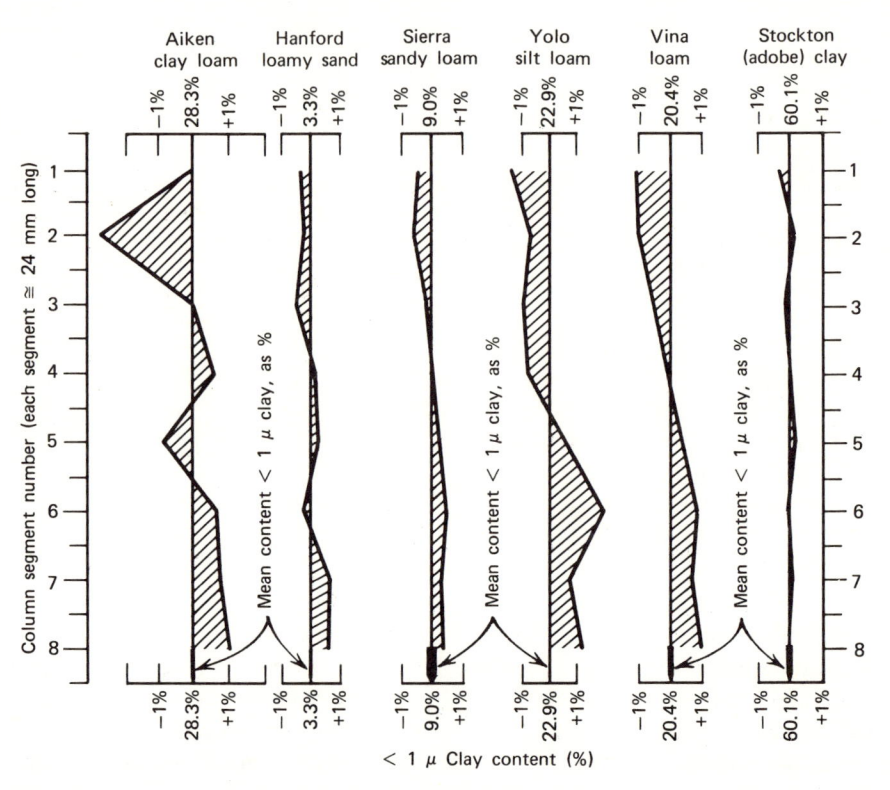

*Fig. 39* Change in clay in soil columns subjected to leaching. Reproduced from ref. 4 by permission of the Soil Science Society of America.

which gives the change in clay content as a function of depth. The Stockton clay showed no significant clay movement. This is readily accounted for in terms of its high montmorillonite content, its very small effective pore diameter, and the consequent minimal rate of water movement. The intensity of clay migration was calculated per gram of soil per centimeter of water movement. The values ranged from $30 \times 100^{-8}$ for the Yolo silt loam (montmorillonitic) to $2.3 \times 10^{-8}$ for the kaolinitic Aiken clay loam. The amounts of water used ranged from 591 to 8167 cm.

Brewer and Haldane studied the movement of an illitic clay (60% illite, 30% kaolin) into prepared mixtures of quartz, sand, and silt, with different concentrations of clay, different exchange cations, and different concentrations of salt (6). They concluded that the nature of the cation had no effect on clay orientation, that low concentrations of salt caused highly oriented clay coatings to form and that, as expected, flocculated clays give poor orientation. The presence of large proportions of silt in the filtering bed greatly reduced orientation.

Hallsworth (19) also used an artificial soil with kaolinitic and montmorillonitic clay suspensions. Clay movement was high for kaolinite suspensions up to 40% clay content, but montmorillonite suspensions blocked the pores at 20% concentration.

In the formation of heavy B horizons by the simple filter-bed mechanism, the early stages would correspond to the greatest penetration of mobile clay. The larger the initial pores, the longer such deep penetration continued. Thus materials high in clay to begin with and having small pores completed the filtering action within a relatively shallow depth. The rate of percolation soon decreased, and the B horizon remained shallow with a very sharply defined boundary between A and B. The relationship of clay peptization and movement in the A horizon to its arrest in the B is clearly a matter of the electrokinetic properties of the soil combined with the percolation characteristics of the profile. Heterogeneity in such properties is naturally to be expected. If only a small proportion of the total clay is mobile the B horizon will be less accentuated and its upper boundary less sharp. Over long periods of time the various surfaces exposed will have had a chance to equilibrate with their aqueous chemical environment. The latter cannot be thought of as constant: it varies through cyclic changes with a certain mean amplitude and period. These cyclic changes are themselves functions of depth in both respects. Thus experiments involving continuous percolation are only a partial model for natural soil systems.

The total distance over which clay movement occurs is difficult to estimate because in many cases the B horizon grades somewhat indistinctly into the C, which has a lower clay content. Brewer (6), as well as

Haseman and Marshall (29), concluded that considerable depths beyond identifiable B horizons may be involved. Mechanical analyses often show this but of course do not enable one to separate movement from formation in place. Then there are numerous cases in which clay content increases with depth beyond the normal range of B horizons. Many soils from limestones close to climatic regions where heavy B horizons normally form, show these characteristics. The whole group of Red-Yellow Podzolic soils, in which kaolinite is the dominant clay mineral, also shares this property. Simonson (47) considers that clay formation occurs in the lowest parts of the profile, but progressively more destruction of clay occurs as the surface is approached. There is X-ray evidence that the clay minerals of the A horizons of these soils shows poorer kaolinite lines than those from deeper layers.

### *Movement of Individual Elements*

Chemical analyses have, in a few cases, supplemented work done by the index mineral method. Barshad (2) quotes data for the A horizon of the Sheridan soil. Complete data for the Grundy profile were obtained by Matthews (36), but net gains in silica and alumina as well as in total mass were found, and these are difficult to account for, except through errors in the original zircon figures or slight nonuniformity of the parent material during deposition of the loess as suggested by Barshad (1,6). Table 7 gives

**TABLE 7** Losses in $A_1$ Horizon of the Sheridan Soil from Granite (2)

|  | Total | $SiO_2$ | $TiO_2$ | $Fe_2O_3$ | $Al_2O_3$ | CaO | MgO | $K_2O$ | $Na_2O$ |
|---|---|---|---|---|---|---|---|---|---|
| Parent material | 100.0 | 57.40 | 1.50 | 14.02 | 16.33 | 4.80 | 2.80 | 0.50 | 2.70 |
| Soil ($A_1$) | 86.6 | 53.15 | 1.06 | 9.64 | 14.10 | 3.84 | 1.88 | 0.26 | 2.61 |
| Loss | 13.4 | 4.25 | 0.44 | 4.38 | 2.23 | 0.96 | 0.92 | 0.24 | 0.09 |

Barshad's figures for the $A_1$ horizon of the Sheridan Prairie soil from granite (0–15.2 cm) calculated on an ignited basis.

It is clear that losses have occurred in all elements, including $Al_2O_3$, $Fe_2O_3$, and $TiO_2$. Even with this relatively mild weathering and consequent slight soil development, it would have been misleading to have chosen any one index element.

Another example was worked out by Mick (36) in the case of the St. Clair profile formed on calcareous till in Michigan. This is particularly interesting because the $A_2$ horizon is carbonate free, the B horizon contains a reduced amount of carbonates, and the C horizon chosen as parent material has 14.5% carbonates. Table 8 gives the results recalculated to

**TABLE 8** Changes per 100 g of Parent Material for the St. Clair Profile from Calcareous Till in Michigan (10)

| Horizon | Depth (cm) | | Change (g) | | |
|---|---|---|---|---|---|
| | | | SiO$_2$ | Fe$_2$O$_3$ | Al$_2$O$_3$ |
| A$_2$ | 5–20 | Original | 68.3 | 4.89 | 8.92 |
| | | Present | 48.2 | 2.39 | 3.84 |
| | | Loss | 20.1 | 2.50 | 5.08 |
| B$_2$ | 36–69 | Original | 68.3 | 4.89 | 8.92 |
| | | Present | 62.5 | 3.99 | 5.86 |
| | | Loss | 5.8 | 0.90 | 3.16 |
| C | >91 | Original | 68.3 | 4.89 | 8.92 |

the basis of 100 g of parent material. The losses of the A$_2$ horizon clearly indicate a podzolic type of process, since the proportion of sesquioxides lost is much greater than that of the silica. In the comparison of the B horizon with the C there may have been gains from the A horizon as well as losses, and the figures available do not enable us to separate them.

An interesting comparison of a Chernozem (Barnes) with an associated Solonetz profile (Cavour) has been made by Westin in South Dakota (50). The soils were less than 2 m apart and were sampled in the same trench. The parent material was glacial till. The ratio of zircon to tourmaline of the very fine sand fraction was almost constant down each profile, and it had the same magnitude for both. The Solonetz soil was in a very slight depression, as is usual in the mosaic pattern of these two components of a landscape. This feature caused alternate ponding and desiccation, whereas the Chernozem drained normally. Analyses were made of the upper horizons and of the assumed parent material at 122–153 cm depth. Zircon of the very fine sand fraction served as index mineral.

The assumed parent material contained soluble salts, carbonates, and exchangeable sodium, some of which were most probably of pedogenic origin. Conversion to the composition of the upper horizons for both profiles involved large losses of carbonates, of soluble salts, and of exchangeable sodium, as well as processes of "lessivage," entailing losses of clay from the A horizons and losses of silt from both A and B horizons. The final results showed clear-cut differences between the two soils. Clay and silt were apparently more mobile in the Solonetz than in the Chernozem. In the Solonetz, exchangeable sodium was appreciable in the A horizon (0.8 meq/100 g) and increased markedly in the B (4.6 and 7.4 meq/100 g), whereas the Chernozem had very little down to 33 cm. Carbonates and soluble salts were absent from the Chernozem profile down to 33 cm but appeared at 18–25 cm in the Solonetz. Some accumulations in the B horizon could be caused by seasonal upward movement of ground water as well as by lessivage and leaching.

Very similar associations of Chernozems with Solonetzes are found in various parts of the world. The author observed examples in northern Kazakhstan, where, in addition, the more accentuated depressions give rise to forest vegetation and the production of an acidic forest solod. Such soils are considered to arise by degradation of the Solonetz. The exchangeable sodium of the latter is much more easily replaced by hydrogen than is the calcium of the Chernozem. Hence this type of pedological change is easily initiated by a shift in vegetation from prairie to forest and is relatively rapid.

## REFERENCES

1. Barshad, I., Factors affecting clay formation. In *Clays and Clay Minerals, Proceedings of the 6th National Conference on Clays and Clay Minerals* (Berkeley, Calif., 1957), Pergamon Press, New York (1958), pp. 111–132.
2. Barshad, I., Soil development. In *Chemistry of the Soil*, F. E. Bear, Ed. American Chemical Society Monograph 126, Reinhold, New York (1955), Chap. 1.
3. Bear, F. E., Ed., *Chemistry of the Soil,* American Chemical Society Monograph 126, Reinhold, New York (1955).
4. Bodman, G. B. and E. F. Harradine, Mean effective pore size and clay migration during water percolation in soils, *Soil Sci. Soc. Am., Proc.*, **3,** 44 (1938).
5. Bourne, W. C. and E. P. Whiteside, A study of the morphology and pedogenesis of a medial Chernozem developed in loess. *Soil Sci. Soc. Am. Proc.*, **26,** 484 (1962).
6. Brewer, R., *Fabric and Mineral Analysis of Soils,* Wiley, New York (1964).
7. Buol, S. W. and F. D. Hole, Clay skin genesis in Wisconsin soils, *Soil Sci. Soc. Am. Proc.*, **25,** 377 (1961).
8. Buol, S. W. and F. D. Hole, Some characteristics of clay skins on peds in the B Horizon of a gray-brown podzolic soil, *Soil Sci. Soc. Am. Proc.*, **22,** 239 (1958).
9. Buol, S. W. and M. S. Yesiloy, A genesis study of a Mohave sandy loam profile, *Soil Sci. Soc. Am. Proc.*, **28,** 254 (1964).
10. Cann, D. B. and E. P. Whiteside, A study of the genesis of a podzol-gray brown podzolic intergrade soil in Michigan, *Soil Sci. Soc. Am. Proc.*, **19,** 497 (1955).
11. Carroll, D., Weatherability of zircon, *J. Sed. Petrol.*, **23,** 100 (1953).
12. Druif, J. H., Aantastung van mineralen in den boden van Deli, Deli Proefstation, Medan; Sumatra, Bull. 38 (1937).
13. Dryden, L. and C. Dryden, Comparative rates of weathering of some common heavy minerals, *J. Sediment Petrol.*, **16,** 91 (1946).
14. Evans, D. D. and S. W. Buol, Micromorphological study of soil crusts, *Soil Sci. Soc. Am. Proc.*, **32,** 19 (1968).
15. Fitzpatrick, E. A., *Pedology, A Systematic Approach to Soil Science,* Oliver and Boyd, Edinburgh (1971).

16. Franzmeier, D. P., E. P. Whiteside, and M. M. Mortland, A chronosequence of Podzols in northern Michigan. III, *Mich. Agr. Exp. Sta. Quart. Bull.,* **46,** 37 (1963).
17. Frei, E. and M. G. Cline, Profile studies of normal soils of New York. II. Micromorphological studies of the grey-brown podzolic-brown podzolic soil sequence, *Soil Sci.,* **65,** 33 (1949).
18. Gerasimov, I. P. and M. A. Glazovskaya, *Fundamentals of Soil Science and Soil Geography,* Israel Program of Scientific Translations, Jerusalem (1965).
19. Hallsworth, E. G., An examination of some factors affecting the movement of clay in an artificial soil, *J. Soil Sci.,* **14,** 360 (1963).
20. Haseman, J. F. and C. E. Marshall, The use of heavy minerals in studies of the origin and development of soils, University of Missouri Agricultural Experimental Station Research Bulletin 387 (1945).
21. Harrison, J. B., The katamorphism of igneous rocks under humid tropical conditions, Imperial Bureau of Soil Science, Harpenden, England (1934).
22. Humbert, R. and C. E. Marshall, Mineralogical and chemical studies of soil formation from acid and basic igneous rocks in Missouri, University of Missouri Agricultural Experimental Station Research Bulletin 359 (1943).
23. Iler, R. K., *The Colloid Chemistry of Silica and Silicates,* Cornell University Press, Ithaca, N.Y. (1955).
24. Jackson, M. L. et al., Weathering sequence of clay size minerals in soils and sediments. I. Fundamental generalizations, *J. Phys. Colloid Chem.,* **52,** 1237 (1948).
25. Joffe, J. S., Lysimeter studies, I, *Soil Sci.,* **34,** 123 (1932); II. The Movement and translocation of soil constituents in the soil profile, *Soil Sci.,* **35,** 239 (1933).
26. Khalifa, E. M. and S. W. Buol, Studies of clay skins in a Cecil (Typio Hapludult) soil. I. Composition and genesis, *Soil Sci. Soc. Am. Proc.,* **32,** 857 (1968).
27. Khangarot, D. S., L. P. Wilding, and G. F. Hall, Composition and weathering of loess mantled Wisconsin and Illinoian-age terraces in central Ohio, *Soil Sci. Soc. Am. Proc.,* **35,** 621 (1971).
28. Köster, H. M., Vergleich einiger Methoden zur Untersuchung von geochemischen Vorgängen bei der Verwitterung, *Beitr. Mineral. Petrogr.,* **8,** 69 (1961).
29. Kubiena, W. L., *Micromorphological Features of Soil Geography,* Rutgers University Press, New Brunswick, N.J. (1970).
30. Marshall, C. E., A petrographic method for the study of soil formation processes, *Soil Sci. Soc. Am. Proc.,* **5,** 100 (1940).
31. Marshall, C. E. and C. D. Jeffries, Mineralogical methods in soil research. Part 1. The correlation of soil types and parent materials, with supplementary information on weathering processes, *Soil Sci. Soc. Am. Proc.,* **10,** 397 (1945).
32. Marshall, C. E. and J. F. Haseman, The quantitative evaluation of soil formation and development by heavy mineral studies: A Grundy silt loam profile, *Soil Sci. Soc. Am. Proc.,* **7,** 448 (1942).
33. Marshall, C. E. and L. L. McDowell, The surface reactivity of micas, *Soil Sci.,* **99,** 115 (1965).
34. Marshall, C. E., M. Y. Chowdhury, and W. J. Upchurch, Lysimetric and chemical investigations of pedological changes. Part 2. Equilibration of profile samples with aqueous solutions, *Soil Sci.,* **116,** 336 (1973).
35. Matelski, R. P. and L. M. Turk, Heavy minerals in some podzolic soil profiles in Michigan, *Soil Sci.,* **64,** 469 (1947).

36. Matthews, B. C., Chemical studies of soil formation and development in the Grundy silt loam, M.A. thesis, University of Missouri (1948).
37. Merrill, G. W., *Rocks, Rock Weathering and Soils*, Macmillan, London (1906).
38. Mick, A. H., The pedology of several profiles developed from the calcareous drift of eastern Michigan, Michigan Agricultural Experimental Station Technical Bulletin 212 (1949).
39. Mickelson, G. A., Mineralogical composition of three soil types in Ohio with special reference to change due to weathering as indicated by resistant heavy minerals, Abstr. doctoral Diss., Ohio State University, **40** (1943).
40. Nettleton, W. D., K. W. Flach, and B. R. Brasher, Argillic horizons without clay skins, *Soil Sci. Soc. Am. Proc.*, **33**, 121 (1969).
41. Nikiforoff, C. C. and M. Drosdoff, Genesis of a clay-pan soil, *Soil Sci.*, **55**, 459 (1943).
42. Oertel, A. C., Pedogenesis of some red-brown soils based on trace element profiles, *J. Soil Sci.*, **12**, 242 (1961).
43. Pallmann, H., Grundzüge der Bodenbildung, *Schweitz. Landwirtsch. Monatsh.* **22**, 1 (1942).
44. Raeside, J. D., Stability of index minerals in soils with particular reference to quartz, zircon and garnet, *J. Sediment. Petrol.* **29**, 493 (1959).
45. Redmond, C. E. and E. P. Whiteside, Some till-derived Chernozem soils in Eastern North Dakota. II. Mineralogy, micromorphology and development, *Soil Sci. Soc. Am. Proc.*, **31**, 100 (1967).
46. Rutherford, G. K., Ed., *Soil Microscopy*, Limestone Press, Kingston, Ontario (1964).
47. Simonson, R. W., Genesis and classification of red-yellow podzolic soils, *Soil Sci. Soc. Am. Proc.*, **14**, 316 (1949).
48. Smith, B. R. and S. W. Buol, Genesis and relative weathering intensity studies in three semiarid soils, *Soil Sci. Soc. Am. Proc.*, **32**, 261 (1968).
49. Upchurch, W. J., M. Y. Chowdhury, and C. E. Marshall, Lysimetric and chemical investigations of pedological changes. Part I. Lysimeters and their drainage waters, *Soil Sci.*, **116**, 266 (1973).
50. Westin, F. C., Solonetz soils of eastern South Dakota: Their properties and genesis, *Soil Sci. Soc. Am. Proc.*, **17**, 287–293 (1953).
51. Whiteside, E. P. and C. E. Marshall, Mineralogical and chemical studies of the Putnam silt loam soil, University of Missouri Agricultural Experimental Station Research Bulletin 386 (1944).
52. Wild, A., Loss of zirconium from 12 soils derived from granite, *Aust. J. Agr. Res.*, **12**, 300 (1961).
53. Yassaglou, N. J. and E. P. Whiteside, Morphology and genesis of some soils containing fragipans in northern Michigan, *Soil Sci. Soc. Am. Proc.*, **24**, 396 (1960).

# 6 Characterization of the pedological environment and its products: Physical and chemical methods as applied to soil profiles

The following list of methods begins with characterization of the external environment as a first step in the understanding of the internal environment (Chap. 6). Field observations and descriptions are presented next, then we discuss the use of thin sections and other observations on the micro scale, and finally standard quantitative physical methods (Chap. 7). In this way a relatively complete physical description of the soil profile can be achieved.

The next group (Chap. 8) comprises mineralogical analysis with its important application through index mineral determination, as already discussed in Chapter 5.

Chemical analyses for mineral elements (Chap. 9) can be applied to the whole soil, to extracts obtained by the use of powerful reagents, and to the clay fraction.

The distribution and characterization of soil organic (Chap. 9) matter is clearly of great importance where the latter dominates the exchange complex.

The ionic properties of the exchange complex (Chap. 10) include the amounts and properties of exchange cations and anions, ionic selectivity, cationic bonding energies, and characterization through small dilute exchanges.

The soil solution in its composition and variation is related both to the exchange complex and to mineral equilibria or approaches to equilibrium.

Finally the electrokinetic properties of soils are of great importance in connection with the movement and arrest of soil constituents in developmental processes. Quantitatively this field of study offers virgin territory for the investigator.

All these categories are set out below with their practical subdivisions. Of course no soils have been investigated using all these methods, but the relevance of each is discussed and illustrated by examples from the literature.

## CLIMATIC FACTORS

### *Temperature*

The general relationship of soil temperature to atmospheric temperature in its diurnal and annual cycles has been discussed in Chapter 4. In comparing conditions in different parts of the world it would be useful to have internal measurements of an integrating character. Pallmann et al. (17) developed chemical methods based on the kinetics of particular reactions, by which the integral effect over a period of weeks or months can be measured. This should give a number roughly proportional to other chemical effects such as the incidence of weathering reactions and the oxidation of organic matter to $CO_2$. The main drawback is that application of the method really applies to continuously humid conditions and does not allow for dry soil in certain periods.

It has frequently been pointed out that the large qualitative differences between upland soils in the tropics as compared with those of temperate regions are difficult to explain in terms of chemical kinetics. It has been mentioned, that in the weathering sequence from feldspars to oxides and hydroxides there might be one step with an unusually high temperature coefficient, corresponding to a very high activation energy. This could even involve a thermodynamically unstable species, since we know that reactions often proceed through such steps. There is, however, no evidence yet available that identifies such a reaction.

Apart from kinetic effects, equilibrium conditions on mineral reactions at different temperatures are sometimes surprisingly sensitive to relatively small changes. Thus the diagrams presented by Helgeson et al. (10) for the system $KCl$-$Al_2O_3$-$SiO_2$-$H_2O$ show that muscovite completely disappears as a stable species below some temperature between 25 and 0°C.

## Precipitation

It has long been realized that total annual precipitation provides only a very crude measure of effectiveness in pedological terms. Water is essential for the continuance of three general processes—mineral transformation, physical transport, and biological activity—which together account for the major part of pedological change in soil profiles.

Various measures of effectiveness have been proposed. Penck used the ratio of precipitation to evaporation; Lang proposed combining precipitation and temperature in a rain factor $P/T$; Meyer suggested the $N \cdot S$ quotient defined by the ratio of precipitation to absolute saturation deficit of the atmosphere. Thornthwaite has shown the advantages of a precipitation·effectiveness index calculated on a monthly basis and summed over the year:

$$I = \Sigma^{12}\left[115\left(\frac{P}{T-10}\right)^{10/9}\right]$$

All these are, of course, external to the soil and mainly useful in relation to a very general definition of major soil zones in relation to climatic zones. As soon as the detail of soil profile–moisture relationships is examined, much more complex considerations become apparent. How do the external climatic cycles reflect themselves in internal cycles? The latter clearly need to be examined in relation to individual soil horizons, and this brings with it the further complexity that soil layers have a strong influence on the moisture regimes of adjacent layers.

Considering a single thin layer of soil, under the most general conditions, three characteristic cycles operate. First, the inactive periods of dryness, frost and drought, can be represented by a certain mean wavelength and amplitude. For instance, the wavelength might be a 100 days and the amplitude 26 days. Second, cyclic variations from the mean moisture content can be characterized by their mean wavelength and amplitude, the former expressed in days and the latter in inches of water or percentage water content above or below the mean. Third, the occasions when moisture passes from the given layer to that below can be described in terms of a mean interval (wavelength) and a mean amount transmitted (amplitude). Passage of water from layer to layer through the whole soil with eventual entry into the drainage is an irreversible process and in many soils can be treated as a rare event.

As layers at increasing depths are considered under humid conditions, the dry periods become less frequent until they may disappear entirely. The cyclic changes diminish in amplitude, and throughput of water becomes smaller in amount and less frequent.

In terms of the soil solution this means that in an initially uniform soil and under moderate rainfall, the layers near the surface have both the longest dry periods and the greatest throughput of water. The time available for mineral interaction with water is a minimum, but losses in solution are greater than for deeper layers. In many cases mineralogical examination of the clay fraction has shown more weathering in the A horizons than at greater depths, indicating that the second factor is of greater importance than the first.

The net movement of water in any particular layer of soil is affected also by evaporation from the surface, or by evapotranspiration, which includes the effects of plant cover. It can be evaluated experimentally by a system of weight lysimeters or by calculation from weather data. In general it is simply included in the total water balance without any attempt to assign its magnitude for different layers of the soil. Clearly it would be expected to diminish with depth. A recent study by Scrivner and Horn (12) in which the moisture regime of the Menfro soil (loess) under forest was examined in detail showed that a logarithmic equation described well the water loss by evapotranspiration as a function of depth. The details on this soil are given in Chapter 7, Figures 48 and 49.

A slightly different emphasis in the evaluation of moisture cycles in their effects upon soil forming processes was used by Scrivner, Baker, and Brees (22). Parent material was kept relatively constant by consideration of deep loess adjoining the Missouri and Mississippi river valleys. From northwest to southeast Missouri, rainfall increases from about 89 to 127 cm (35–50 in.). The potential evapotranspiration was calculated from climatic data extending over 40 years, and by subtraction from the total rainfall the excess was obtained. The number of months out of 500 total in which the soil was completely recharged with water was also determined. This represents the total period in which possible drainage might occur. Table 9 gives the main results of this study. Because of the rather uniform precipitation distribution, with a weak minimum in the winter at all sites,

**TABLE 9a** Months of Complete Moisture Recharge out of 500 total; Excess Moisture Over Potential Evapotranspiration (22)

| Station (Missouri) | Mean Annual Rainfall (cm) | % Excess Moisture Pot. Evapotrans. | Months of Recharge |
|---|---|---|---|
| Tarkio | 86 | 10.1 | 75 |
| Brunswick | 94 | 10.6 | 81 |
| Columbia | 94 | 17.7 | 115 |
| Sikeston | 119 | 34.8 | 130 |

**TABLE 9b**  Soil Characteristics for Soils Derived from Loess

| Station | Soil Type | Classification | Clay (max %) | Depth to $B_2$ (cm) |
|---|---|---|---|---|
| Tarkio | Marshall | Mollisol | 31 | 46 |
| Brunswick | Grundy | Mollisol | 45 | 38 |
| Columbia | Mexico | Alfisol | 55 | 30 |
| Sikeston | Memphis | Alfisol (fragipan) | 35 | 23 |

the months of recharge are almost all in the spring of the year. A 40% increase in rainfall is reflected in a 225% increase in excess moisture and a 73% increase in months of recharge. Soil-forming processes should reflect this increase in throughput of water, and according to Table 9*b* they clearly do so.

Evaporation and transpiration cause upward movement of mineral elements, the former by capillarity over relatively short distances, the latter by accumulation through the action of roots to considerable depths and movement into the tops of plants, with later release into the soil solution and the exchange complex. This action is distinctive because of the high ionic selectivity of plants, especially with regard to the alkali cations sodium and potassium. In most soils at depth exchangeable sodium and potassium are comparable in amount, the sodium commonly somewhat exceeds the potassium. Upward movement by evaporation would carry distinctly more sodium than potassium into the topmost layers, because of its greater quantity in the equilibrium soil solution. On the other hand, transpiration by plants with subsequent decomposition will release at the soil surface much more potassium than sodium. Hence it is common to find that the ratio of exchangeable potassium to exchangeable sodium is relatively high in the A horizon, even though the expressed soil solution may contain more soluble sodium than potassium. This reflects the relatively high bonding energy of potassium in the exchange complex as compared with sodium, especially in soils containing 2:1 clay minerals. Illustrations of these effects are given below.

## GEOLOGICAL FACTORS

Geological factors can be divided into two groups, those relating to topography and those concerning parent materials. Since topography is already recognized as an independent soil-forming factor, it is dealt with separately.

## Parent Materials

The importance of parent materials in determining the characteristics of soils was recognized long before the Russian investigators placed climatic factors in the forefront of discussion. There are still many areas recognized in which factors originating in the parent materials are dominant. This is true, for instance, of limestone soils in many parts of the world. Juvenile soils in general still bear the stamp of the original parent rocks.

Some discussion, ably summarized by Brewer (5), has been concerned with the question of whether the original rock itself (consolidated or not) should be taken as parent material, or whether the C horizon in the form of incoherent saprolitic material should be used pedologically as the material on which soil-forming processes have operated. The present author agrees with Brewer and with Whiteside (24) that because of the impossibility of separating weathering reactions of the original material from pedological processes, one should refer back to the rock itself. Unconsolidated materials in the unweathered condition are taken as rocks for this purpose.

The recognition of the part played by parent materials depends on evaluations of their original physical and chemical properties and of the manner in which these have been modified in the formation of soil horizons. Thus comparisons between soil horizons and with underlying rocks form the central experimental theme. The methods are generally physical, mineralogical, and chemical, as described below.

The highly detailed study of the heavy minerals, which is necessary in deciding difficult cases of provenance, has been dealt with in Chapter 5 as a means of distinguishing between geological, depositional, and pedological differences.

The modern view of the reversibility of weathering reactions, with its emphasis on the immediate chemical environment, carries important connotations in pedology. These can well be illustrated using limestone soils as examples.

Assuming that the coherent rock has undergone no further change since its final consolidation, we now consider its change in environment. First, there is a moderate to large reduction in pressure to atmospheric, as compared with tens to hundreds or thousands of atmospheres at final consolidation. This affects chiefly the volatile components, water and carbon dioxide, through the thickness of adsorbed films, the equilibria with hydrous species of minerals, and the equilibria with carbonates.

An important equilibrium involving muscovite and kaolinite can be written:

$$2KAlSi_3O_6Al_2O_4(OH)_2 + 2H^+ + 3H_2O \rightleftarrows 3Si_2O_3 \cdot Al_2O_2(OH)_4 + 2K^+$$

Then

$$K_r = \frac{[K^+]^2}{[H^+]^2 [H_2O]^3}$$

Under atmospheric conditions the activity of water in the dilute aqueous system is taken as unity, but at high pressures it is higher, roughly proportional to the pressure. The free energy difference per atmosphere is about 0.43 cal/mole $H_2O$. Thus the value of $K_r$ decreases as the cube of the pressure increases, and the reduction in pressure from high values to atmospheric favors muscovite as compared with kaolinite, but the free energy differences are small fractions of a kilocalorie. However exposure to rainfall changes conditions even more drastically. The consolidation of a limestone involves the "locking in" of small quantities of seawater and their attendant salts. The first change experienced by the limestone upon exposure to rainwater is the loss of these soluble salts. Thus $[K^+]$ in the foregoing equation will fall drastically, probably by a factor of hundreds to tens of thousands; $[H^+]$ will remain constant because of the carbonate buffer system. Hence kaolinite will be favored. Thus the saprolitic rock in the presence of water tends to contain more kaolinite and less muscovite or illite than the original limestone. A further drastic change occurs when the limestone buffer system disappears by solution of the rock. The pH falls from 8.4 to much lower values; that is, $[H^+]^2$ rapidly increases. This again pushes the equilibrium toward the kaolinitic side. According to Figure 40, muscovite has a very narrow range of stability in terms of variation in silicic acid activity ($10^{-4.1}$ to $10^{-4.2}$), the ratio $a_{K+}/a_{H+}$ being fixed at $10^{5.7}$. Below this muscovite is not stable. Thus weathering, which tends to decrease $a_K/a_H$, promotes kaolinite. This passes into gibbsite independently of the $a_K/a_H$ ratio when the silicic acid in solution drops below $10^{-4.2}$. Theoretically it cannot do this in the presence of quartz, which tends to maintain a silicic acid activity of $10^{-4.0}$ at 25°C. However since this a very sluggish system, mixtures of quartz and gibbsite may persist for long periods.

The actual situation in limestones usually involves micas in the compositional range illite-glauconite, which lose silica upon production of kaolinite. The mica-kaolinite boundary is no longer horizontal, and the area for illite stability is greater than that for muscovite. The $K^+/H^+$ activity ratio is about $10^{3.5}$ (Reesman), considerably lower than for muscovite.

Thus even under mild conditions of weathering it is possible for a profile from uniform material to contain different proportions of the clay minerals in different horizons.

The literature contains numerous examples of different geological materials producing different soils under the same climatic conditions. This is of course commoner under mild conditions than under extreme weath-

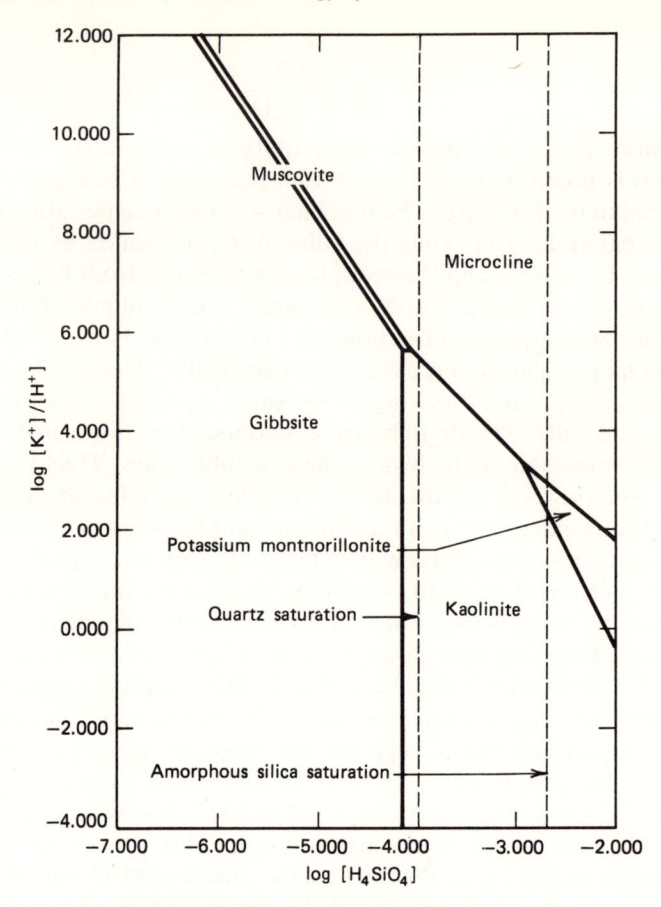

**Fig. 40** Garrels and Christ diagram for the system $HCl$-$H_2O$-$Al_2O_3$-$K_2O$-$SiO_2$ at 25°C (10). Used by permission, Freeman, Cooper and Co.

ering, but even in the humid tropics the parent material can make itself felt. Some extremely well-documented cases are given in Harrison's monograph (9). The whole tenor of this impressive work is that both the parent material and the detailed conditions of drainage and rainfall affect the weathering of rocks and the production of soils. For instance at Issorora Hill in Guyana, where the rainfall is about 297 cm (117 in.) annually, very distinct differences were found between a hornblende schist (5% quartz, 40% feldspar, 55% hornblende) and an epidiorite (amphibolite) (0 quartz, 55% feldspar, 44% hornblende). The surficial soil was somewhat similar, being kaolinitic lateritic earth containing quartz in both cases; but the epidiorite, as the more basic rock, gave more primary laterite (gibbsite) near the weathering rock contact. The hornblende schist showed more kaolinization of the primary laterite and much greater amounts of quartz in the soil layers. Harrison's whole outlook on rock weathering is very much in line with modern concepts.

Under less extreme conditions, although differences in soils from different rocks can persist, similarities also show over long periods. Humbert and Marshall compared the Ashe stony loam in southeastern Missouri with an unnamed soil developed on a diabase boss in the same area. The rainfall is about 112 cm, well distributed through the year. Both soils are relatively old, having been produced near the tops of ridges by weathering of Precambian rocks well to the south of glaciated areas in Missouri. These soils are compared in Figures 35 and 36 (Chapter 5) and in Table 10. The nature of the clays was deduced mainly from the cation

**TABLE 10** Properties of Two Soils from Igneous Rocks in Southeastern Missouri

| Depth (cm) | Proportion $<2\mu$ Clay (%) | Exchange Capacity of Clay (meq/100 g) | Exch. [Ca + Mg]/ [Na + K] | Equiv. Rock [Ca + Mg]/ [Na + K] |
|---|---|---|---|---|
| \multicolumn{5}{c}{Granitic Soil (Ashe Stony Loam)} |
| 0–10 | 14.5 | 60 | 3.16 | |
| 10–25 | 23.2 | 51 | 7.17 | |
| 25–46 | 39.9 | 60 | — | |
| 46–61 | 35.9 | 59 | 16.9 | |
| 61–76 | 24.1 | 53 | 6.85 | |
| Rock | — | — | — | 0.265 |
| \multicolumn{5}{c}{Diabase Soil} |
| 0–13 | 13.4 | 49 | 3.45 | |
| 13–23 | 26.1 | 57 | 4.83 | |
| 23–41 | 32.0 | 59 | 10.70 | |
| 41–53 | 29.2 | 50 | 27.06 | |
| 53–64 | 26.7 | 49 | 11.13 | |
| 64–84 | 33.0 | 46 | 62.35 | |
| 84–119 | 32.7 | 45 | 38.28 | |
| 119–152 | 15.9 | 54 | 8.83 | |
| Rock | — | — | — | 9.00 |

exchange capacities, refractive indices, and appearance in electron micrographs. At the time of this study X-ray methods had not reached their present sensitivity.

The results indicate considerable similarity in the clays formed from the two rocks. They are probably either mixtures of beidellite-nontronite smectites with hydrous micas, or mixed layer clays of similar composition. The exchange capacities are intermediate between those of beidellites and illites, and they vary relatively little down the two profiles. The mineralogical composition of the coarser fractions has been presented in Figures 35 and 36. In the diabase profile quartz and chalcedony appear as secondary minerals in approximately equal amounts.

The ratios of divalent to monovalent cations in the exchange complexes and the original rocks are interesting. In spite of large differences in the two rocks, the exchange complexes near the surface have reached very similar ratios that widen in the subsoil, then decrease again near the rock surface. A very intense loss of monovalent cations takes place in the subsoil so that the ratio of divalent to monovalent becomes large. Nearer the surface, the decomposition of plant residues exerts an influence; their high content of potassium causes a reduction in the ratio of divalent to monovalent cations. This ratio varies for different crops, but it is often in the range 2–4, similar to that of the surface soils.

## BIOTIC FACTORS

### Vegetation

Ordinary vegetation exerts its effects on soil formation and development in three distinct sets of ways, one internal, one external, and one that begins externally but ends internally. They are, respectively, root action, canopy effects, and leaf fall.

*Internal effects of roots.* The most obvious effects of roots are physical. That is, roots actively penetrate the soil, thus establishing a new set of pathways for the transmission of water through active transpiration. This is mainly one-way traffic, carrying water from the soil reserves to the tops and so to the atmosphere. The original capillary pores of the soil are not one way, although downward movement of water dominates under most conditions. Upon death of the roots, their one-way function ceases completely. Processes of decomposition occur, soil organic matter may or may not remain, and the channels occupied by the roots now become ordinary capillary conductors. They may then become especially prominent in providing surfaces for the deposition of translocated clay, as has frequently been demonstrated in the last 30 years (Brewer 5).

The roots of most species are highly concentrated in and near the topmost layers of soil. Roots that penetrate more deeply tend to be larger in diameter and less branched. Their channels after decomposition are more easily recognized than the smaller and more numerous channels near the surface. A mat of roots near the soil surface effectively breaks up the original soil structure and imposes a distinctly granular character. This physical action is strongly reinforced by the action of aqueous exudates while the root is still living, and of polymeric carbohydrates, pectins, lignins, and humic compounds produced when the root decomposes. All the compounds named tend to stabilize the granular structure initiated by root elongation and ramification. These effects do not persist indefinitely after removal of the original vegetative cover. Thus after the plowing of a virgin prairie the surface plow layer changes to a denser and much more easily eroded mineral soil within very few seasons.

The figures given by Troughton (23), part of which are quoted in Table 11, indicate something of the physical dominance of grass roots near the

**TABLE 11** Percent Dry Weight of Roots in Soil Layers (23)

| Species | 0–10 | 10–20 | 20–30 | 30–40 | 40–50 |
|---|---|---|---|---|---|
| *Holars lanetus* | 51 | 16 | 18 | 11 | 4 |
| *Grastis vulgaris* | 59 | 25 | 9 | 5 | 1.5 |
| *Festuca rubra* | 73 | 13 | 5 | 4 | 5 |
| *Poa trivialis* | 83 | 9 | 5 | 2 | 1 |
| *Lolium perenne* | 80 | 9 | 5 | 3 | 2 |
| *Poa pratensis* | 62 | 29 | 4 | 4 | 1 |
| *Trifoleum repens* | 82 | 6 | 8 | 3 | 1 |

Depth (cm)

surface of pastures. In the 0–10 cm layer they constitute 50–80% of the dry mass. The diminution with depth varies considerably according to the species. In such swards the root biomass considerably exceeds that of the tops.

Forest conditions are considerably different, but again a large proportion of the fine roots are concentrated near the soil surface. The annual amount of leaf fall can be very high. The mass of roots per unit mass of soil tends to be lower near the surface than for pastures or natural prairie. The ratio of biomass of roots to that of tops is low for trees of mature height; a figure around 0.25 is often used. Thus in forest soils the vegetative factor tends to express itself chiefly through the leaffall and through canopy effects, whereas in prairie soils direct effects of roots on the soil are dominant.

*Canopy effects.* The physical effects of plant cover take rather obvious forms. The shading of the soil reduces its daytime temperature near the surface and diminishes the daily spread between extremes. Shade-loving species are encouraged to grow.

Light rains are largely intercepted by the canopy, and a considerable part of their rainfall never reaches the soil. Thus evapotranspiration is effectively increased, and throughput of water decreased. These effects vary considerably according to the species involved, and with deciduous plants they are more evident in summer than in winter.

Plant canopies can also exert strong chemical effects through the dissolution from the leaves of organic compounds having strong complexing action. Such leaf drip can initiate marked processes of soil development. The classic example of this is the Kauri pine, under whose canopy marked podzolization occurs in areas where the soils are not podzolic under other species. Thus each tree of this species produces its own Podzol shadow in a relatively short time.

Bloomfield (3) has pointed out the importance of such leaf drip. It is not necessarily acidic, although the most striking example, that of the Kauri pine (*Agathis australis*), gives highly acidic leaf extracts with very strong complexing for iron and aluminum. The resulting Podzol is directly under the canopy, being most highly developed near the trunk and thinning out toward the perimeter. The characteristic shape of the soil features under each tree has given rise to the name Basket Podzol.

Complexing compounds are present in the leaves of certain species as neutral salts and may give rise to eluviation and deposition of iron and aluminum very similar in appearance to the classical Podzol. Bloomfield demonstrated that decomposition of the leaf fall gave rise to humus that showed a weaker complexing action than the aqueous extract of fresh leaves. Thus in the formation of Podzols it is very difficult to separate the effects of humus from those of fresh leaf drip and leachates from undecomposed leaf litter.

A canopy effect of a different kind—namely, the special incidence of water that flows down the stem of a tree—has been investigated by Gersper and Holowaychuk (8). These authors group stem flow, leaf drip, and canopy interception together as the biohydrologic factor in soil development. The example chosen was a beech tree (*Fagus grandiflora*) growing on the Miami soil in Ohio. They showed that in accord with the increased throughput of water near the stem, the soil there exhibited greater podzolization tendencies than at greater distances. The base saturation in particular was markedly lower near the stem, and the organic matter was higher.

*Leaf fall.* The chemical composition of the leaf fall has long been regarded as a controlling factor in soil development. Grasses and legumes contain higher percentages of potassium plus calcium and magnesium than do the leaves of trees. Among the latter, sharp distinctions exist. Thus beech forests in northern Europe normally give rise to "brown forest" or weakly podzolized soils, whereas pines produce more advanced degrees of podzolization. Beech litter is much less acidic and is higher in bases than pine litter. Differences of these kinds show themselves most strongly in temperate climates where there is persistence of humus in the soil profile. If biochemical processes of breakdown are rapid, as in the tropics, they become less apparent.

From what has been said of canopy effects, it is clear that leaves giving neutral water extracts can still exert effective complexing action on iron and aluminum. In recent years examples have been studied in which apparent podzolization has proceeded without appreciable development of soil acidity.

The general influence of vegetative factors under temperate conditions can well be seen in Figure 41 and Table 12. This compilation gives the characteristics of three soils formed from loess in a region of gradation from the Brunizem to the Gray-Brown Podzolic profile. In the table

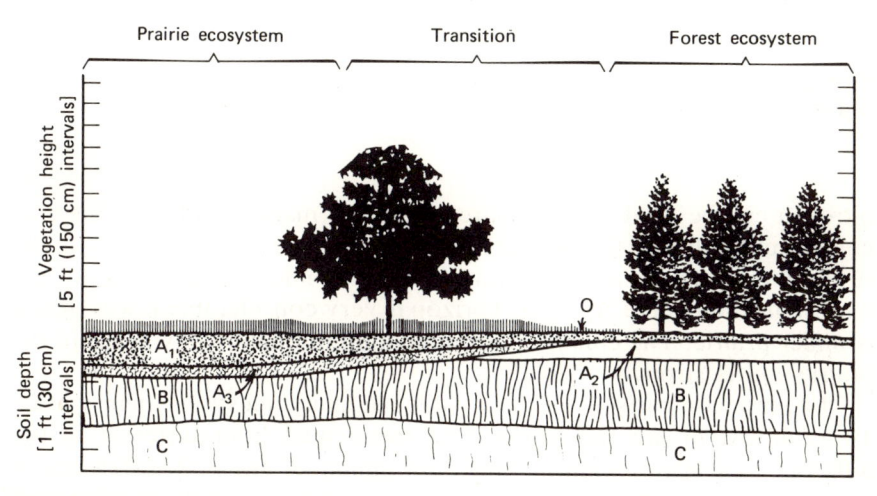

*Fig. 41* Diagram of organism-soil relationships at a prairie-forest border. Reprinted by permission from ref. 6, Iowa State University Press, Ames, Iowa.

**TABLE 12** Some Properties of Soils Developed Under Prairie, Prairie-Border, and Nearby Deciduous Forest Cover in the North Central Region of the United States (6)

| Soil Property | Soil Formed Under Prairie | Soil Formed Under Prairie-Border Conditions | Soil Formed Under Deciduous Forest |
|---|---|---|---|
| Clay content (%) | | | |
| A horizon | 28 | 23 | 21 |
| B horizon | 34 | 36 | 36 |
| Ratio, B:A | 1.21 | 1.57 | 1.71 |
| Total nitrogen (%) | | | |
| $A_{11}$ (0–7 cm) | 0.35 | 0.35 | 0.35 |
| $A_{12}$ or $A_2$ (20–30 cm) | 0.21 | 0.11 | 0.08 |
| $B_2$ (55–65 cm) | 0.09 | 0.06 | 0.05 |
| Soil reaction (pH) | | | |
| $A_{11}$ (0–7 cm) | 5.1 | 6.1 | 6.5 |
| $B_2$ (55–65 cm) | 5.2 | 5.0 | 5.2 |
| C (115–130 cm) | 6.2 | 6.0 | 5.2 |
| Base saturation (%) | | | |
| $A_{11}$ (0–7 cm) | 90 | 68 | 80 |
| $A_{12}$ or $A_2$ (20–30 cm) | 68 | 80 | 42 |
| $B_2$ (55–65 cm) | 78 | 87 | 78 |
| Silt-size opal phytoliths (mT/Ha) in solum (largely in A) | 20 | 10 | 5 |

Based on representative data from Daniels, Brasfield, and Riecken (1962); Jones and Beavers (1964); Riecken (1965); and White and Riecken (1955).

nitrogen can be used as a measure of organic matter. Although the clay content in the B horizon changes relatively little, its ratio with that in the A horizon shows highly significant variation. Under forest conditions the base saturation in the $A_{12}$ or $A_2$ horizon is very considerably lower than it is under prairie. The enhanced accumulation of opaline phytoliths in the prairie soil is strongly visible.

Of course comparisons of this kind leave many questions unanswered. To what extent are the *qualities* of the annual litter different in the different vegetative regimes, and how much of the soil difference is caused by different *amounts* of litter? To what extent do the different rooting systems play a part?

*Microbiological and Other Regimes*

The effects on soil formation and development of the microbiological population of soils are exceedingly difficult to deal with quantitatively. This population is very sensitively affected by climatic factors and by the chemical differences between horizons which these have caused. Taken as a whole, its most important functions are the breakdown of plant and animal remains, the production of humus, and the final decomposition of the latter to give simple substances in solution. For obvious quantitative reasons its functions are chiefly concentrated near the soil surface.

The internal factors affecting the microbiological activity at any point in a soil profile are (1) moisture, (2) content and nature of organic matter, (3) soil pH, (4) temperature, and (5) presence and availability of nutrients.

Animals of various families exert important influences on soils. The activity of earthworms in mixing different soil layers near the surface, stressed by Darwin, has been commented on frequently since. Large numbers of earthworms are only found where there is abundant digestible organic matter. The factors that tend to increase microbiological activity also favor earthworms.

Larger burrowing animals such as moles and mice also effect mixing and in so doing tend to obscure differentiation of soil horizons near the surface.

Ants and termites produce extensive changes in soil used for their hills and mounds. In the tropics an appreciable proportion of the landscape may be covered by these features. They form, in regions of poor soils, islands of enhanced productivity, corresponding to the concentration in them of nutrient elements from a wider area. In regions of poor soils in the tropics these mounds are notably enriched in calcium and phosphorus.

## TOPOGRAPHIC FACTORS

Factors connected with topography or landscape relief were recognized as very important from the earliest days of soil classification. Relief was given independent status as a soil-forming factor in Jenny's book (13).

Of course if the *internal* climatic environment of soils could be satisfactorily determined, it would be less important to deal with topography as a separate factor in soil development—it would simply be one of the determinants of this internal climatic environment. However its present status has the great practical advantage that since soil differences are shown over relatively small distances, the external macro climate can be taken as constant. Over these small distances parent material differences are also often minimal, therefore very clear-cut effects of slope, aspect, drainage conditions, and depth of water table can be demonstrated. Soils related to one another through these factors form catenas (Milne, 16). Many studies of soils related through topographic factors have been made all over the world since the 1930's.

*Slope and Aspect*

For any soil, the latitude, slope, and aspect determine the amount of direct radiation per unit area received from the sun. Thus northerly slopes correspond to cooler conditions than southerly, a common observation in the northern hemisphere. The difference in terms of vegetation, soil formation, and so on, should be less in practice than the fraction of maximum radiation received would indicate. This is because of winds and the convectional mixing of air over hilly topography, which tends on the average to equalize air temperatures.

In addition to these effects, slope has an important relationship to the net moisture penetrating the soil surface. Depending on the permeability of the soil, more or less of the rainfall may be lost by surface runoff. This factor is most effective near the tops of uniform slopes. Lower down the slopes, more moisture is absorbed, especially where the slope diminishes. Thus a broad complex of factors determines the internal environment of the soil in undulating or hilly regions. Numerous studies have testified to this.

*Drainage and Depth of Water Table*

Although drainage can be regarded both as an internal property of soils and an external factor, the depth of the water table is mainly an external factor determined by the topographic position. However perched water tables that persist for only part of the year can exist alongside permanent water tables. The former are commonly a result of the formation of impervious subsoils. Thus a description of variation in the water table throughout the year is an exceedingly important part of the assessment of moisture factors in their effect on soil development. As mentioned earlier, saturated soil layers differ greatly from unsaturated layers both in respect to reactions involving constituents of the air and in respect to their microbiological and consequently chemical properties. In absence of the free access of oxygen, soil organic matter imposes reducing conditions that bring iron and manganese into solution, thus rendering them more mobile.

## DURATION

The time scale for pedological changes is clearly shorter than that for cyclic geological processes such as the deposition and uplift of a sedimentary rock. From common observation we know that it is longer than is required for laboratory models (e.g., tubes of soil subjected to leaching) to show signs of pedological changes. Historical evidence, discussed in Jenny's book (13), using the example of a ruined castle in eastern Europe, indicates an order of magnitude of hundreds to thousands of years. This has been corroborated by studies of soils formed on exposed glacial moraines during recession of the glacier. Archaeological evidence can also be used.

One experimental method of considerable value is the determination in preserved organic matter of $^{14}C$. The radioactive isotope approach has been widely used to date Pleistocene deposits up to about 50,000 years from the present. In several published studies soil organic matter, extracted from different horizons, has been dated.

Gerasimov (7) has summarized our present knowledge in a broad way by the statement that forest ("podzolic") soils are characterized by apparent ages (radiocarbon dates) of hundreds of years, whereas prairie soils (Chernozem) show values of thousands of years. He ascribes this to differences in the rates of soil development.

Two obvious questions arise. First, are all the fractions of soil organic matter of the same apparent age? Second, what can be learned through variations horizon by horizon in the soil profile? These practical questions are related to one's view of soil carbon as a pedogenic entity. Should determinations be confined to humic materials, as distinct from the total carbon? If determinations could be made separately on soil charcoal, coal fragments, pedogenic soil carbonates, and soil humic matter, how would they be interpreted? Even if we confine ourselves to the truly humified material, differences of viewpoint are found. A number of investigators have identified the radiocarbon age as the mean residence time of the humus. It is generally agreed that this value does not represent the total age of the soil horizon. The meaning to be ascribed to it varies in different cases. At one extreme we have the buried profile (Paleosol), in which the age since burial is large compared with the duration of the original soil development (18). In this case the Paleosol can be approximately dated. At the other extreme is a modern Podzol in which there is differential movement of different humic fractions acting alongside the net resultant of gains and losses of humus per horizon. In such cases the nature of the average is different in different horizons.

Scharpenseel (19, 20) has concluded that the humus is best represented by the portion that accompanies the clay in sedimentation by centrifugal means. Using this method he has examined and reported on a considerable number of profiles. He also investigated the $^{14}C$ ages of different fractions of soil organic matter—fulvic acid, hymatomelanic acid, humic acid (brown and grey), and humin. Although in most cases the values increase as expected with increasing molecular weight or degree of polymerization, there are exceptions. Thus it does not seem possible consistently to separate the young, biologically active organic matter from the old, biologically inert material.

The variation of $^{14}$C age with depth within profiles of different great soil groups has yielded the following results.

*Podzols.* A and B horizons give values in the range 0–3000 years. There is poor correlation of $^{14}$C age with depth. This is in agreement with the vertical mobility of humus, particularly that of the soluble fulvic acid fraction.
*Brown earths* (Alfisoils). The range of values extends to 15,000 years; $^{14}$C age is well correlated with depth in the profile.
*Chernozems* (Mollisols). The range extends to about 12,000 years. The correlation with depth is even better than that of the Alfisols.
*Tropical soils.* Very little information is available. Herrera and Tamers (11) have found good correlation of $^{14}$C age with depth for a series of profiles from alluvial parent materials in Venezuela. Most of the values were in the range 0–3000 years. Deep Australian Krasnozems on basalt shown similar characteristics irrespective of their age as inferred from site description. Thus Beckmann and Hubble (2) conclude that biological decomposition operates eventually on subsoil humus, implying limits to the $^{14}$C ages. The maximum found for these lateritic soils on basalt was 4000 years at about 1 m depth.
*Histosols* (peats). McDowell et al. (15) have examined peats resting on limestone in the Everglades region of Florida. Both humic acid and humin fractions were used and gave similar values. Figure 42 gives the $^{14}$C ages in relation to distance upward from the limestone. The linear relationship between $^{14}$C age and depth holds to within a short distance of the soft rock.
*Limestone soils.* Ballagh and Runge (1) have determined the $^{14}$C ages of layers within the heavy clay horizons that rest on limestone. In one case the lower layer gave 4270 and the upper 9330 years. This strongly suggests that organic matter is moving from the loess soil above into the heavy clay and is precipitated near the weathering surface of the limestone. Thus the most recent limestone layer to weather out should contain the youngest organic matter. The deposition of loess began 12,000–20,000 years ago, but the organic matter of the heavy clay horizon is much younger than this.

Pedogenic carbonates can be dated in the same way as humus in areas of limited rainfall. Bowler and Plach (4) have found that $^{14}$C ages of

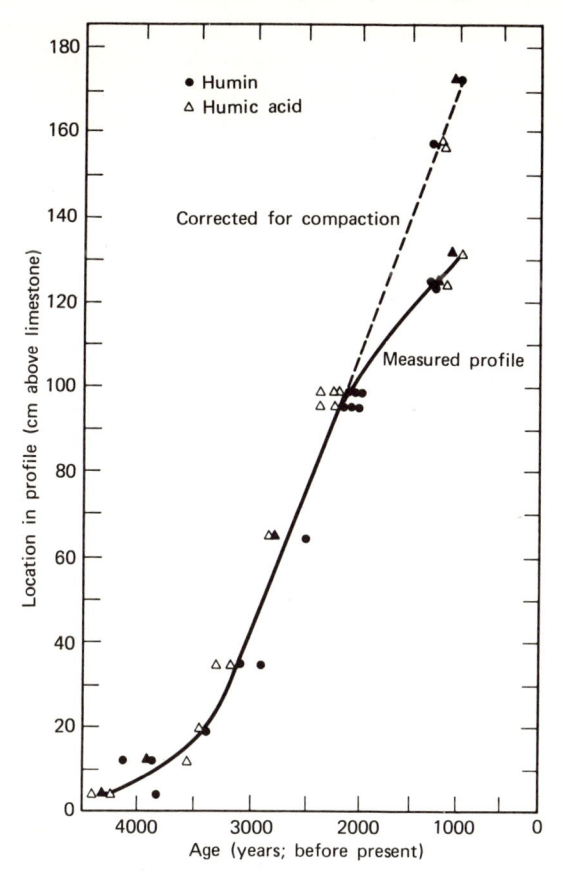

*Fig. 42* Radiocarbon ages of humin and humic acid fractions of Florida Everglades peat. Reproduced from ref. 15 by permission of the Soil Science Society of America.

carbonates are younger than the sediments in which they occur. The $^{14}C$ ages tend to be smaller, the more humid the environment. This indicates that active interchange of these porous carbonates with the soil solution and the atmosphere continues after their initial precipitation.

## REFERENCES

1. Ballagh, T. M. and E. C. A. Runge, Clay-rich horizons over limestone-illuvial or residual, *Soil Sci. Soc. Am. Proc.*, **34**, 534–536 (1970).
2. Beckmann, G. G. and G. D. Hubble, "The significance of radiocarbon measurements of humus from Krasmozems (ferralsols) in subtropical Australia, *Trans. 10th Int. Congr. Soil Sci.*, **VI**, 362–371 (1974).

3. Bloomfield, C., Organic matter and soil dynamics. In *Experimental Pedology*, E. G. Hallsworth and D. V. Crawford, Eds., Butterworth, London (1965).
4. Bowler, J. M. and H. A. Polach, Radiocarbon analyses of soil carbonates: An evaluation from paleosols in southeastern Australia. In *Paleopedology: Origin, Nature, and Dating of Paleosols*, D. H. Yaalon, Ed., International Society of Soil Science and Israel University Press, Jerusalem (1971), pp. 97–108.
5. Brewer, R., *Fabric and Mineral Analysis of Soils*, Wiley, New York (1964).
6. Buol, S. W., F. D. Hole, and R. J. McCracken, *Soil Genesis and Classification*, Iowa State University Press, Ames (1973).
7. Gerasimov, I. P., Nature and originality of paleosols. In *Paleopedology: Origin, Nature and Dating of Paleosols*, D. H. Yaalon, Ed., International Society of Soil Science and Israel University Press, Jerusalem (1971), pp. 15–27.
8. Gersper, P. L. and N. Holowaychuk, Effects of stemflow water on a Miami soil under a birch tree. I. Morphological and physical properties. II. Chemical properties, *Soil Sci. Soc. Am. Proc.*, **34**, 779; 786 (1970).
9. Harrison, J. B. The katamorphism of igneous rocks under humid tropical conditions, Imperial Bureau of Soil Science, Harpenden, England (1934).
10. Helgeson, H. C. et al., *Handbook of Theoretical Activity Diagrams*, Freeman, Cooper, San Francisco (1969).
11. Herrera, R. and M. A. Tamers, Radiocarbon dating of tropical soil associations in Venezuela. In *Paleopedology: Origins, Nature, and Dating of Paleosols*, D. H. Yaalon, Ed., International Society of Soil Science and Israel University Press, Jerusalem (1971).
12. Horn, F. E., The prediction of amounts and depth distribution of water in a well-drained soil, M. S. thesis, University of Missouri, Columbia (1971).
13. Jenny, H., *Factors of Soil Formation*, McGraw-Hill, New York (1941), Chap. III.
14. Leamy, M. L., The use of pedogenic carbonate to determine the absolute age of soils and to assess rates of soil formation. *Trans. 10th Int. Congr. Soil Sci.*, **IV**, 331–339 (1974).
15. McDowell, L. L., J. C. Stephens, and E. H. Stewart, Radiocarbon chronology of the Florida Everglades peat, *Soil Sci. Soc. Am. Proc.*, **33**, 743 (1969).
16. Milne, G., Composite units for the mapping of complex soil associations, *Trans. 3rd Int. Congr. Soil Sci.*, **I**, 345 (1935).
17. Pallmann, H., E. Eichenberger, and A. Hasler, Principles of a new method for temperature measurements in ecological or soil investigations, *Soil Res.*, **7**, 53 (1940).
18. Runge, E. C. A., K. M. Goh, and T. A. Rafter, Radiocarbon chronology and problems in its interpretation for quaternary loess deposits—South Canterbury, New Zealand, *Soil Sci. Soc. Am. Proc.*, **37**, 742–746 (1973).
19. Scharpenseel, H. W., Radiocarbon dating of soils-problems, troubles, hopes. In *Paleopedology: Origin, Nature and Dating of Paleosols*, D. H. Yaalon, Ed., International Society Soil Science and Israel University Press, Jerusalem (1971), pp. 77–88.
20. Scharpenseel, H. W., Radiocarbon dating of soils, *Sov. Soil Sci.*, **3**:1, 76–83 (1971).
21. Scharpenseel, H. W., Natürliche Radiokohlenstoffmeassungen als Mittel zur Untersuchung von Bodenprozessen und deren Dynamik, *Trans. 10th Int. Congr. Soil Sci.*, **VI**, 315–330 (1974).
22. Scrivner, C. L., J. C. Baker, and D. R. Brees, Combined daily climatic data and dilute solution chemistry in studies of soil profile formation, *Soil Sci.*, **115**, 213 (1973).

23. Troughton, A., Studies on the roots of leys and the organic matter and structure of the soil, *Emp. J. Exp. Agr.,* **29,** 165 (1961).
24. Whiteside, E. P., Some relationships between the classification of rocks by geologists and the classification of soils by soil scientists, *Soil Sci. Soc. Am. Proc.,* **19,** 164 (1955).

# 7 Physical description of soil profiles

## FIELD OBSERVATIONS AND DESCRIPTIONS

Modern descriptions of soil profiles include a much more thorough set of field observations than was usual 50 years ago. The *Soil Survey Manual* of 1951 and later revisions should be consulted for details. The modern description is designed to make differentiations possible between related soils, although it is recognized that final decisions may rest on results of laboratory determinations. The reader is referred to Chapters 2, 3, and 4 of *Soil Genesis and Classification* (4) for a clear account of the present position.

In field observations we find that important aspects of soil heterogeneity take the center of the stage. Chemical properties can sometimes be inferred in the field. For instance, color sometimes but not invariably gives a clue to the content of organic matter. Minerals of the sand fraction can sometimes be identified by eye, thus providing important evidence on parent material. Carbonates are easily recognized by their effervescence with acid.

In general the main horizonation of the soil is recognized by field observation, although as Hallsworth (7) has pointed out, unseen horizonations may also be present.

## MICROSCOPIC OBSERVATIONS AND DESCRIPTIONS

Examination of soils under the microscope is an old technique, but its potentialities were first developed by Kubiena (12), who devised a special microscope for the observation of soil profiles in the field. This instrument was equipped with advanced illuminating systems using incident light. He also utilized and improved the thin section techniques of the sedimentary petrologist. On the basis of this work he introduced a classification of the soil fabric, that is, of the soil structure, as shown by relationships of the solid constituents to voids.

More recently, mainly on the basis of thin section work, Brewer has developed a different and more elaborate classification of soil fabric (2). At the same time the study of thin sections has been made more quantitative. At present this provides the most direct method for quantitative determination of illuviated clay. The relationship of the terms structure and fabric is defined by Brewer as follows:

*Soil structure.* The physical constitution of the soil material as expressed by the size, shape and arrangment of the solid particles and voids, including both the primary particles to form compound particles and the compound particles themselves; fabric is the element of structure which deals with arrangement.

In the light of microscopic observations, the horizonation of soils can clearly be seen in its relation to structure. Beyond this, the evidence often suggests mechanisms that have operated in the heterogeneous material of the soil to produce the final result (Figures 43 to 45).

The versatility, sophistication, and power of microscopic techniques applied to thin sections of soils from many parts of the world are well shown in *Soil Microscopy,* the proceedings of a conference held in 1973 (16). The details of soil heterogeneity, which throw much light on soil formation and development, complement and frequently refine the conclusions drawn from laboratory examination of soils in bulk with which this book is chiefly concerned. Such data also are likely to improve the teaching of soil science right down to the elementary level, as particularly demonstrated by Fitzpatrick (6). He has also improved techniques for making extra-large thin sections, which is advantageous because pore size distributions can be reliably determined by microscopic measurements.

The use of the scanning electron microscope, an instrument that provides the resolution necessary to show changes in the colloidal realm (1–0.01 $\mu$) is already carrying us to a further stage. Very intimate crystallizations or flocculations of mixtures are thereby made apparent (Figures 46 and 47).

The use of thin section methods to study quantitatively the illuviation of clay has been strongly recommended by Brewer. Increasingly in recent years, however, processes have been recognized that cause disorientation of clay films. It is then difficult to decide how much clay has been moved and redeposited. Adding to this uncertainty, we have little direct informa-

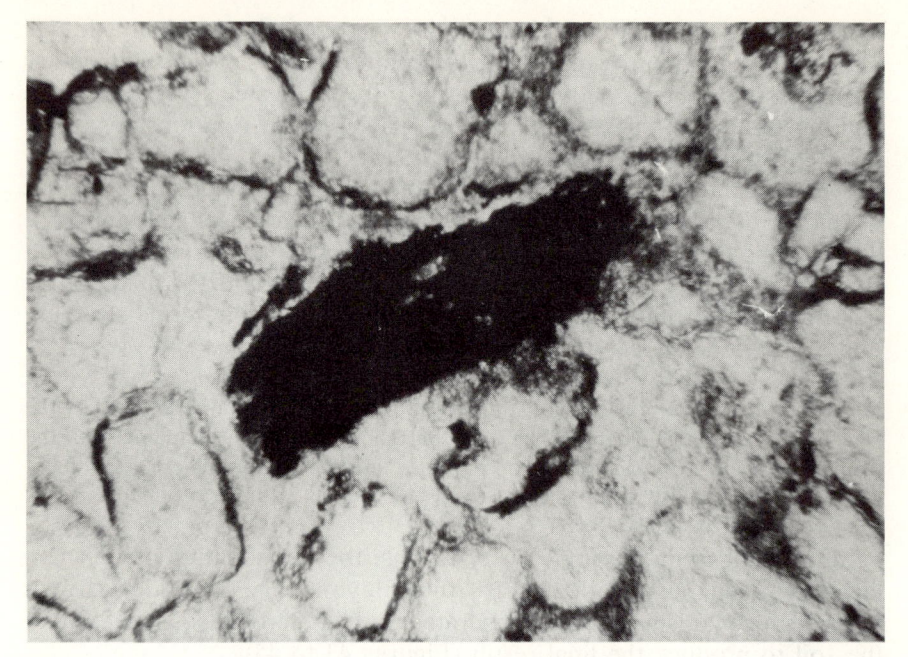

*Fig. 43* Thin section of $A_2$ horizon (10–38 cm) of an Alfisol (Cryoboralf, McVickers very fine sandy loam) from northern Arizona: precipitation, 50–55 cm; mean annual temperature, 7.2°C; parent material, sandstone. Note mass of humus in center surrounded by fairly clean sand grains. Courtesy of Prof. S. Buol.

tion on the electrokinetic situation either for the initial release of clay in upper horizons or for its deposition by illuviation. The following summary is an attempt to clarify the present position.

*Colloid-chemical aspects of clay eluviation and illuviation (lessivage).* We omit from consideration the two extreme processes of passage of clay into true solution and of dispersion and washing down by turbulent flow of water. The former process is grossly nonstoichiometric; that is, the small amount that passes into true solution is highly deficient in aluminum as compared with soil clay. The latter process is known to occur in self-mulching soils (Vertisols) such as the Houston clay, which develop deep cracks when drying. Rain washes topsoil directly into such cracks.

Clay deposited in B horizons by nonturbulent illuviation should have the chemical composition of fine clay of the A horizon. Characteristically it has a higher content of the finest clay than either the A or the C

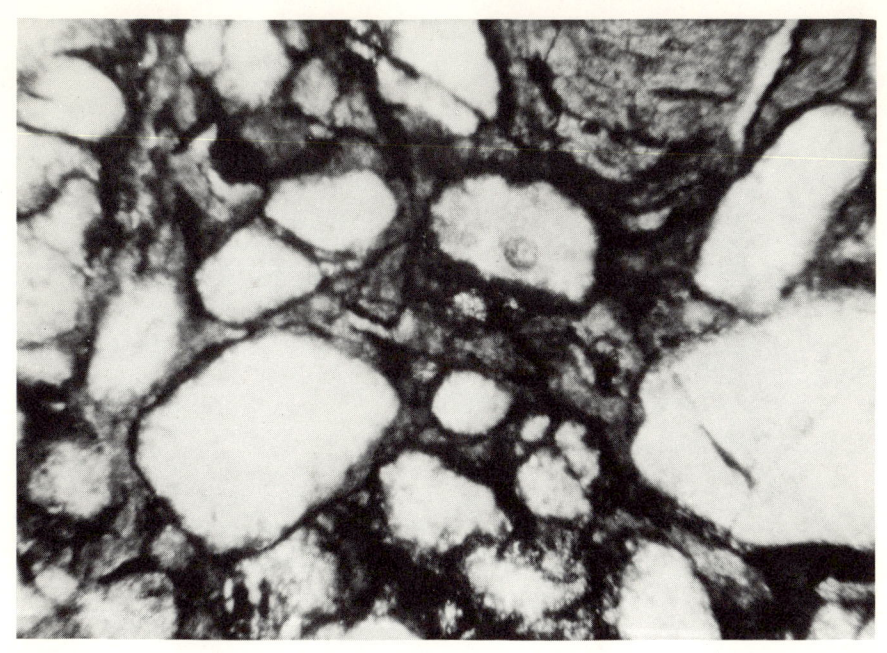

*Fig. 44* Thin section of $B_2$ horizon (50–71 cm) of soil illustrated in Figure 43. Note illuviated clay filling pores and oriented around sand grains. Courtesy of Prof. S. Buol.

horizons. It is frequently accompanied by organic matter in moderate to small amounts.

Thus the first process to consider is a differential dispersion or peptization of fine clay. It has received extraordinarily little study. From general colloid-chemical considerations two factors favor ready dispersibility, namely, high hydration and high zeta potential. External forces available for detachment of particles from each other are obviously extremely small when turbulent flow of liquid is excluded.

The work of Lutz (13) indicated that relatively slight differences in the zeta potential of clay particles might be correlated with large differences in dispersibility or capacity to erode. The Iredell soil was erosive, the Davidson not; but the zeta potentials with monovalent cations averaged 40.3 and 38.6, respectively. With calcium the values were 35.8 and 36.3 mV, the nonerosive Davidson samples have slightly the higher value. Because of those very slight differences, Lutz placed particular emphasis on clay hydration as an important factor. The Iredell clay was appreciably more hydrated than the Davidson.

The values for zeta just quoted were obtained on extremely dilute

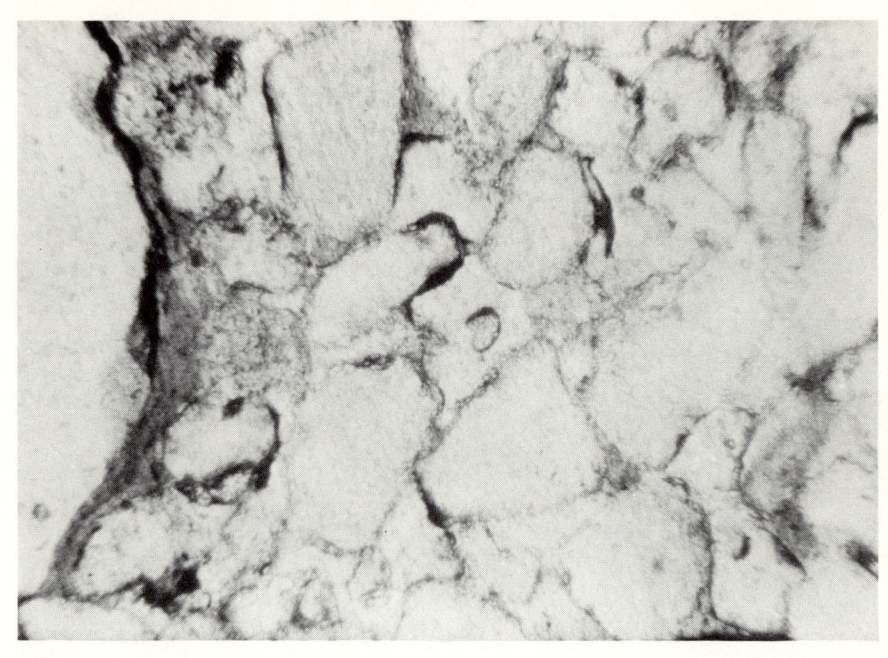

*Fig. 45* Thin section of C horizon (102–152 cm) of soil illustrated in Figure 43. Clean sand grains with a few clay bridges, and illuviated clay on edge of large pore. Courtesy of Prof. S. Buol.

suspensions. What would have been the value for a concentrated clay aggregate in presence of a thin film of water? Unfortunately, information on the mean zeta potential of soil profiles is lacking. It is very probable that the individuality of different clays would show itself much more clearly in electroendosmosic movement of water in soil plugs than in the cataphoresis of suspended particles.

As mentioned in Volume I, Chapter 9, the calculated zeta potential for very small particles is greater than it is for large particles of the same cataphoretic velocity. Hauser and Le Beau found that the finest fraction ($\sim$20 m$\mu$) of sodium montmorillonite had a higher cataphoretic velocity than coarser fractions. Thus there are good colloid-chemical reasons for greater ease of peptization of the very finest clay.

These considerations lead to the general prediction that the process of "lessivage" should produce ratios of coarse to fine clay having their lowest values in illuvial horizons where the clay content is highest. This conclusion is well supported in the work of Whiteside and Marshall (18) on two claypan soils from Missouri and Illinois. These loessial soils are dominated by beidellitic clay, with some illite. Similar conditions were

**Fig. 46** Scanning electron micrograph of sand grains coated with spodic material from $B_{21}$ horizon of Croghan soil, New York, × 100. Reprinted from ref. 14 by permission of the Soil Science Society of America.

also found by Khalifa and Buol (11) in the Cecil soil dominated by kaolinite.

However the weathering of primary minerals to give clay would also be expected to produce a predominance of fine particles, at least in the early stages. Thus in profiles in which both lessivage and weathering are active factors, but dominant in different horizons, more complicated relationships can be expected. Examination by the author of the relationship of clay content to the ratio of coarse clay to fine clay for a series of soils from materials of igneous origin in southeastern United States (5) showed a number of cases in which C horizons gave coarse-to-fine ratios equal to or

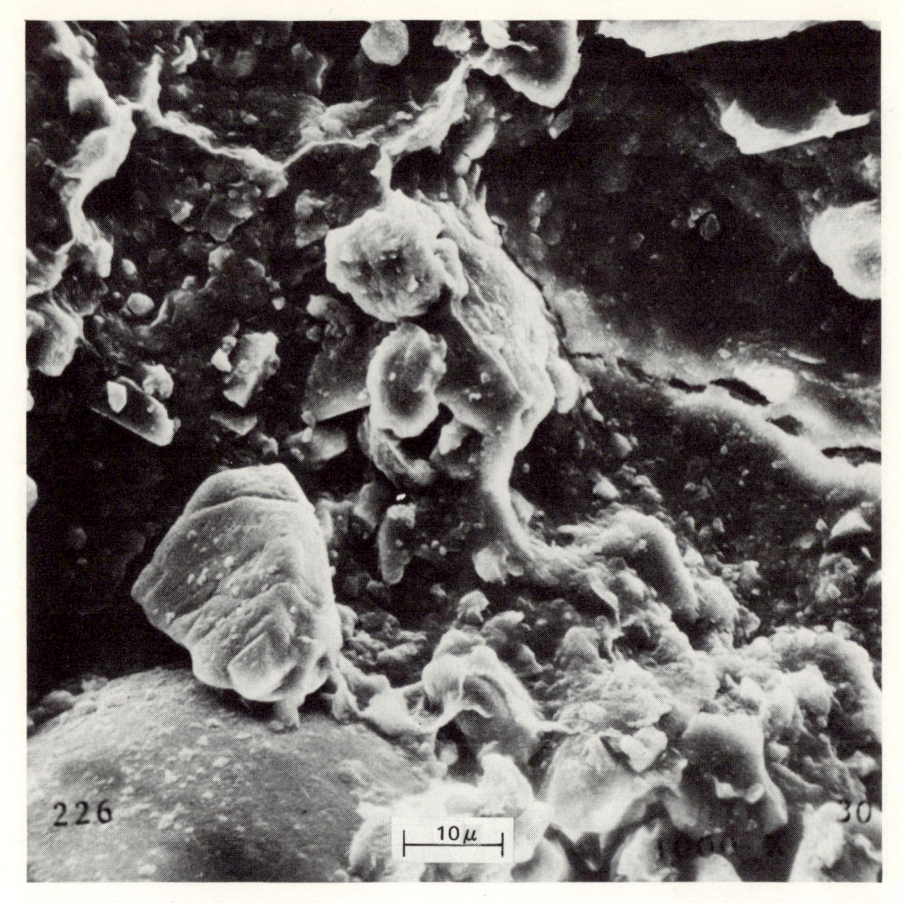

**Fig. 47** Scanning electron micrograph of fragipan, B×2 horizon of Caddo soil, Texas, ×1000. Reprinted from ref. 14 by permission of the Soil Science Society of America.

lower than the respective B horizons. Others, however, showed the usual minimum in the B. The data available do not indicate why some soils are exceptional. There may be cases in which fine clay moves farther down than the assumed base of the B horizon, but differences in the internal environment for weathering and clay synthesis may be more important.

## PARTICLE SIZE DISTRIBUTION

A central feature of all pedological studies involving soil development is a quantitative mechanical analysis of the mineral matter. The techniques employed vary somewhat for soils of differing degrees of cementation.

The general objective is to obtain dispersion to the ultimate particle units in as mild a manner as possible.

Pedologically the data are important in a number of ways, their value depending to a considerable extent on the detail of subdivision by particle sizes. The long established International Method gives very broad classes, the ranges being 2–0.2 mm coarse sand, 0.2–0.02 mm fine sand, 0.02–0.002 mm silt, and less than 0.002 mm clay. For deciding questions of origin, more detailed subdivision of the sand fractions is advantageous. Questions of movement within soil profiles also require more detailed silt and clay values.

The possible use of whole sand fractions as immobile indicators of gains and losses has been reviewed in Chapter 5. Limitations are clearly serious, and additional evidence justifying its use in particular cases should be sought.

The question of differential movement of different fractions within soil profiles is usually discussed with reference to clay. However in the Grundy profile studied by Haseman and Marshall (8), movement of the silt fractions also was apparent, extending up to the 0.05 mm size.

In claypan soils containing swelling clay such as beidellite, the content of the very finest particles in the B horizon can rise to high values. Thus in the Putnam and Cisne profiles studied by Whiteside and Marshall (Table 6, Chap. 5) this clay fraction contains 40% or more of particles less than 0.02 $\mu$ in equivalent spherical diameter, corresponding to more than 20% of the B horizon in its maximum development.

Few particle size distribution studies of the clay fractions of profiles dominated by nonswelling clays or by oxide and hydroxides are available. A fractionation at 0.2 $\mu$ is now becoming widely used because the two fractions 2–0.2 $\mu$ and 0.2 $\mu$ often exhibit distinctive X-ray characteristics. The coarser fraction usually contains a greater variety of clay minerals than the finer. It reflects the parent material more than the latter. The finer fraction affords evidence on chemical transformations and physical translocations. Thus the change from beidellite to kaolinite by progressive weathering has shown itself in a small increase in the kaolinite content of the fine clay.

The ratio of coarse clay (2–0.2 $\mu$) to fine clay (< 0.2 $\mu$) provides a useful index in the study of soil profiles, especially in conjunction with mineralogical studies. Fine clay is more easily moved than coarse clay of the same composition; hence the ratio provides a measure of illuviation. As mentioned earlier, if this is the dominant factor, the coarse-to-fine ratio should be highest in the A horizon and lowest in the B, the parent material being intermediate. Some profiles, but not all, among Grey-Brown Podzolic, Red-Yellow Podzolic, and Lateritic soils in the southeastern United States show this relationship.

An additional relationship is that mineralogical composition of fine clay is shifted toward products of weathering as compared with coarse clay. This is simply a restatement of one line of evidence that led Jackson (10) to his concept of a weathering sequence. It is a natural consequence of the character of soil systems in relation to secondary mineral formation that surface films and very small particles should be favored. It could hardly be otherwise, considering the heterogeneous nature of the solid phases and the very variable chemical environment in solution that follows cyclic changes in moisture content.

## AGGREGATE ANALYSIS

The study of soil aggregation begins with the field description, in which the visible and palpable properties of natural units (peds) are placed in relation to one another, horizon by horizon. In this description of soil structure both the arrangement of soil units and the forces holding them together find quantitative expression.

In aggregate analysis the soil is subjected to very mild forces of disruption. It can be carried out by sieving dry, in a nonpolar liquid, or in an aqueous system—in each case with gentle agitation. In this way one obtains a quantitative particle size distribution that can be compared with the results of ultimate mechanical analysis.

Variation in aggregation in different horizons is a well-marked feature of many soils. Table 13 illustrates this for two Chernozems, a Lateritic clay and a Podzol.

## BEHAVIOR TOWARD WATER

### *Moisture in Relation to Depth*

In practical agriculture and forestry few factors are more important than the type of aqueous reservoir presented to plants by the soil profile. This is established partly by the nature of the soil as a porous body and partly by the intensity and distribution of rainfall.

As an illustration, Figures 48 and 49 based on a thesis by Horn (9) are presented. The soil under investigation was a Typic Hapladulf, the Menfro silt loam, under oak and hickory forest cover. Such soils, with heavy swelling subsoils, behave quite differently from tropical soils dominated by nonswelling clays.

Figure 49 gives for two dates in 1970 the percentage of water by volume at depths down to 300 cm. On May 26 the soil was essentially at field capacity down to 280 cm. It received 5.23 in. of rain between May 26 and July 22, when the moisture profile had decreased fairly uniformly down to

**TABLE 13** The Structure Capacity or Extent of Aggregation of Various Soil Profiles (1)

| Soil Type | Depth (in.) | Mechanical Separates Larger than 0.05 mm (%) | State of Aggregation: Aggregates Larger than 0.05 mm (%) | Degree of Aggregation: Mechanical Separates smaller than 0.05 mm aggregated into units Larger than 0.05 mm (%) |
|---|---|---|---|---|
| Chernozem | 0–3 | 41.1 | 44.2 | 75.5 |
| (fine sandy | 3–9 | 55.3 | 30.4 | 68.0 |
| loam) | 9–15 | 55.2 | 32.5 | 72.1 |
|  | 18–28 (lime) | 37.1 | 33.6 | 53.4 |
| Chernozemlike | 0–6 (A$_1$) | 13.3 | 43.4 | 50.0 |
| silt loam | 7–12 (A$_2$) | 7.6 | 70.9 | 76.5 |
| Lateritic clay | 0–5 | 29.1 | 67.1 | 94.6 |
|  | 5–30 | 14.1 | 70.6 | 82.1 |
|  | 30–60 | 28.7 | 37.7 | 52.9 |
| Podzol | 0–4 (A$_1$) | 72.8 | 8.3 | 24.5 |
| (fine sand) | 4–10 (A$_2$) | 64.8 | 3.9 | 11.1 |
|  | 10–16 (B$_1$) | 77.5 | 10.9 | 51.2 |
|  | 16–28 (B$_2$) | 85.7 | 7.7 | 53.6 |
|  | 28–36 (B$_3$) | 85.6 | 8.1 | 35.8 |
|  | 36+ (C) | 74.5 | 0.7 | 26.8 |

300 cm, corresponding to a potential evapotranspiration loss of 10.03 in. The physical makeup of this soil in terms of the relative volumes of water, air and solids is shown in Figure 48. The upper limit of the b horizon is very well marked by the rapid increase in clay and the corresponding increase in unavailable water (> 15 atm tension). The available water actually decreases appreciably. Changes farther down the profile are gradual. There was almost no runoff during the period considered, and no water passed completely through the profile.

In the zone of maximum clay content at about 63 cm depth, the soil lost 4.2% moisture between May 21 and July 22. But the total available water at this depth was only 7.7%. A further loss of only 3.5% would bring this layer to the wilting point.

At 25 cm depth the soil lost 15.6% moisture out of a total available of 24%, leaving 8.4%. Thus the heaviest clay horizon acts as a strongly limiting layer in the delivery of water.

## *Pore Size Distribution*

Pore size distributions are generally calculated from moisture tension curves that relate moisture content to the tension applied through a water column or by air pressure applied above a column of wet soil. Under such experimental conditions the soils are in the wet portion of the moisture range, and therefore are fully or partially swollen. Similar measurements

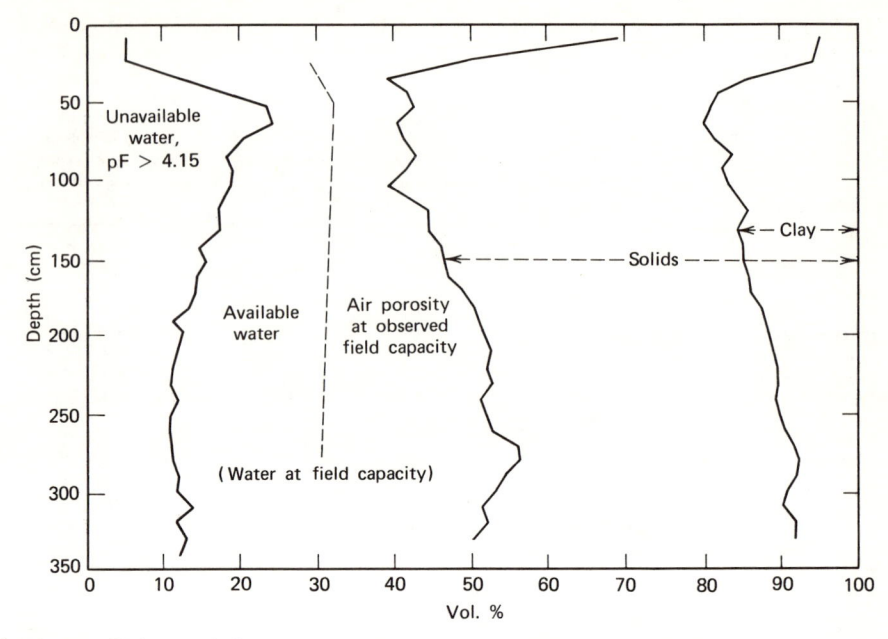

*Fig. 48* Volume relationships in the Menfro silt loam (typic hapladulf) under forest cover (9).

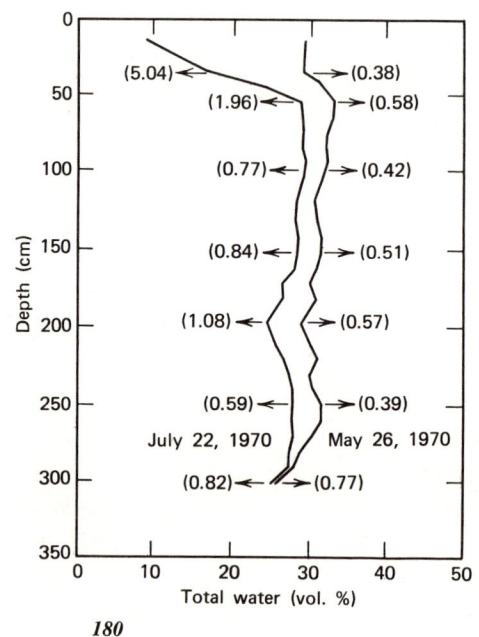

*Fig. 49* Relationship of water content and depth at field capacity (May 26) and under moisture stress (July 22) for the Menfro soil. Figures in parentheses indicate tensions in bars at designated points (9).

can also be carried out with nonpolar liquids, to eliminate swelling, but few are available. The use of mercury offers advantages in convenience and speed of equilibration. The pore sizes are generally calculated using Maxwell's equation for cylindrical tubes:

$$h = \frac{2\gamma \cos \theta}{\rho g r}$$

where $h$ is the height of rise in a capillary tube of radius $r$ (this is exactly equivalent to the tension applied by a water column that just causes emptying of the pore of radius $r$); $\theta$ is the angle of contact (usually taken as 0 for water in contact with nongreasy mineral grains), $\gamma$ is the surface tension of the liquid against air, $\rho$ is the density, and $g$ is the gravitational constant. This simple relationship breaks down for very small pores (the micron range and smaller), because eventually the surface tension $\gamma$ is no longer a constant but is affected by the solid wall.

Marked hysteresis occurs between wetting and drying, in the curves connecting tension or pF with moisture content. Clearly if a pore size distribution is to be deduced from the results, the drying curve, beginning with complete saturation, is the one to use. The characteristics of such curves vary with the horizonation of the profile. Since in deducing pore sizes a cylindrical shape is assumed, the agreement between such data and those derived through microscopic examination of thin sections can only be approximate. Even more important in such comparisons, thin sections represent dry conditions.

Pore size distributions have been used to calculate permeability and hydraulic conductivity, properties that may differ very strikingly for different soil horizons. As we show below, variations in these properties can cause very great changes in soil development.

## Behavior at High pF Values (Adsorption)

As described in Volume I, Chapter 5, the adsorption of vapors can be used as a measure of total accessible surface. It is necessary to begin with dry materials and to follow the uptake through the adsorption of successive molecular layers. Thus the surface measured by the use of inert gases is "external"; it does not involve internal surfaces of the swelling clays. If small polar molecules such as ethylene glycol are used, however, the majority of reactive surfaces can participate.

Measurements of surface area are of course related to the particle size distribution, to the mineralogical nature of the clay minerals and amorphous minerals present, and to the content of organic matter. Thus the method would seem to be sensitive in relation to profile characteristics, but not highly discriminating with regard to specific compounds and mechanisms.

## Permeability; Infiltration Capacity

The permeability or conductivity of a porous body for fluids can be expressed as a single characteristic resistance, provided one deals with saturated flow. Unsaturated flow involves constants that vary with the degree of unsaturation. Because in natural soils the latter is the rule, the overall capacity to take up water is expressed in terms of what is called the infiltration capacity. This quantity is fairly easily measured in the field under standardized conditions. Field measurements can be made on the whole soil, or various horizons from the surface downward can be removed for measurements at successive depths. Infiltration rates can also be derived from watershed data on rainfall and runoff.

As Baver (1) points out, the infiltration capacity is a very dynamic property. It is affected by the intrinsic permeabilities of the different horizons, by the condition of the soil surface, and by the soil moisture situation. Nevertheless, maximum and minimum infiltration capacities have fairly definite values at a given season. Various equations reviewed in Baver's book relate infiltration rate to time for homogeneous soils. However the main pedological interest lies in the very strong effects exerted by soil horizonation on the entry of water as a function of time. These show themselves chiefly in (1) surface effects and (2) subsurface horizon effects.

1. The formation of a surface crust by the impact of raindrops is a strong characteristic of silty soils containing easily dispersed clays. Silt and clay are both dispersed, then deposited in pores and depressions to form a thin surface layer with very low pore space and permeability. On drying this hardens to a cracking or noncracking crust.

2. Dense subsurface horizons are extremely effective in reducing infiltration once the advancing water front begins to encounter them. Both siltpans and claypans act in this way. In such layers the larger pores, which are the chief conductors of water, are greatly reduced, and conduction becomes chiefly a function of very small pores. But these, although numerous, are of small effect because of the fourth power dependence of flow on radius.

Accordingly the relationship between pore size distribution and infiltration can be formulated in terms of groupings of the larger pores into categories, with elimination of the fine pores at an arbitrary cutoff point. This approach was fruitfully explored by Smith, Browning, and Poehlman (17), who found, comparing equal volumes, that pores draining at tensions between 40 and 100 cm water were only about one-tenth as effective as pores draining at tensions below 10 cm water. They therefore used a cutoff point at 100 cm tension.

The general relationship between moisture content, tension curves, and conduction for water has also been explored through various equations based on partial models of soil-water systems (T.J. Marshall, 15). Alongside these are purely empirical relationships describing infiltration. They are reviewed in Chapter 10 of Baver's book (1).

The importance of infiltration in pedological processes can hardly be overestimated, but its quantitative expression is far from adequate. Certain striking cases come readily to mind. Under moderate rainfall, well distributed through the year, claypan and siltpan soils may drain to the permanent water table very rarely or not at all. This situation is vastly different from that in sandy soils under similar climate. In extreme cases of impermeable subsoils we may have perched water tables existing for part of the year, with great influence on oxidation-reduction conditions in the profile.

## REFERENCES

1. Baver, L. D., W. H. Gardner, and W. R. Gardner, *Soil Physics*, 4th Ed., Wiley, New York (1972).
2. Brewer, R., *Fabric and Mineral Analysis of Soils*, Wiley, New York (1964).
3. Buol, S. W. and F. D. Hole, Clay skin genesis in Wisconsin soils, *Soil Sci. Soc. Am. Proc.*, **25,** 377 (1961).
4. Buol, S. W., F. D. Hole, and R. J. McCracken, *Soil Genesis and Classification,* Iowa State University Press, Ames (1973).
5. Cooperative Regional Research Project S-14, Certain properties of selected southeastern United States soils and mineralogical procedures for their study, Southern Regional Bulletin 61, Virginia Agricultural Experimental Station, Blacksburg, (1959).
6. Fitzpatrick, E. A., *Pedology, A Systematic Approach to Soil Science,* Oliver & Boyd, Edinburgh, (1971).
7. Hallsworth, E. G., *In Experimental Pedology,* E. G. Hallsworth and D. V. Crawford, Eds., Butterworth, London (1965).
8. Haseman, J. F. and C. E. Marshall, The use of heavy minerals in studies of the origin and development of soils, University of Missouri Agricultural Experimental Station Research Bulletin 387 (1945).
9. Horn, F. E., The prediction of amounts and depth distribution of water in a well-drained soil, M.S. thesis, University of Missouri, Columbia (1971).
10. Jackson, M. L. et al., Weathering sequence of clay size minerals in soils and sediments. I. Fundamental generalizations, *J. Phys. Colloid Chem.*, **52,** 1237 (1948).
11. Khalifa, E. M. and S. W. Buol, Studies of clay skins in a Cecil (typic hapludulf) soil. I. Composition and genesis, *Soil Sci. Soc. Am. Proc.*, **32,** 857 (1968).
12. Kubiena, W. L., *Micropedology,* Collegiate Press, Ames, Iowa (1938).
13. Lutz, J. F., The physico-chemical properties of soils affecting soil erosion, University of Missouri Agricultural Experimental Station Research Bulletin 212 (1934).
14. Lynn W. C. and R. B. Grossman, Observations of certain soil fabrics with the scanning electron microscope, *Soil Sci. Soc. Am. Proc.*, **34,** 645 (1970).
15. Marshall, T. J., A relation between permeability and size distribution of pores, *J. Soil Sci.,* **9,** 1 (1958).
16. Rutherford, G. E., Ed., *Soil Microscopy,* Limestone Press, Kingston, Ontario (1974).

17. Smith, R. M., D. R. Browning, and G. G. Poehlman, Laboratory percolation through undisturbed soil samples in relation to pore size distribution. *Soil Sci.*, **57,** 197 (1944).
18. Whiteside, E. P. and C. E. Marshall. Mineralogical and chemical studies of the Putnam silt loam soil, University of Missouri Agricultural Experimental Station Research Bulletin 386 (1944).

# 8 *Mineralogical analysis*

## MINERALS OF THE CLAY FRACTION

### General Considerations

In relation to the dynamics of pedological processes, the clay minerals, together with the oxides and hydroxides, follow closely behind soil organic matter in reactivity and in sensitivity to varying factors of soil formation. The methods used for identification and quantitative determination have steadily been improved over the last 30 years. They are not discussed in detail here, since monographs describing X-ray methods (11), differential thermal methods (36), and electron-optical methods (23) are available. Considerable information on the clay mineralogy of soil horizons has now accumulated. However it must be recognized that the quantitative estimation of individual minerals in clay fractions varies greatly in precision. It depends to a considerable extent on the nature of the mineral mixture present. The presence of amorphous and mixed layer materials in quantity reduces precision. Subjective factors are also present, particularly in the preparation and manipulation of oriented films. Hence quantitative comparisons between results from different laboratories require great care in interpretation. For these reasons, semiquantitative estimates of abundance are widely used.

As pointed out in Chapter 7, a preliminary quantitative division of the < 2 $\mu$ clay into fractions is a great advantage. A separation at 0.2 $\mu$ is convenient. Some workers have used three fractions instead of two. The coarsest fraction usually provides a more complex mixture of minerals than the finest, which contains more of the poorly crystallized components. In the interpretation of results in terms of pedological processes, it is exceedingly important to consider what is inherited from the parent materials, hence what the results of pedological change have been.

The Kittrick type of formulation (Figure 4, Chap. 2) in which pH − ⅓pAl$^{3+}$ of the solution phase is plotted against pH$_4$SiO$_4$ provides a quick review of weathering conditions in a profile, provided that the activities corresponding to ½pMg$^{2+}$ and pK$^+$ are at very low levels or remain constant. The comparison of this diagram with weathering data on a worldwide basis is also very illuminating. Extreme lateritic conditions are represented on the right, where gibbsite is the dominant phase. This represents the highest equilibrium value for Al$^{3+}$ in solution and operates over a range of silica values from Si(OH)$_4$ 10$^{-4.7}$ $M$ at the gibbsite kaolinite junction downward. Less extreme weathering conditions correspond to kaolinite stability, which extends to 10$^{-2.8}$ $M$ for Si(OH)$_4$, at which point montmorillonite appears. Finally at 10$^{-2.7}$ $M$ and up we have amorphous silica. These data typify rock weathering in absence of organic matter, and of course where sodium, potassium, magnesium, and calcium in solution are very low. The actual figures may be modified as more accurate free energy values become available.

However pedological conditions often involve the participation of organic matter. Its effects are dramatic, primarily because of the complexing action of humus or humus constituents on aluminum. A complexing mechanism that reduces pH − ⅓pAl$^{3+}$ by less than one unit (see Figure 4) could bring the equilibrium from one dominated by gibbsite to one dominated by montmorillonite. But this is exactly the kind of change needed to explain the recurrent observations that in many Podzols the A$_3$ horizon contains smectite clays.

A secondary factor that operates in the same direction arises from the occurrence of opal phytoliths in many plants. They remain in the soil throughout the humification process and tend to raise the soluble silica level towards the equilibrium value pSi(OH)$_4$ = 2.7. This trend also favors smectite clay formation. The same can be said for any mechanism that favors the retention of soluble silica. For instance, if throughput of water by leaching becomes a rare event, the dominant pedological environment will correspond to higher silica in solution, provided there are still minerals available that on balance release silica by weathering.

Other secondary factors that may become important in the right circumstances are those which favor accumulation of magnesium, potassium, calcium, and sodium in solution. In this way the dominant equilibria may come to favor minerals such as illite, which contains both potassium and magnesium, or zeolites, which contain calcium and sodium.

Variations in pH are also effective in changing the conditions for mineral reactions. In some cases the aqueous pH is determined by a specific aluminosilicate reaction, but other modifying factors such as carbonic acid–carbonate equilibrium, the acidity of soil humic matter and its decomposition products, and variations in oxidation-reduction equilibria, are also effective. Kittrick's diagram (Figure 4) illustrates the sensitivity of equilibria to pH. At constant ⅓pAl$^{3+}$ a pH change of about 1.5 units encompasses the whole range from gibbsite to amorphous silica.

These considerations clearly indicate that the study of aqueous equilibria, along with determination of the mineral composition layer by layer in the profile, is likely to prove exceedingly important in defining pedological processes in chemical terms. The role of organic matter is also critical and will require quantitative evaluation in respect of three factors: (1) its complexing of cations, particularly aluminum and iron; (2) its acidic function; and (3) its function as a reducing system towards iron, manganese, and so on. The fulvic acid fraction, being water soluble, has already received partial characterization in these terms (Schnitzer and Skinner, 49).

This section, however, summarizes clay minerals in soils as related to pedological factors, using data from the best-defined parent materials. Igneous rocks and loess deposits provide widely studied parent materials of relatively uniform character.

## Specific Groups of Soils

*Laterites.* The powerful combination of chemical rock and ground water analysis with the petrographic study of thin sections enabled J. B. Harrison (28) to demonstrate the main features of laterization long before quantitative X-ray methods became available. His interpretations were based on the prevailing climatic factors in a manner that makes it possible to link his work with modern diagrams (Kittrick, Garrels and Christ) illustrating mineral stability relationships. He showed that under conditions of high rainfall (391 cm per year at Eagle Mountain) and rapid drainage, gibbsite accumulates as the first recognizable product from basic igneous rocks. This is within millimeter distances of the unweathered material. No intermediate zone containing secondary silicates was found. This production of what Harrison called primary High Level Laterite was characterized petrographically in three ways: (1) by the disappearance of original aluminosilicates (chiefly labradorite and augite), (2) by the appearance of gibbsite, and (3) by a reduction in the amount of crystalline quartz, which was a minor constituent of the dolerite rock.

Two studies of similar rock weathered under conditions of somewhat lower rainfall (254–260 cm.) with restricted drainage ("Low Level Laterite") indicated that Primary Laterite was again formed within millimeter distances of the rock surface, but that at larger distances halloysite and kaolinitic clays took the place of gibbsite. However the formation of the Primary Laterite was accompanied by an increase in quartz. Thus it appears as though quartz and gibbsite were formed as primary weathering products of the dolerite.

In terms of the Kittrick and Garrels and Christ diagrams, the simultaneous presence of these two minerals can indicate only the prevalence of nonequilibrium conditions. They are incompatible at equilibrium. Quartz, however, participates very sluggishly in equilibration with aqueous solutions. Kittrick (32) summarizes the situation as follows:

> The very slow dissolution rate and the negligible precipitation rate of quartz at room temperature resulting from the high activation energy of the Si–O–Si bond have two very important consequences in the weathering of quartz. The very slow dissolution rate means that quartz will be relatively resistant to weathering and the negligible precipitation rate means that quartz is unlikely to control any silica equilibria.

With this in mind the interpretation of Harrison's results can be based on processes involving the climatic cycles in different localities. At Eagle Mountain, the High Level Laterite is formed under a somewhat irregular annual cycle consisting of 8½ months of wet weather averaging 40 cm per month and 3½ months of drier weather averaging 14.5 cm per month. Under these conditions leaching is interrupted insufficiently to cause the buildup of silica to produce either quartz, aluminosilicates, or amorphous silica. At Tumatumari, where Low Level Laterite was formed, the wet season averaged 30.5 cm per month and the dry 11.2 cm; total rainfall 287 cm. The layers of Primary Laterite contained about 12% quartz, whereas the original rock contained only 1.6%. Thus for part of the year the silica in solution near the surface of the rock had risen above $1.8 \times 10^{-4}$ mole/liter, the normal solubility of quartz, for so long that the subsequent wet period did not completely dissolve what had crystallized out. Reference to Kittrick's diagram (Figure 4, Chap. 2) indicates that at this content of soluble silicic acid, kaolinite could also be present. Why was it not found in the primary laterite? The answer would seem to lie in relative rates of equilibration. If kaolinite or halloysite equilibrates more rapidly than quartz, small amounts precipitated during the dry season could redissolve more or less in the wet. In the outer layer of the Primary Laterite a little halloysite was found. Quartz, however, accumulated much more.

At Tumatumari the layers of Primary Laterite extended about 70 mm from the fresh rock. They were separated by a thin (2 mm) crust from a very thick layer (5–6 m) of lateritic earth. This consisted of about 12% quartz, 46% kaolin, 10% bauxite, and about 30% hydrous iron oxide and ilmenite.

At Eagle Mountain the layers of Primary Laterite were developed to a considerable thickness and were overlaid by a ferruginous bauxite or in places by lateritic ironstone. In neither of these was there evidence of secondary aluminosilicates, and both were extremely low in quartz. However migration and redeposition of gibbsite and of hydroxides or iron was evident. Titanium oxide was found to be relatively immobile in the mineral ilmenite.

In the formation of Primary Laterite close to the rock surface it is evident that water movement in bulk must operate to remove almost continuously the alkalis, alkaline earths, ferrous iron, and silica. The process is clearly akin to a surface dialysis in which these products of decomposition diffuse through relatively thin layers of stationary water into the moving layers. This process is clearly consistent with the finding that when water movement becomes restricted, silica is the first element to show evidence of accumulation. The rate of self-diffusion of monomeric silicic acid (molecular weight 96) and of its polymers, is appreciably less than that of the dissociated hydroxides, carbonates or bicarbonates of $Na^+$, $K^+$, $Ca^{2+}$, $Mg^{2+}$, or $Fe^{2+}$ in dilute solution. Near the rock surface, where primary minerals are still present, the reaction of the liquid layer is slightly alkaline. Thus $pH - \frac{1}{3}pAl^{3+}$ may be sufficiently high (i.e., 2.7 or higher) and silica sufficiently low ($<10^{-4.7}M$ $Si(OH)_4$ so that gibbsite is the stable phase. With slower or stationary water layers, $Si(OH)_4$ is the first constituent to attain a sufficient concentration; thus new phases such as kaolinite and halloysite can appear and even quartz may increase in amount. According to Kittrick's curves (32), for a given value of $pH - \frac{1}{3}pAl^{3+}$ halloysite appears at lower values of $pSi(OH)_4$ than kaolinite. Thus with restricted silica, kaolinite is favored; as silica increases, halloysite may appear as an intermediate; as the position of the curves indicates kaolinite will be the stable phase.

Outside the narrow zone of Primary Laterite formation the products formed clearly depend largely on the silica in solution, which may vary greatly according to the rainfall distribution. At Eagle Mountain Harrison showed that the ferruginous bauxite and lateritic ironstone contained almost no free quartz and less than 1% combined silica. At Tumatumari, with somewhat restricted drainage the thick layer of lateritic earth contained quartz, kaolinite, bauxite, and hydrous iron oxides, as mentioned. It would thus appear that silicic acid in solution varied at least from $10^{-3.7}$ (quartz) to $10^{-4.7}$ (gibbsite-kaolinite boundary) $M$. At Tumatumari Harrison nowhere mentions amorphous silica, which of course would only appear at $10^{-2.7}$ $M$ and lower.

Kittrick's ideas on the sluggishness of equilibria involving silica thus fit well with Harrison's observations, especially when consideration is given to the detail of the climatic cycles at different localities. However secondary quartz does appear to accumulate.

Another type of weathering characterizes the formation of crusts on rock surfaces directly exposed to the atmosphere. Harrison comments as follows on the surficial crust found on dolerite at Tumatumari. "The weathered crust has no resemblance to primary laterite. It is a complex mixture of quartz, titaniferous iron (largely ilmenite) with hydrated titania, hydrated peroxide of iron, and apparently a hydrated silicate of iron and alumina, possibly nontronite."

He also showed that Primary Laterite, subsequently exposed to atmospheric weathering, loses all its gibbsite, with concentration of quartz, combined silica, hydrous iron oxides, and titania. Thus in most respects atmospheric weathering, with its discontinuous wetting and drying and high exposure to carbon dioxide, is different from the production of Primary Laterite by weathering at depth.

In general it is unnecessary to postulate resilicification of gibbsite to form kaolinitic earth as Harrison did. Milne (41) criticized Harrison's interpretations in this respect, arguing against upward movement of siliceous waters and pointing out two other possibilities of confusion: namely, that residues of other strata might form part of the surface layers now exposed, and that climatic variations from the present regime may well have occurred in the past. The general experience of soil scientists who have worked with tropical soils under high rainfall is that acid igneous rocks can give rise to gibbsite, often mixed with kaolinite. This almost brings one to the point of considering quartz to be an inert constituent, but numerous observations of the surface weathering of quartz crystals make this risky. At the same time the production of secondary quartz, which was a prominent feature of Harrison's Low Level Laterite, must be taken into account. In this material Harrison found a complex mixture: residual minerals, mainly feldspar and mica from the diabase, together with gibbsite, secondary quartz, and iron minerals. Further out from the rock surface, kaolinitic products were observed.

Seasonal changes thus seem to impose two different chemical environments. During the period of highest rainfall the conditions are those for the production of Primary Laterite. Alkalis, alkaline earths, and silicic acid are very effectively removed. Under moderate rainfall the dialysis process is less effective, particularly with regard to silicic acid. This combines with released alumina to form kaolinite or halloysite, and it may even give rise to quartz. In the subsequent wet season, however, the kaolinite or halloysite reequilibrates readily under the very low silica conditions to give gibbsite, but the quartz is so sluggish in this respect that it largely remains. Thus in successive cycles quartz slowly increases in amount. Higher up in the profile the dialysis conditions favor direct production of kaolinite, halloysite, and quartz more and more; thus gibbsite diminishes and eventually disappears.

The surface and subsurface layers described by Harrison are characterized by much higher quartz contents than the underlying lateritic earth. Milne thought that in some cases residues from different strata were present (41). However an appreciable source of soluble silica is present in the plant residues. Many tropical plants are high in silica, which can be recognized as plant opal in the surface soil. There exists also the possibility that aluminum and iron are weakly complexed and carried down to some extent without being redeposited in a definite B horizon as in typical Podzols. Also external sources of silica deposited from the atmosphere and stratosphere need to be considered, such as dust from distant volcanic eruptions and from local dust storms. Thus the work of Harrison, highly detailed though it was, left some pedological questions unanswered.

Hardy and Rodriguez (25,26) and Hardy and Follet-Smith (27) attempted to verify Harrison's conclusions by the development of quantitative chemical methods for the determination of gibbsite and iron hydroxides. They confirmed the relationships he found, but they believed that resilication by upward movement of siliceous waters during the dry season was improbable or at least infrequent as a method of formation of lateritic earths overlying Primary Laterite. Several alternative ideas for the predominance of kaolin minerals in the lateritic earths were discussed.

The mineralogy of true Laterites (Oxisols with plinthite) is dominated by oxides and hydroxides of iron and aluminum. Minerals of the kaolin group are frequently present, but 2 : 1 lattice minerals such as smectites have never been found and hydrous micas are rarely present, even relatively close to the parent rock. Amorphous aluminosilicates such as allophane, however, may account for an appreciable part of the total silica.

The distinctions between laterites and lateritic soils may be roughly equated with those between Oxisols and Ultisols in the American system of classification. By definition, this rests on the relative dominance of oxides and hydroxides of iron and aluminum in the Oxisols and of kaolin group minerals in the Ultisols.

The past usages of the word laterite are ably discussed by Sivarajasingham et al. in their valuable review article of 1962 (54). They restrict the term to profiles containing the material now known as plinthite, which is capable of hardening upon exposure and drying. This hardening process is extensively discussed. The authors conclude that hardening is related to the iron content, specifically to a greater degree of crystallinity of goethite or hematite or a greater continuity of the crystalline phase. Thus the low iron materials of a bauxitic character would not necessarily be included. However they are included in the group of Oxisols.

The transition, by virtue of climatic variation, from Oxisols to Ultisols has often been commented on but rarely demonstrated in absence of interfering factors. Beinroth et al. (7) have recently followed this transition in a region of Hawaii where the parent rock is a Pleistocene lava and the rainfall varies from 1000 to 3000 mm. On the lava surface of low slope a change from Eutrustox→Eutrorthox→Gibbsiorthox→Gibbsihumox with increasing rainfall is apparent. This sequence implies increasing dominance of gibbsite. Areas of greater slope are characterized by Ultisols, which are distinguished from the Oxisols by the presence of argillic horizons. They contain more kaolinitic clays than the Oxisols. Oxides and hydroxides of iron and aluminum are much less readily translocated as such in soil profiles than are kaolinitic clays. However in tropical soils translocation can occur with minimal production of birefringent clay skins. Variably oriented clay is found filling up former voids in the soil fabric and forms turbicutans, which are revealed by thin section studies.

In the formation of Laterites from basic igneous rocks, the completeness of the changes within a few millimeters from fresh rock to "Primary Laterite" is remarkable. Hence later changes are very limited in scope, because of the inertness of the products to subsequent weathering. There is minor translocation downward of finely divided material as revealed by mechanical analysis. Kaolinitic clays if present might be expected to move more readily than gibbsite, goethite, or hematite, but there seems to be little mineralogical evidence that such a differential movement is significant in true Laterites (Oxisols). Harrison, in his interpretation of thin sections, repeatedly mentions evidence of recrystallization of gibbsite at some distance from the rock surface.

Regarding the iron minerals in Laterites and lateritic soils, we have to consider primarily amorphous ferric hydroxide ($Fe(OH)_3$), lepidocrocite ($FeO \cdot OH$). hematite ($Fe_2O_3$), and goethite ($FeO \cdot OH$). Study of the free energy situations for the various equilibria indicates (43) that (1) goethite is stable with respect to lepidocrocite, (2) amorphous ferric hydroxide is unstable with respect to hematite plus water or goethite plus water, and (3) goethite is stable with respect to hematite plus water. However the free energy difference for well-crystallized hematite plus water to give well-crystallized goethite is only $-0.2$ k. This means that for equilibrium at atmospheric temperatures, the vapor pressure of water would correspond to about 70% relative humidity. At higher relative humidities goethite is the stable phase; at lower values hematite is stable. Thus changes in relative humidity that are common in soil systems might greatly influence the dominant species. However, as already noted, the hydration of hematite to give goethite is exceedingly sluggish. Fischer and Schwertmann (20) believe that in soils goethite is never formed by hydration of hematite but from iron in complexed form. Unfortunately, little information is available on the rates of these reactions, or on the stability of aluminous goethites, which are also found in Ultisols and Oxisols.

In reviewing the literature on true Laterites (i.e., Oxisols) with plinthite, it becomes evident that quantitative X-ray determinations of the clay-size minerals present are difficult to obtain and infrequently presented. Where complete chemical analyses are available, calculations of the "normative" mineralogical composition can be performed. Rock weathering has proceeded sufficiently far that relatively simple mixtures of known products remain. Hence these values have significance. However assumptions used, such as mineralogically ideal compositions, need careful consideration. *Tropical Soils* (43) contains many of these calculated values.

One example from the work of Satyanarayana and Thomas in India (ref. 98: T. S. profile 2, pp. 206–208) is given below. This profile is derived from basaltic rock. The present dry climate involves a rainfall of 300–375 mm, which occurs predominantly from May to November. Table 14 indicates that gibbsite is not a dominant constituent. The chief minerals are kaolins and goethite. The latter dehydrates to hematite near the surface. A rather unusual feature is the mica content (calculated from the total potassium).

The essential characteristics of Laterites remain when the climate has changed to a drier regime. Köster (34) studied an Indian Laterite from Kot (originating from a coarse granite), which contained calcite in cracks and voids in several horizons. No gibbsite was found. Kaolinite and small amounts of mica were present in all horizons. The appearance and physical properties were those of Laterite with hard concretionary material at the surface.

An example of a relatively youthful Laterite is found in Hawaii (51). Table 15 gives the clay mineralogy of the Naiwa soil, originally described

**TABLE 14** Normative Mineralogical Composition of Laterite (South Kanara, India) from Basalt (48)

| Depth (cm) | Character | Quartz | Illite | Kaolin | Gibbsite | Goethite | Hematite |
|---|---|---|---|---|---|---|---|
| 0–25 | Red clay soil | 4.6 | 4.0 | 52 | — | 23 | 10 |
| 25–350 | Hard Laterite | 0.6 | 5.0 | 43 | — | 46 | — |
| 350–490 | Soft plinthite | — | 2.4 | 41 | 19 | 32 | — |
| 490–610 | Weathered rock | — | 4.0 | 55 | 3 | 26 | — |

# TABLE 15

**Name:** Naiwa; humic ferruginous Latosol
**Locality:** Maui, Hawaii (51)
**Parent Mat.:** Basalt
**Climate:** Alternating wet and dry; annual rainfall 88–200 cm

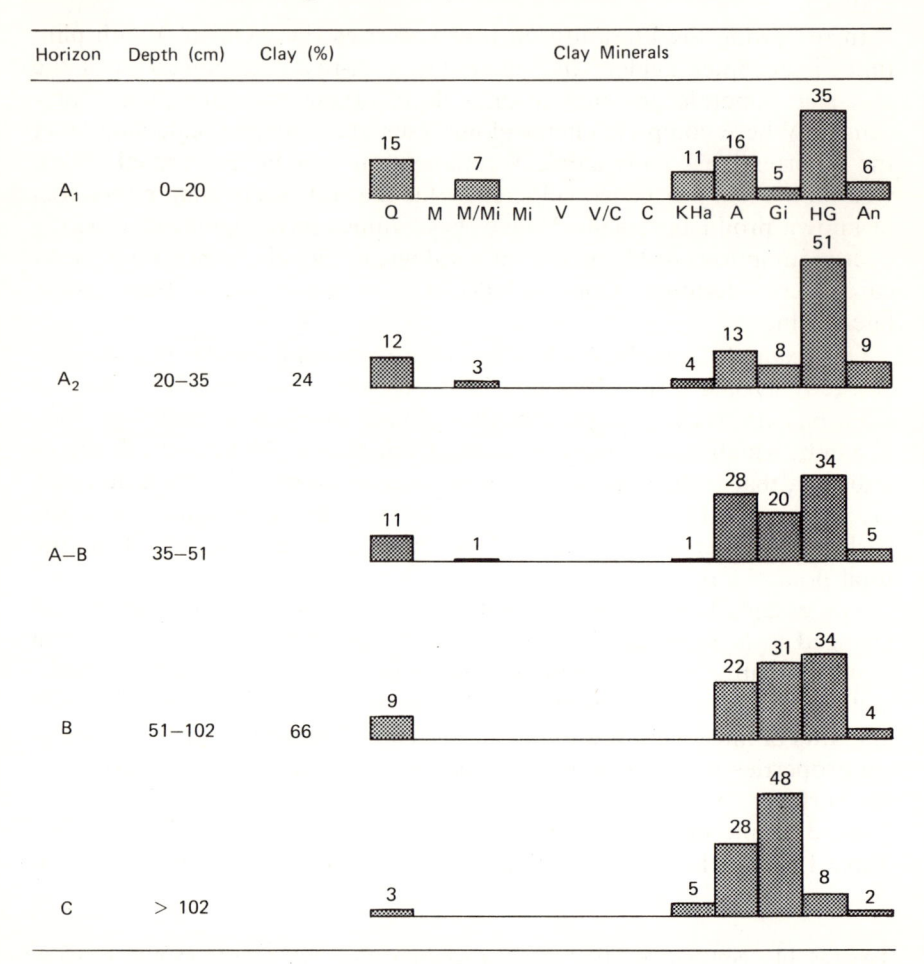

| Horizon | Depth (cm) | Clay (%) | Clay Minerals (Q, M, M/Mi, Mi, V, V/C, C, KHa, A, Gi, HG, An) |
|---|---|---|---|
| A₁ | 0–20 | | Q 15, M 7, KHa 11, A 16, Gi 5, HG 35, An 6 |
| A₂ | 20–35 | 24 | Q 12, M/Mi 3, KHa 4, A 13, Gi 8, HG 51, An 9 |
| A–B | 35–51 | | Q 11, M/Mi 1, C 1, KHa 28, A 20, HG 34, An 5 |
| B | 51–102 | 66 | Q 9, A 22, Gi 31, HG 34, An 4 |
| C | >102 | | Q 3, A 5, Gi 28, HG 48, An 8, (2) |

**Key:** Q, quartz; M, montmorillonite group; M/Mi, interstratified montmorillonite-mica; V, vermiculite; V/C, vermiculite-chlorite; C, chlorite; KHa, kaolinite + halloysite; A, allophane; Gi, gibbsite; HG, hematite + goethite; An, Anatase. Also for Tables 16, 17, 18, 20, 21, 22, 23. In Table 15 only, percentages are given above the respective columns.

as a humic ferruginous Latosol, which is derived from basalt under alternating wet and dry seasons.

*Lateritic Soils, Ultisols.* The broad grouping of soils described at various times as Lateritic Soils, Tropical Red Earths, Ferrisols, Ferri-allitic Latosols, as well as some of the Red-Yellow Podzolic soils, fall mainly, though not exclusively in the category Ultisols of the American system of classification. They have been extensively characterized by mineralogical and chemical methods. They occur not only in the tropics but in broad adjacent regions. They are formed from diverse parent materials under moderately high rainfall and temperature that lead to weak accumulation of organic matter. They are found on old, unglaciated landscapes, originally under forest.

The conditions for their formation involve extensive leaching for a major part of the year, with temperatures high enough for rapid decomposition of added organic matter. Clay illuviation is well marked by clay skins in the B horizon. The B horizon extends to relatively great depths, and within it the clay content increases with depth. Simonson (52) believes that the leached $A_2$ horizon increases in thickness at the expense of the upper B horizon and that clay formation by weathering in place is more important than illuviation and redeposition. However the index mineral method has not been applied strictly to these profiles and uncertainties regarding Simonson's arguments remain. Barshad (4) used his chemically based procedure on several Ultisols (Cecil, Appling, Davidson, and some South African soils). Many examples from the southeastern United States have been studied physically, chemically, and mineralogically (57, 58).

Mineralogically, the Ultisols show greater diversity than the Oxisols. However the tendency for the clay fraction to be dominated by kaolinite plus halloysite is very strong. Mineralogical differences between horizons are sometimes well marked. Tables 16 and 17 summarize some of the results from the literature.

In general Ultisols from upland sites are less complex mineralogically than those from valley and lowland sites. The latter usually show the influence of more diversified and mixed parent materials, as well as marked fluctuations in water table relatively near to the surface. Table 18 gives an example of the Aripo fine sand from North Trinidad, where some important observations on the formation of iron concretions were made. It was possible to identify concretions of different ages, up to 30 years. Very marked increase in iron was found with increasing age (Table 19). Manganese was very low throughout, indicating that strongly reducing conditions characterized the occurrence of the high water table.

The differences and relationships of the Red-Yellow Podzolic soils in a

## TABLE 16

**Name:** Cecil sandy loam
**Locality:** Mecklenberg County, Virginia, No. 1 (57)
**Parent Mat.:** Mica gneiss
**Climate:** Annual rainfall 117 cm; mean annual temperature 14.7°C

| Horizon | Depth (cm) | Clay (%) | Clay Minerals (Q, M, M/Mi, Mi, V, V/C, C, KH, A, Gi, HG) |
|---|---|---|---|
| $A_1, A_2$ | 0–15 | 13 | Q, M/Mi, V/C, KH |
| $A_3$ | 15–22 | 13 | Q, Mi, C, KH |
| $B_1$ | 22–32 | 35 | Q, V/C, KH |
| $B_{21t}$ | 32–45 | 72 | C, KH |
| $B_{22t}$ | 45–62 | 71 | C, KH |
| $B_{23t}$ | 62–80 | 65 | V/C, KH |
| $B_{24t}$ | 80–92 | 50 | V/C, KH |
| $B_{31t}$ | 92–114 | 50 | V/C, KH |
| $B_{32t}$ | 114–147 | 35 | V/C, C, KH |
| $B_{33}$ | 147–183 | 32 | V/C, C, KH |
| $C_1$ | 183–239 | 21 | V/C, C, KH |

given climatic region have been summarized by McCaleb (38) for North Carolina. Figure 50 shows that a clear relationship exists with the mineralogy of the parent material. Basic igneous rocks, low in quartz, give rise to Reddish-Brown Lateritic soils. Typical granites and sedimentary rocks produce Yellow Podzolic soils. In all B horizons and in most C horizons the dominant clay mineral is kaolinite or halloysite.

## TABLE 17

**Name:** Hayesville fine sandy loam (57)
**Locality:** Oconee County, South Carolina, No. 1
**Parent Mat.:** Mica gneiss, low in mica
**Climate:** Annual rainfall 152 cm; mean annual temperature 15.6°C

| Horizon | Depth (cm) | Clay (%) | Clay Minerals (Q  M  M/Mi  Mi  V  V/C  C  KH  A  Gi  HG) |
|---|---|---|---|
| $A_1$ | 0–4 | 9 | |
| $A_2$ | 4–20 | 10 | |
| $B_{1t}$ | 20–35 | 22 | |
| $B_{2t}$ | 35–56 | 38 | |
| $B_{31t}$ | 56–81 | 36 | |
| $B_{32}$ | 81–94 | 26 | |
| $C_1$ | 94–140 | 19 | |
| $C_{21}$ | 140–165 | 10 | |

*Alfisols (Brown forest soils and some claypan soils under grass).* In passing from tropical to temperate climates it becomes convenient to separate the two chief vegetative regimes, forest and grassland. The resultant soil profiles show large differences in the amount and distribution of organic matter. The Alfisols represent chiefly forest soils (Brown forest soils and related groups such as claypan soils); the Mollisols,

## TABLE 18

**Name:** Aripo fine sand (1)
**Locality:** North Trinidad
**Parent Mat.:** Alluvium from mica schist
**Climate:** Annual rainfall 250 cm; mean annual temperature 26°C

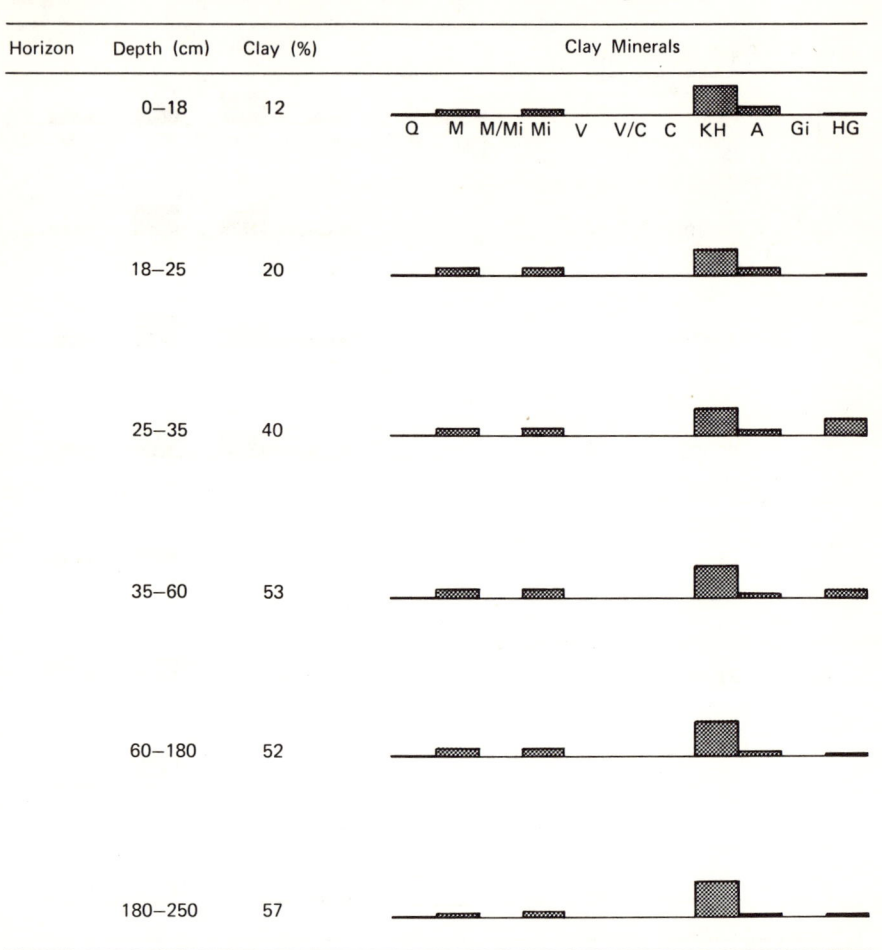

| Horizon | Depth (cm) | Clay (%) | Clay Minerals |
|---|---|---|---|
|  | 0–18 | 12 | Q  M  M/Mi  Mi  V  V/C  C  KH  A  Gi  HG |
|  | 18–25 | 20 |  |
|  | 25–35 | 40 |  |
|  | 35–60 | 53 |  |
|  | 60–180 | 52 |  |
|  | 180–250 | 57 |  |

chiefly grassland soils (Prairie soils and related groups) of the middle latitudes.

The Alfisols are characterized by a moderate base status, low organic matter, a moderate abundance of 2 : 1 lattice clays and the presence of clay accumulation in the B horizon. Weathering is only moderate. Accordingly the clay mineralogy is mixed in all horizons, but in any given

**TABLE 19** Chemical Data on Concretions Formed from Mottled Areas in Exposed Clay Subsoil of the Aripo Soil from North Trinidad (1)

| Age (years) | Range of $Fe_2O_3$ Content (%) | $Mn_2O_3$ (%) |
|---|---|---|
| 1 | 6–18 | 0.01 |
| 5 | 13–41 | 0.01 |
| 30 | 23–44 | 0.01 |

profile, qualitative and even quantitative similarity in the suite of clay minerals is found. Many examples derived from loess and glacial deposits in the American Middle West have been studied. The great power of the combination of clay mineralogy with studies of light and heavy minerals of the sand fractions has been well demonstrated by White, Bailey, and Anderson (65).

The mineralogical uniformity characteristic of these soils is demonstrated in the Saylesville (Wisconsin) and Fayette (Illinois) profiles (Tables 20, 21). The former is dominated by clay mica and the latter by a montmorillonite, probably beidellite. Yet although the proportion of montmorillonite remains about the same down the profile, its character is somewhat different in the A horizons. Much of it does not swell beyond 14 Å in these layers. The Illinois group considers that partial aluminum hydroxide interlayering is responsible for this feature. It is commonly found in A horizons of loessial soils and glacial soils of the midwest,

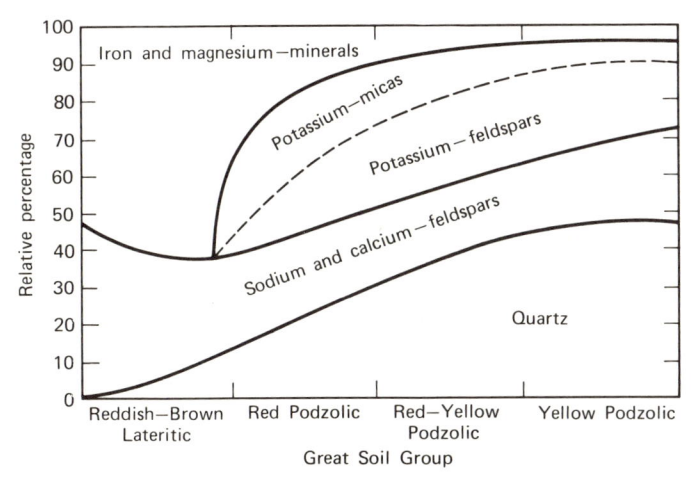

*Fig. 50* Schematic average mineral composition of slightly altered parent material of acid crystalline and sedimentary origin. Reproduced from ref. 38 by permission of the Soil Science Society of America.

## TABLE 20

**Name:** Saylesville silt loam
**Locality:** Jefferson County, Wisconsin (65)
**Parent Mat.:** Glacial outwash, silts, and clays
**Climate:** Rainfall 83 cm; mean annual temperature 7.5°C

| Horizon | Depth (cm) | Clay (%) | Clay Minerals (Q M M/Mi Mi V V/C C KH A Gi HG) |
|---|---|---|---|
| $A_1$ | 0–15 | | |
| $A_{22}$ | 23–28 | | |
| $B_{21}$ | 28–36 | | |
| $B_{22}$ | 36–43 | | |
| $C_1$ | 46–51 | | |
| $C_2$ | 51–81 | | |
| $C_3$ | 81–168 | | |
| $C_4$ | 168–203 | | |

including many Mollisols. It increases with increase in the degree of weathering.

*Mollisols.* In the American system the Mollisols include soils previously described as Prairie soils, Chernozems, Chestnut soils, Brunizors, Brunigra, Humic Gleys, and Wiesenböden. They contain more organic matter than the Alfisols. In the American Middle West they

## TABLE 21

**Name:** Fayette silt loam No. 6
**Locality:** LaSalle County, Illinois (65)
**Parent Mat.:** Peorian loess
**Climate:** Rainfall 82 cm; mean annual temperature 11°C

| Horizon | Depth (cm) | Clay (%) |
|---|---|---|
| A₁ | 0–15 | 15 |
| A₂₁ | 15–23 | 17 |
| B₁ | 33–41 | 27 |
|  | 41–56 | 32 |
| B₂₂ | 56–81 | 31 |
|  | 82–103 | 28 |
| B₃₂ | 103–126 | 27 |
| B₃₂ | 126+ | 24 |

Clay Minerals columns: Q  M  M/Mi  Mi  V  V/C  C  KH  A  Gi  HG

are largely derived from glacial deposits, loess, and various secondary rocks.

Beavers, Johns, Grim, and Odell (6) made a careful comparative study of loess and glacial deposits as parent materials of Illinois soils, all under grassland vegetation. Two examples (Tables 22 and 23) show a high degree of soil development for the prevailing climatic conditions. This is a useful comparison because the loessial soils studied were all dominated by montmorillonitic clay, whereas the glacial soils were highly illitic.

Clearly the Clarence soil from heavy glacial till has undergone little

## TABLE 22

**Name:** Clarence silt loam (Brunizem)
**Locality:** North central Illinois (6)
**Parent Mat.:** Glacial till
**Climate:** Rainfall 84 cm; mean annual temperature 10.6°C

| Horizon | Depth (cm) | Clay (%) | Clay Minerals |
|---------|------------|----------|---------------|
| A | 2–15 | 33 | Q  M  M/Mi  Mi  V  V/C  C  KH  A  Gi  HG |
| B | 28–38 | 70 | |
| C | 61–81 | 60 | |

## TABLE 23

**Name:** Cisne (Planosol)
**Locality:** South central Illinois (6)
**Parent Mat.:** Peorian loess
**Climate:** Rainfall 97 cm; mean annual temperature 12.8°C

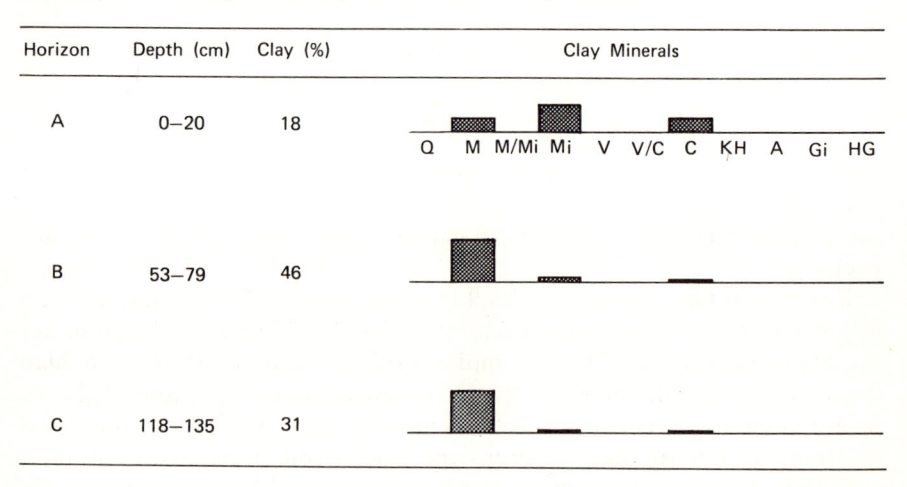

| Horizon | Depth (cm) | Clay (%) | Clay Minerals |
|---------|------------|----------|---------------|
| A | 0–20 | 18 | Q  M  M/Mi  Mi  V  V/C  C  KH  A  Gi  HG |
| B | 53–79 | 46 | |
| C | 118–135 | 31 | |

change in mineralogy. Mica dominates all horizons, and only a small amount of montmorillonitic clay appears to have been formed. The main process is clay illuviation.

The Cisne soil from loess has lost much montmorillonite from the A horizon, with marked gains in mica and chlorite. In neither profile is there evidence of the formation of kaolin group minerals.

The Chernozems form part of the Mollisol order. Several studies of their mineralogy have been published, one of the most complete being that of Redmond and Whiteside (47). Although not expressed quantitatively, their X-ray results indicate that montmorillonitic clay was dominant throughout the three profiles studied. The amount present was considerably less in the A horizons than the B. Mica was more prominent in the A horizons than the B. Only small amounts of kaolinite were found in all horizons. There was a slight accumulation of clay in the B horizons. Carbonates were present in the clay fractions of the C horizons but were absent from the A. In some instances they were found in the lower B.

St. Arnaud and Mortland (59) studied Chernozems formed on glacial till in Saskatchewan. Illite and a montmorillonite (beidellite) were dominant in the coarse clay; the former was most abundant in the A horizon. In the fine clay, montmorillonitic clay dominated, but the sharpness of the 18° Å peaks diminished in the horizon order C > B > A. They concluded that illitic clay was formed in the upper horizons.

Russian investigators of Chernozems also discovered a relative concentration of micaceous clay in the A horizon. Loss of montmorillonitic clay to the B horizon accounts for only part of this increased proportion. It is uncertain to what extent enhanced potassium fixation of the upper layers is caused by the cycling of potassium through vegetation. There is also the possibility that recent accessions of loess high in micas serve to increase the potassium content.

A comparison of compact (heavy texture) Chernozems with normal calcareous Chernozems (Bolyshev, 8) indicated that whereas clay fractions of the latter were dominated by beidellitic clay with a little quartz, the compact Chernozems contained nontronite in the $B_1$ horizon at 80–90 cm, together with halloysite and a little montmorillonite. The pedogenic production of nontronite clay is ascribed to mobility of iron caused by periodic occurrence of high water tables.

The *Guide Book, Tour 5, Northern Kazakhstan* of the 10th International Congress of Soil Science, 1974, gives clay mineral data on an ordinary Chernozem, a Brown forest soil, and a forest Solod. Hydrous micas were usually dominant, followed by chlorites and kaolinite. Montmorillonitic clay was interstratified with mica in the forest Solod, but was not found in the Chernozem. In the Brown forest soil it was interstratified with chlorite. Feldspars in small amount were present in the clay fraction in all horizons of these three soils. Hence changes caused by pedogenic weathering are minimal.

A Solonetz in this region, high in clay content throughout, showed a dominance of kaolinite in the A horizon, with montmorillonites and chlorites in the B, increasing in amount with depth. This distribution may not be pedogenic because the sand fractions showed evidence of depositional variation, and the surface horizon had a pH of 7.6, which is high for the formation of kaolinite under the existing conditions.

*Soils of Semiarid and Arid Regions.* The soils that provide a transition from Chernozems to Desert soils, with decreasing rainfall and increasing temperature, are a classical sequence in the history of pedology. They were studied early by Russian investigators beginning with Dokuchaev. However detailed studies of their mineralogy, especially that of the clay fraction, are very recent. The overall view of the relation of these soils to weathering is simple—namely, the sequence demonstrates decreasing weathering and increasing accumulation of weathering products. Accordingly, secondary carbonates are found closer and closer to the surface as the rainfall decreases.

A somewhat more sophisticated view is readily formulated in terms of the frequency and duration with which water passes from each horizon of the profile to the next downward or upward. Thus the relative depletion or accumulation of weathering products layer by layer could be qualitatively expressed. In most of these soils there is no complete throughput to a water table. Thus all products of weathering are arrested somewhere and accumulate at different depths according to solubility and equilibrium relationships. Thus there will be one depth function for carbonates, another for calcium sulfate, and others for sodium sulfate and sodium chloride. Where rainfall is seasonal, these depth functions will vary accordingly, especially those of the most soluble salts.

The secondary aluminosilicates are exceedingly interesting when viewed in the light of the Garrels and Christ type of diagram. Since soluble monomeric silicic acid is likely to vary relatively little, the main interest lies in ratios of cation to hydrogen ion and their variation. Moderately high sodium and high pH (which generally go together) favor the formation of sodium montmorillonite or sodium beidellite according to the level of magnesium in solution. The simultaneous presence of high sodium and high magnesium has been shown to give analcime plus trioctahedral smectite in synthesis at 200°C. Very high sodium alone gives analcime, as was earlier shown in Noll's syntheses. The mineral illite is readily formed under conditions of moderate $K^+/H^+$ and low to moderate magnesium, the area of stability increasing with the magnesium ion activ-

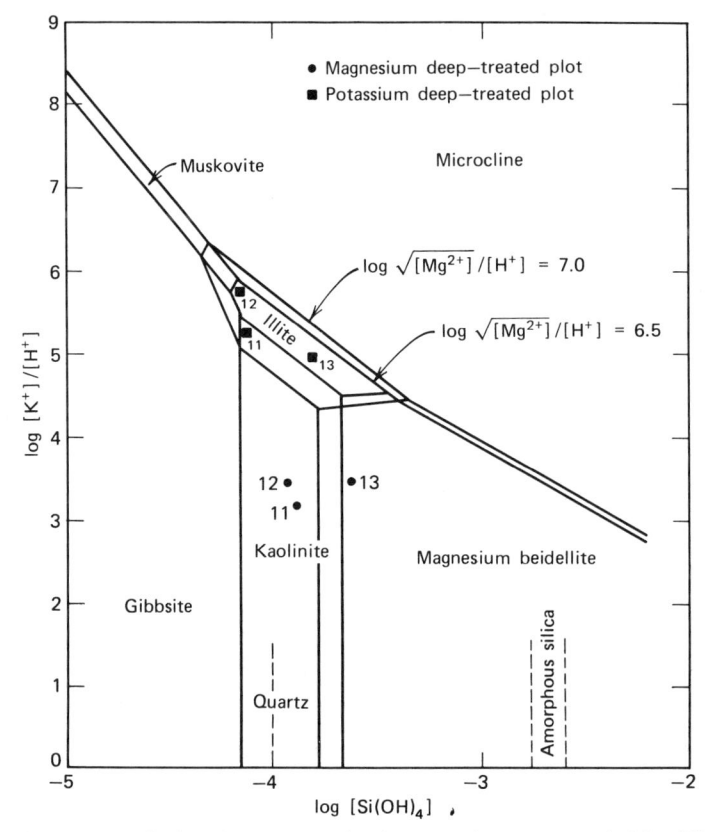

*Fig. 51* Garrels and Christ diagram showing increase in area occupied by illite as log $\sqrt{[Mg^{2+}]}/[H^+]$ increases. Note corresponding reduction in area for kaolinite and increase for beidellite (42).

ity (Figure 51). Very high magnesium might favor the formation of palygorskites (i.e., attapulgites). Thermochemical data to place them in Garrels and Christ diagrams have only recently become available.

Thus montmorillonites, members of the beidellite-nontronite series, illite, analcime, and possibly attapulgites and sepiolite might be formed by synthesis from the weathering products of mixed feldspars, micas, and ferromagnesian minerals. They have all been recorded in different soils of arid regions.

Three semiarid soils studied by Smith and Buol (55) were dominated by illitic clay in the upper horizons. Montmorillonite was found with calcium carbonate below 74 cm in the White House soil (a reddish brown soil now classified as a Mollic Haplargid). It was present in limited amount in the slightly calcareous $B_3$ horizon (112–127 cm) of the Sonoita soil (a red desert soil now classified as a typic Haplargid). It was also present in the $C_3$ horizon (53–86 cm) of the Hathaway soil (a Calcisol, now classified as a typic Calciustoll). All these soils showed some lithological discontinuities, making it uncertain that the montmorillonite was formed pedogenically. The authors conclude that clay movement in the profiles is present in combination with clay formation in the B horizons. Maximum weathering occurred in the A horizons.

At the other extreme, with high salt contents, we have the highly sodic alkali soils studied in California (2), in which analcime is a dominant mineral of undoubted pedogenic origin. The Pond clay loam (classified as a Paleargid) derived from granitic alluvium in a poorly drained basin, under seasonal rainfall of less than 10 cm per year gave the following data. The content of analcime, which was found predominantly in the fine salt and coarse clay fractions, was estimated at 22% of the soil in the 0–15 cm layer. The subsoil layer at 15–46 cm contained approximately the same amount. Decreasing amounts were present at 46–72 cm and 72–122 cm, and none was found below 122 cm. From the diffractograms of the coarse clay, 10 Å mica was present throughout the profile, with a distinct accentuation in the surface layer. A peak at about 14.7 Å appeared for all layers except the surface.

Data collected from related soils in the same area indicated that analcime was not found if the pH of a saturation extract was below 9. The surface layer of the Pond soil gave a pH of 9.7, and the saturation extract contained 1.66 moles/liter sodium, mainly as chloride and sulfate, with some carbonate. These results would indicate a value of about 9.7 for pH-pNa. Reference to the Garrels and Christ diagram (Figure 5, Chap. 2) indicates that at this value analcime would be stable for all values of log $[H_4SiO_4]$ between $-4.7$ and $-2.7$. The triple point analcime–montmorillonite–amorphous silica occurs at pH-pNa = 8.3 and pSi(OH)$_4$ = 2.7.

In contrast to the clear evidence for pedogenic analcime in California, the presence of attapulgite-sepiolite minerals in soils of the Near East has been ascribed to the parent limestones (44). The surface soils contain less of the attapulgite-sepiolite minerals and more of the montmorillonite groups than the lower horizons and parent limestones. However Singer and Norrish (53) have found occurrences in the lower horizons of some Australian arid soils indicating that palygorskites (attapulgites) are pedogenic, since they form coatings on soil peds. In these horizons the composition of the saturated soil extract in respect of $Mg^{2+}$, $Al(OH)_4^-$, $H^+$, and $Si(OH)_4$ correspond closely to that derived by the equilibration of pure palygorskite with water. From the latter, the free energy of formation was calculated. The authors conclude that at the pSi value 3.6, common in many soils, palygorskite would require pH values above 7.7 and pMg$^{2+}$ values below 4.0 for precipitation. Such values are most likely to be found in salt lakes and playas in areas of rocks rich in magnesium.

The *Guide to Soil Excursion, Tour 7, Central Asia, Uzbekhstan,* of the 10th International Soils Congress, 1974, gives clay mineral data on a number of Desert soils, some under irrigation, including Sierozems, Gray-Brown Desert soils, a takyr (heavy textured Desert soil with hard crust and very little vegetation), and a Cinnamon-Brown soil. Some of these contained detectable palygorskites associated with carbonates in the lower horizons and some had 7 Å serpentines throughout. The dominant mineral in almost all cases was hydrous mica, followed by chlorites and kaolinite. Montmorillonites were minor constituents of irregular occurrence. These clay minerals were chiefly inherited from the parent material, which was loess or fine alluvium in most of the profiles.

*Spodosols (Podzols and podzolization).* Although the word Podzol implies the presence of an ashy layer in the upper part of the soil, the actual process of podzolization involves not only the differential loss of iron and aluminum from this layer but also their accumulation together with that of humus, in the B horizon immediately below. This marked difference in the composition of the colloidal fraction of the soil between the bleached A$_2$ and the B horizon suggests that mineralogical relationships should prove exceedingly interesting.

It has long been known that Podzols form most readily on sandy parent materials and that certain dominant species of plant cover favor their formation. The climatic and drainage conditions are usually taken to be such that acidic organic matter can accumulate at the surface. These considerations all point to crucial interactions between soil organic matter and mineral matter. Four sets of ideas have been used to explain how these interactions might lead to Podzol formation: (1) the protective colloid theory, (2) the isoelectric precipitation theory, (3) the complexing of iron and aluminum, and (4) the reduction of trivalent iron to divalent, its mobilization in this form, and its subsequent reoxidation and deposition. These ideas are not mutually exclusive. They have been utilized to explain that the A horizon is more siliceous than the B and that the latter may be subdivided according to the dominant accumulation of humus, iron, and aluminum.

Of the chemical environment for mineralogical change we can say that the A horizon should be distinctly different from the B, being more siliceous, lower in pH, and possibly, through complexing by organic matter, lower in the activity of ionized aluminum and iron species. If free hydroxides are present in the B horizon, this should be revealed through the constancy of the appropriate ionic products and through chemical solubility relationships.

Modern mineralogical methods applied to the clay fractions from Canadian Podzols have shown some fairly consistent relationships (Kodama and Brydon, 33). Comparison of the A and C horizons for five Podzol profiles in New Brunswick indicated that the illite and chlorite dominating the C horizons had been changed to a randomly interstratified mica-vermiculite-smectite with no trace of chlorite in the A horizons. Kaolinite was only a minor constituent. It was concluded that the hydrated components in the interstratified clay of the A horizons increased in proportion with the degree of weathering.

The B horizons were not reported on in this study because of the dominance of amorphous or poorly crystalline material. A later study (Brydon and Shimoda, 14) of a Podzol from Nova Scotia indicated that this B horizon was dominated by allophanelike material with small amounts of partially chloritized vermiculite and a smectite. In this soil, as in those from New Brunswick, the A horizon contained interstratified mica-vermiculite-smectite. The C horizon had illite, vermiculite, and a small amount of amorphous material, but no chlorite.

Thus the Canadian work definitely points to a pedogenic origin of smectite and vermiculite in A horizons, and of allophane in B horizons, with a hint that chlorites may also be formed there. Conditions in A horizons are not compatible with long-continued presence of chlorites, nor do they appear to favor the formation of halloysite or kaolinite. Whether the latter minerals would eventually crystallize in B horizons remains to be determined.

In northern Michigan a series of Podzols that form a chronosequence of 3000–10,000 years in their pedological development from sandy glacial till was studied by Franzmeier, Whiteside, and Mortland (22). The Blue Lake profile No. 1, of about 10,000 years development, showed montmorillonitic clay as dominant in the $A_2$ horizon (46%) with appreciable quartz (22%) and illite (13%). The $B_h$ horizon gave 19% montmorillonite, 17% quartz, 14% illite, and in addition, 12% chlorite and 19% vermiculite. The C horizon had only 4% montmorillonite with 34% illite, 20% quartz, 19% chlorite, and 3% vermiculite. Kaolinite, potassium feldspars, and allophane were present in small amounts in all three horizons. Thus in the A horizon there is clearly pedogenic montmorillonite probably derived from illite and chlorite. The B horizon has pedogenic vermiculite, but the chlorite content is below that in the C horizon.

A type of Michigan Podzol underlain by a fragipan was very thoroughly investigated by Yassoglou and Whiteside (66). The upper Podzol $A_2$ horizons contained abundant montmorillonite derived from illite, chlorite, and interstratified clays in the parent glacial till material.

Podzols in the Alps studied by Bouma et al. (10) showed beidellite in the A horizons, evidently derived from chlorites and micas in the parent materials.

A mountain Podzol formed under heath-type vegetation from feldspathic sandstone in eastern Tennessee (McCracken, Shanks, and Clebsch, 39) was found to be dominated by vermiculite with some kaolinite and a little gibbsite and vermiculite-chlorite in the B horizon. The weathering and soil development processes seem to center around a decrease in feldspars and chlorite and an increase in vermiculite in the B and A horizons. The presence of gibbsite seems to point to the incidence of nonpodzolic processes at some stage in the history of this quite old soil. The soil mineralogy is very similar to that of several "sols bruns acides" (acidic Brown forest soils) in the same area.

Another old Podzol predating the last glaciation in Scotland has been studied by Stevens and Wilson (60). The parent material contained micas and chlorites and these also dominated the B horizons, but in the $A_2$ horizons the chlorite content was much reduced and kaolinite was found. No smectites were found. This and the preceding example seem to indicate that eventually podzolic conditions change so that montmorillonites are unstable and the chemical environment approaches that found in Ultisols.

The prevalence of smectites or vermiculites in the $A_2$ horizon of Podzols from Scandinavia led Gjems (24) to study the variations in X-ray spacing with exchange cation in some detail. He concluded that the evidence supported the presence of a mixed layer dioctahedral smectite–vermiculite.

Thus the leached horizons can be dominated by smectites or dioctahedral vermiculites or both, probably depending on the parent material. There is no evidence of the formation of kaolinite, halloysite, or chlorites. Illites are sometimes present, but as a residue from the parent material.

*Limestone Soils*

*1. General.* The pedological problems of limestone soils are now seen to be much more interesting and complex than was at first realized. Here are cases in which soils are formed from more or less minor residues of the original rocks after the major carbonate components have passed into true solution. These minor residues are very finely divided silicates and aluminosilicates, often with aggregates of amorphous or cryptocrystalline silica (chert, etc.). They are thus highly reactive when the environment changes. Two sets of changes were discussed in Chapter 2: loss of soluble salts and complete dissolution of carbonates. The products are then subject to the usual factors of soil formation and development.

Limestones may contain a variety of clay minerals. Weaver (64) originally considered that the clays deposited largely remain as such in the limestone. Other authors have contended that transformations occur after deposition (diagenesis). In particular Burst (15) has shown that deep burial produces definite increase in illitic clays at the expense of smectites. The general conclusion from X-ray studies of clay minerals in limestones is that kaolinite is only occasionally a major constituent; illites are generally dominant; smectites are major constituents in limestones influenced by volcanic dust; chlorites, vermiculites and interstratified clays also occur in secondary quantities, as does quartz.

Pedologically, limestone soils have long been divided into two broad groupings—the red soils and the black soils. The former include the Terra Rossa of the Mediterranean region. The latter include black self-mulching clays (sometimes called Grumosols) now classified in the United States system as Vertisols, such as the Houston series of Texas; also the thin soils of the chalk uplands in England (Rendzinas). Clearly the incidence of climatic factors is wide-ranging, once the carbonates with their buffering action have been lost by leaching.

As already discussed in Chapter 2, initial contact of fresh limestone with water greatly reduces the soluble salt content and with it the effective $K^+/H^+$ ratio. Whether this change is drastic enough to bring the system from the illite stability field to that of kaolinite depends on the localized water regime and its cyclic fluctuations, and also on the bonding energy of potassium to the particular illite. If the latter is very high, very small values of $K^+/H^+$ will be effective in maintaining the stability of the illite. In the Hagerstown soil studied by Brydon and Marshall (13), little change in clay minerals was found even after complete removal of the solid carbonates.

This situation may be contrasted with that found by Scrivner (50) in the lower horizons of the Lebanon soil in Missouri, which was formed on a cherty dolomite. A change involving rapid reduction in illite and increase in kaolinite and in interstratified illite-montmorillonite occurred in the zone of disintegrating rock (Table 24). This change had been noted earlier by Carroll and Hathaway (18) in a soil formed from the cherty, dolomitic Lenoir limestone in Virginia.

The contact between weathering limestone and four pedogenic profiles in the same Missouri Ozark area was examined further by Miller (40). All four soils showed increasing clay content with depth down to the contact with weathered limestone. In all cases rapid increases in kaolinite and in interstratified 2 : 1 clays occurred close to the limestone where the illite content diminished. Higher up in the soil profiles the content of illite diminished further, and interstratified minerals were found. Near the soil surface a 14 Å mineral with characteristics somewhat different from chlorites or vermiculites was found.

**TABLE 24** Properties of Clay in Lebanon Profile, Missouri (50)

| Distance Above Limestone (cm) | Fine clay (%) | Coarse/Fine | pH$_s$ | Interstratified Montmorillonite-Illite (%) | Illite (%) | Kaolinite (%) |
|---|---|---|---|---|---|---|
| | | | Fine Clay (<0.2 μ) | | | |
| 0 (Dolomite) | 25 | 1.2 | 7.5 | — | 100 | — |
| 0–1.3 | 22 | 1.8 | 7.4 | — | 95 | 0.5 |
| 1.3–6 | 30 | 0.93 | 7.4 | 25 | 60 | 15 |
| 6–13 | 47 | 0.57 | 7.4 | 0 | 20 | 50 |
| 16–20 | 65 | 0.35 | | 30 | 15 | 55 |

| Depth from Surface (cm) | | | | Montmorillonite-Illite-Vermiculite (%) | | |
|---|---|---|---|---|---|---|
| 351–366 | 46 | 0.46 | 6.1 | 0 | 35 | 60 |
| 152–168 | 46 | 0.56 | 3.5 | 10 | 30 | 60 |
| Surface–104 | 8–25 | 1.02–0.71 | 4.4–3.3 | 40–70 | 0 | 20–40 |

| Distance above Limestone (cm) | Coarse Clay (%) | Illite (%) | Kaolinite (%) | Quartz (%) | Feldspar (%) | Montmorillonite (%) |
|---|---|---|---|---|---|---|
| | | | Coarse Clay (2–0.2 μ) | | | |
| 0 | 23 | 70 | 5 | 5 | 5 | |
| 0–1.3 | 26 | 70 | 5 | 5 | 5 | |
| 1.3–6 | 29 | 25 | 5 | 5 | 5 | |
| 6–13 | 26 | 5 | 5 | 5 | 5 | |
| 16–20 | 23 | 5 | 5 | 5 | 5 | 5 |

Similar results have been reported by Miller and Springer for three similar limestone profiles in Tennessee. Further work will be needed to establish the relative roles of illuviation, weathering from primary minerals, and clay transformations in a changing environment. It is clear, however, that the transition from calcareous to acidic environment greatly favors the disappearance of illite and the appearance of kaolinite, as would be expected from the stability diagrams. At the same time some silica becomes available (illite is more siliceous than pure muscovite), making reactions favoring 2:1 silicates feasible, as well.

The transformed residue after loss of carbonates is subsequently the parent material for pedogenic change, which will be subject to the usual variations as the factors of soil formation dictate. The remarkable feature of the red limestone soils considered previouly is that in a climatic zone where heavy subsoils and claypans are common, the limestone soils show increases in clay content right through the B and C horizons to the uppermost calcareous layer near the limestone. Since this increase in clay corresponds to an increased proportion of the fine clay fraction (see Table 24), the ratio of coarse to fine clay decreases in the same manner as is shown by heavy clay B horizons. However mineral transformations as well as illuviation could contribute to this type of change. The content of primary minerals, such as feldspars, in residues from limestones is variable. Scrivner's study of the Lebanon profile demonstrated that the acidic layers of the C horizon were almost free of silt-size feldspar, whereas the residues from dolomite and weathered dolomite contained 16–24% microcline in the 0.025–0.005 mm fraction, 25–43% in the 0.005–0.002 mm fraction, and less than 10% in the coarse clay fraction. The conversion of this feldspar to kaolinite or to 10 or 14 Å clay could supply an appreciable fraction but not the major part of clay found in the lower acidic horizons. However concentration of clay by loss of carbonates is more than sufficient to account for the total clay found. The particle size distribution and X-ray identification of clay minerals clearly indicates that kaolinite and interstratified clays produced from illite are dominant in the fine clay fraction.

We know from a $^{14}$C age study of the organic carbon present that illuviation plays an important part in the formation of very heavy clay layers near the limestone. Organic matter carried down with clay proved to be younger than that present in the acidic layers above (Ballagh and Runge, 3). Thus it is clear that we still need detailed studies of these limestone soils to evaluate quantitatively the relative roles of illuviation and mineral transformations.

In addition to mineral transformations and illuviation of clay, transport in and deposition from true solution need consideration. This is strikingly effective for phosphate in limestone soils. For instance, the layer 5–11 cm above the limestone underlying the Talbott soil contains 1700 ppm phosphorus, whereas none was detected in the fresh rock. Thus there has been translocation and precipitation near the contact.

Similar processes may operate for other elements. Chert probably partially dissolves and sluggishly redeposits as quartz. The stability diagrams show that the solubility of quartz lies within the kaolinite and illite regions. The influence of acidic organic matter in complexing iron and aluminum will show itself by losses in the A horizons and accumulations in the B as normally found for Podzols; and indeed Podzols can be formed on limestone residues. However the soils here considered show no evidence of organic matter accumulation at depth. Nonsilicate iron ("free iron") closely parallels total clay in its variation down the profile.

These limestone soils of the middle latitudes bear many resemblances to red soils from granites, and so on, earlier classified as Red-Yellow Podzolic soils and now included in the Ultisol order. Using the degree of base saturation as the differentiating criterion, some limestone soils fall with the Ultisols and others with the Alfisols. Apart from this, their similarities are great.

In the A and upper B horizons kaolinite is usually accompanied by mixed layer clays somewhat different from those found in the decomposing limestone. The X-ray diffractiograms are characterized by poorly defined peaks and broad bands. Chlorites and vermiculites can sometimes be identified, but frequently complex interstratifications occur. Thus the X-ray diffraction curves are difficult to interpret, and additional evidence is sought.

*2. Carbonates in Calcareous Soils.* The weathering of calcite and dolomite from materials containing both does not correspond to the much greater solubility of calcite in acidic solutions. As shown by Stumm and Morgan (61), the solubilities reverse at pH 7.8 for solutions containing $5 \times 10^{-4}$ $M$ $Mg^{2+}$ and $2 \times 10^{-3}$ $M$ total carbonates. Read and Protz (45) have demonstrated that dolomite disappears proportionately more rapidly than calcite from a series of soil profiles in Ontario, Canada. They plotted the ratios of dolomite at depth to dolomite in a given horizon (dolomite index) and calcite at depth to calcite in a given horizon (calcite index), against depth. Five stages were found. In unweathered materials both ratios remain close to unity at all depths. Next, near the surface, calcite begins to increase at the expense of dolomite. The calcite index becomes less than unity and the dolomite index greater. In the third stage this difference becomes more accentuated and progresses to greater depths. In the fourth stage both indices become greater than unity, the dolomite index more than the calcite index. In the fifth stage both indices are very high at the surface and even higher in the subsoil. They then diminish with depth back to unity. These stages were demonstrated on a series of soils from glacial till corresponding to different landforms.

*3. Terra Rossa.* Soils produced on limestones in the Mediterranean region have long been known as "Terra Rossa." The climatic conditions provide a moist, cool winter and a dry, warm summer. Rather shallow soils have resulted, neutral at the surface and increasingly calcareous in the subsoil, with no clearly defined B horizon. Thus pedological processes and weathering of limestone proceed in close vertical proximity, and accumulation of organic matter is slight. The dry summer season tends to preserve hematite, whether residual or transformed from other iron minerals. In Palestine a heavy textured Terra Rossa soil studied by Barshad et al. (5) contained about 65% montmorillonite and 35% kaolinite in the clay fraction. Other limestone soils in the same area contained in addition attapulgite inherited from limestone. Its amount decreased with increasing vertical distance from the rock. Thus it is not a pedogenic product in these soils and is slowly transformed even in a neutral or calcareous environment, presumably when the ratio of $Mg^{2+}$ to $[H^+]^2$ falls below a fixed value.

*4. Gray and Black Limestone Soils.* These were described as Rendzinas in earlier soil literature. Two broad groups are now distinguished. Upland soils from soft limestones form one group. They are neutral or calcareous at the surface and calcareous below the A horizon. The other group forms the Vertisol order in the new American system. These are deep calcareous black soils, very high in swelling clays. They crack extensively in the dry summers. Surface soil then falls into the cracks, giving rise eventually to a hillocky topography knwon as "gilgai." They are dominated by a high content of montmorillonitic clay. The Houston clay of Texas is a classic example.

The shallower upland group does not exhibit this extreme swelling. The clays present are a mixed assemblage derived from the limestone.

Profile features are weakly expressed in both groups, as would be expected. Thus such soils can be regarded as immature under the prevailing climates, which make for very slow development.

*Volcanic Ash Soils.* Considerable work has been done since 1950 on soils derived from volcanic ash. This parent material varies in its chemical and mineralogical composition and also physically in particle size distribution. In some cases the date of its deposition is known. Thus we have possibilities for very interesting pedological studies. Clearly this highly fragmented material, with its content of silicate glass, should be exceedingly reactive under conditions of weathering.

In comparison with loess, which is also airborne and even more finely divided, several differences should prevail. Volcanic ash is deposited over very short periods associated with single or multiple eruptions. Considerable depths accumulate before weathering begins. By contrast, loess deposition stretches over long periods. The material carried by the wind has already been subject to weathering, which continues as it is deposited in small amount year by year. The initial reaction of volcanic ash with water is highly vigorous, because the extensive mineral surfaces are fresh and mineralogically unstable species, including glass, are present. It will be an atmospheric type of weathering, different from that of igneous rocks at depth as demonstrated by Harrison. Thus it is not surprising that noncrystalline aluminosilicates dominate in the early stages under quite a wide variety of climatic conditions. Then as different moisture regimes make themselves felt, one would expect the corresponding secondary minerals to appear, and eventually to dominate. Thus the special properties associated with volcanic ash soils persist to different degrees in different climatic and environmental regions.

These special properties are associated with the presence of the amorphous aluminosilicate allophane and of its crystalline variety imogolite. In allophane the atomic ratio of silicon to aluminum varies over a range from about 1:1 to 1:2. This amorphous mineral forms very stable associations with soil humic matter. Hence in regions of relatively high temperature and precipitation, the soils are blacker and contain more organic matter than would be expected. These "Ando" soils were first described in Japan. Related soils have been studied in New Zealand, Hawaii, Alaska, Indonesia, South America, and other areas of volcanic activity.

Because of the great reactivity of the original ash and also of the first weathering product allophane, the persistence of the latter should be relatively low. This has been well documented. The sequence volcanic ash → allophane → halloysite ($4H_2O$) → halloysite ($2H_2O$) has been demonstrated under moderate weathering and over time spans of several thousand years. Under such circumstances it might be expected that soil horizonation would show itself in the clay mineral distribution in soil profiles. Study of Andosols in Colombia by Calhoun et al. (17) indicated that such differences exist to some extent both in the qualitative and quantitative distribution of minerals with depth.

Earlier work (51) on volcanic ash soils of Hawaii had shown that under the highest rainfall (hydrol humic soils) the profiles were dominated by gibbsite, allophane, and goethite, but no kaolin group minerals, whereas the low humic soils under moderate rainfall gave soils dominated by kaolins but also containing allophane and a little gibbsite.

Although thermodynamic data for allophane are not available, the position of sodium and potassium glasses with an $SiO_2/Al_2O_3$ ratio of 3 (alkali feldspar composition) has been used by Rai and Lindsay (46) in the Kittrick type of diagram to demonstrate the wide variety of products that might come from volcanic ash. Indeed their diagram suggests that feldspar might be a product under conditions of high cation activity. This would simply be a crystallization of albite or microcline from glass of the same composition. However Figure 5 (Chap. 2) indicates that very high values for soluble silicic acid are required. Even for alkali soils containing sodium carbonates, such concentrations would be unusual.

## SILTS AND SANDS

The examination of sand and silt fractions by microscopic methods dates back to the nineteenth century. It originally served two purposes: to provide evidence on geological origin, and to correlate with soil fertility relationships. The latter were related to content of particular minerals such as calcic and potassic feldspars or apatite. These tend to disappear under weathering, and in decomposing they provide certain essential elements for plant growth.

It is now possible to identify and estimate individual minerals of silt size by X-ray methods. Sands can be ground to silt size for examination. In spite of these advances in technique, relatively few studies of sands and silts in soil profiles are available. Attention has been focused on the products of weathering and their movement. The appearance of these products should be related to the loss of coarser minerals.

The evidence available shows that weathering is usually most intense in the A horizon or upper B horizon. The A horizon may be dry or frozen for part of the year, which limits the extent of its reaction with water. Offsetting this factor is its greater content of organic matter which favors weathering processes.

The work of Jeffries and White (31) provides a good example of detailed mineralogical study of a soil profile, in which heavy liquid separations were combined with microscopic identification and counts. The soil chosen was the Hagerstown soil from dolomitic limestone in central Pennsylvania. Study of the sand fractions and the heavy minerals indicated that transported sediments were present in the upper horizons. Detailed study of the feldspar composition showed that microline had remained unweathered but calcic plagioclases were almost completely lost. The steady increase of clay with depth (Figure 52) did not correspond quantitatively to loss of feldspar and must be ascribed to other causes, such as illuviation and production from other minerals such as hydrous micas.

The identification of mineral grains by refractive index determinations was extended by Marshall (37) to the finest silt and to coarse and fine clay fractions. It was later used in the study of soil profiles from acid and basic rocks in Missouri (30). It led, in combination with electron microscope studies, to the identification of the clay as chiefly a member of the beidellite-nontronite series, mixed with small amounts of quartz, kaolinite, halloysite, and hydrous micas, with possible chlorites near the basic igneous rock. The changes in the quartz and chalcedony, feldspar, and mica of the sands and coarse silt, and those of the total clay are represented in Figures 35 and 36 (Chapter 5) as a function of depth. A minimum in feldspar content occurs in the 9–16 in. layer and is associated with a high clay content. Micas reach a maximum in the 33–47 in. layer. They are formed mainly by decomposition of the hornblendes in the original rock and virtually disappear in the surface and subsurface horizons.

A good example of a quantitative study extending through the whole range of particle sizes is that of Barshad on the Sheridan soil from granite in California (4). He used a separation of clay from nonclay at 5 $\mu$ and employed quartz + albite of the fine sand as the immobile indicator.

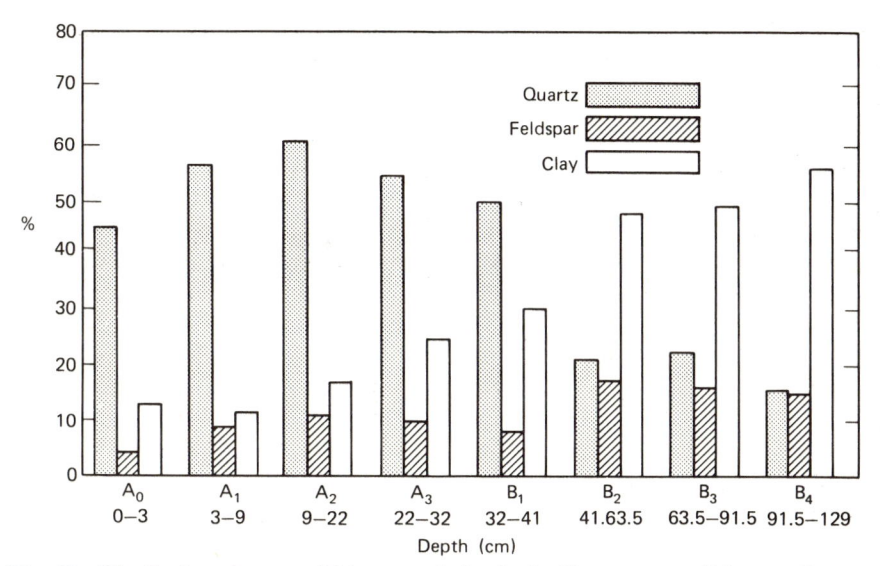

*Fig. 52* Distribution of quartz, feldspar, and clay in the Hagerstown soil from sediments and limestone. Reproduced from ref. 31 by permission of the Soil Science Society of America.

Calculations then yielded both the clay formation and its downward movement. In this case of relatively slight change, formation of clay was confined to the $A_1$, $A_2$ and $B_1$ horizons down to 50 cm. Translocation of clay extended to 75 cm.

## REFERENCES

1. Ahmed, H. and R. L. Jones, A plinthaquult of the Aripo Savannas, North Trinidad. II. Mineralogy and genesis, *Soil Sci. Soc. Am. Proc.*, **33**, 765 (1969).
2. Baldar, N. D. and L. D. Whittig, Occurrence and synthesis of soil zeolites, *Soil Sci. Soc. Am. Proc.*, **32**, 956–958 (1973).
3. Ballagh, T. M. and E. C. A. Runge, Clay-rich horizons over limestone—Illuvial or residual? *Soil Sci. Soc. Am. Proc.*, **34**, 534 (1970).
4. Barshad, I., Factors affecting clay formation. In *Clays and Clay Minerals: Proceedings of the Sixth National Conference on Clays and Clay Minerals* (Berkeley, Calif., 1957), Pergamon Press, New York, 1958, pp. 110–132.
5. Barshad, I., E. Halevy, H. A. Gold, and J. Hagin, Clay minerals in some limestone soils in Israel, *Soil Sci.*, **81**, 423–433 (1956).
6. Beavers, A. H., W. D. Johns, R. E. Grim, and R. T. Odell, Clay minerals in some Illinois soils developed from loess and till under grass vegetation. *Clays and Clay Minerals: Proceedings of the Third National Conference on Clays and Clay Minerals*, National Academy of Science—National Research Council, Publication 395, 356–372 (1954).
7. Beinroth, F. H., G. Uehara, and H. Ikawa, Geomorphic relationships of oxisols and Ultisols on Kauai, Hawaii, *Soil Sci. Soc. Am. Proc.*, **38**, 128–131 (1974).
8. Bolyshev, M. V., Genesis of compact soils in the Chernozem and Chestnut zones, *Sov. Soil Sci.*, No. 6, 654–663 (June 1965).
9. Borchardt, G. A., F. D. Hole, and M. L. Jackson, Genesis of layer silicates in representative soils in a glacial landscape of southeastern Wisconsin, *Soil Sci. Soc. Am. Proc.*, **32**, 399–403 (1968).
10. Bouma, J., J. Hoeks, L. Van der Plas, and B. Van Scheurenburg, Genesis and morphology of some Alpine Podzol profiles, *J. Soil Sci.*, **20**, 384–398 (1969).
11. Brown, G., Ed., *The X-ray Identification and Crystal Stucture of Clay Minerals*, 2nd ed., Mineralogical Society, London (1961).
12. Brydon, J. E., H. Kodama, G. J. Ross, Mineralogy and weathering of the clays in orthic Podzols and other podzolic soils in Canada, *Trans. 9th Int. Congr. Soil Sci.*, **III**, 41 (1968).
13. Brydon, J. E. and C. E. Marshall, Mineralogy and chemistry of the Hagerstown soil in Missouri, University of Missouri Agricultureal Experimental Station Research Bulletin 655 (1958).
14. Brydon, J. E. and S. Shimoda. Allophane and other amorphous constituents in a Podzol from Nova Scotia, *Can. J. Soil Sci.*, **52**, 465–475 (1972).
15. Burst, J. F., Jr., Post-diagenetic clay mineral environmental relationships in the Gulf Coast eocene. In *Clays and Clay Minerals: Proceedings of the Sixth National Conference on Clays and Clay Minerals* (Berkeley, Calif., 1957), Pergamon Press, New York, pp. 327–341 (1958).

16. Cady, J. G., Mineral occurrence in relation to soil profile differentiation, *7th Int. Congr. Soil Sci. Trans.*, **4**, 418 (1960).
17. Calhoun, F. G., V. W. Carlisle, and C. Luna Z., Properties and genesis of selected Colombian Andosols, *Soil Sci. Soc. Am. Proc.*, **36**, 480 (1972).
18. Carroll, D. and J. C. Hathaway, Clay minerals in a limestone soil profile. In *Clays and Clay Minerals: Proceedings of the Second National Conference on Clays and Clay Minerals* (Columbia, Mo., 1953), National Academy of Science–National Research Council, Publication 327 (1954), pp. 171–182.
19. El-Nahal, M. A. and L. D. Whittig, Cation exchange behavior of a zeolitic sodic soil, *Soil Sci. Soc. Am. Proc.*, **37**, 956–958 (1973).
20. Fischer, W. R. and U. Schwertmann, The formation of hematite from amorphous iron III hydroxide, *Clays Clay Miner.*, **23**, 33 (1975).
21. Franzmeier, D. P. and E. P. Whiteside, A chronosequence of podzols in northern Michigan. I. Ecology and description of pedons, *Mich. Agr. Exp. Sta. Quant. Bull.*, **46**, 2–20 (1963); II. Physical and chemical properties, **46**, 21–36 (1963).
22. Franzmeier, D. P., E. P. Whiteside, and M. M. Mortland, A chronosequence of podzols in northern Michigan. III. Mineralogy, micromorphology and net changes occurring during soil formation, *Mich. Agr. Exp. Sta. Quant. Bull.*, **46**, 37–57 (1963).
23. Gard, J. A., Ed., *The Electron-Optical Investigation of Clays*, Mineralogical Society, London (1971).
24. Gjems, O., A swelling clay mineral from the weathering horizon of Podzols, *Clay Miner. Bull.*, **5**, 159–160 (1962).
25. Hardy, F. and G. Rodriguez, Soil genesis from andesite in Grenada, British West Indies, *Soil Sci.*, **48**, 361–384 (1939).
26. Hardy, F. and G. Rodriguez, The genesis of Davidson clay loam, *Soil Sci.*, **48**, 483–495 (1939).
27. Hardy, F. and R. R. Follett-Smith, Studies in tropical soils. II. Some characteristic igneous rock profiles in British Guiana, South America, *J. Agr. Sci.*, **21**, 739–761 (1931).
28. Harrison, J. B., The katamorphism of igneous rocks under humid tropical conditions, Imperial Bureau of Soil Science, Harpenden, England (1934).
29. Hervel, R. C. van den, The occurrence of sepiolite and attapulgite in the calcareous zone of a soil from Las Cruces, New Mexico, *Clays Clay Miner.*, **13**, 193 (1966).
30. Humbert, R. P. and C. E. Marshall, Mineralogical and chemical studies of soil formation from acid and basic igneous rocks in Missouri, University of Missouri, Agricultural Experimental Station Research Bulletin, 359 (1943).
31. Jeffries, C. D. and J. W. White, Some mineralogical and chemical characteristics of a Hagerstown soil profile, *Soil Sci. Soc. Am. Proc.*, **2**, 133–141 (1937).
32. Kittrick, J. A., Soil minerals in the $Al_2O_3$-$SiO_2$-$H_2O$ system and a theory of their formation, *Clays Clay Miner.*, **17**, 157–167 (1969).
33. Kodama, H. and J. E. Brydon, A study of clay minerals in podzol soils in New Brunswick, eastern Canada, *Clay Miner.*, **7**, 295–309 (1968).
34. Köster, H. M., Beitrag zur Kenntnis indischer Laterite, *Heidelberger Beitr. Mineralog. Petrograph.*, **5**, 23–64 (1955).
35. Köster, H. M., Vergleich einiger Methoden zur Untersuchung von geochemischen Vorgängen bei der Verwitterung, *Beitr. Mineralog. Petrograph.*, **8**, 69 (1961).
36. Mackenzie, R. C., Ed., *The Differential Thermal Investigation of Clays*, Mineralogical Society, London (1957).

37. Marshall, C. E., Mineralogical methods for the study of silt and clays. *Z. Krystallogr. (A)*, **90**, 8–34 (1935).
38. McCaleb, S. B., The genesis of the red-yellow podzolic soils, *Soil Sci. Soc. Am. Proc.*, **23**, 164 (1959).
39. McCracken, R. J., R. E. Shanks, and E. E. C. Clebsch, Soil morphology and genesis at higher elevations of the great Smokey Mountains, *Soil Sci. Soc. Am. Proc.*, **26**, 384–388 (1962).
40. Miller, B. J., A characterization of four limestone-derived soils from the Missouri Ozarks, M.S. thesis, University of Missouri, Columbia (1965).
41. Milne, G., A report on a journey to parts of the West Indies and the United States for the study of soils, Government Press, Dar es Salaam (1940).
42. Misra, U. K., Mineral equilibria in a soil system under natural and modified conditions as shown by the physical chemistry of the aqueous phase, Ph.D thesis, University of Missouri, Columbia (1973).
43. Mohr, E. H. C., F. A. Van Baren, and J. Van Schuylenborgh, *Tropical Soils*, Van Hoeve, The Hague (1972).
44. Muir, A., Notes on Syrian soils, *J. Soil Sci.*, **2**, 163–183 (1951).
45. Raad, A. T. and R. Protz, Calcite and dolomite as developmental criteria in young soil derived from dolomitic materials, *Soil Sci. Soc. Am. Proc.*, **38**, 807–812 (1974).
46. Rai, D., and W. L. Lindsay, A thermodynamic model for predicting the formation, stability and weathering of common soil minerals, *Soil Sci. Soc. Am. Proc.*, **39**, 991 (1975).
47. Redmond, C. E. and E. P. Whiteside, Some till-derived Chernozem soils in Eastern North Dakota. II. Mineralogy, micromorphology and development, *Soil Sci. Soc. Am. Proc.*, **31**, 100 (1967).
48. Satyanarayane, K. V. S. and P. K. Thomas, Studies on laterites and associated soils. I. Field characteristics of Laterites of Malabar and South Kanara, *J. Indian Soc. Soil Sci.*, **9**, 107–118 (1961); II. Chemical composition of Laterite profiles, **10**, 211–222 (1962).
49. Schnitzer, M. and S. J. M. Skinner, Stability constants for $Cu^{++}$, $Fe^{++}$ and $Zn^{++}$–fulvic acid complexes, *Soil Sci.*, **102**, 361 (1966).
50. Scrivner, C. L., Morphology, mineralogy and chemistry of the Lebanon silt loam, Ph.D. thesis, University of Missouri, Columbia (1960).
51. Sherman, G. D., Z. C. Foster, and C. K. Fujimoto, Some of the properties of the ferruginous humic latosols of the Hawaiian Islands, *Soil Sci. Soc. Am. Proc.*, **13**, 471–476 (1948).
52. Simonson, R. W., Genesis and classification of red-yellow podzolic soils, *Soil Sci. Soc. Am. Proc.*, **14**, 316–319 (1949).
53. Singer, A. and K. Norrish, Pedogenic palygorskite in Australia, *Am. Mineralog.*, **59**, 508–517 (1974).
54. Sirarajasingham, S., L. T. Alexander, J. G. Cady, and M. G. Cline, Laterite, *Advan. Agron.*, **14**, 1–56 (1962).
55. Smith, R. B. and S. W. Buol, Genesis and relative weathering intensity studies in three semiarid soils, *Soil Sci. Soc. Am. Proc.*, **32**, 261–265 (1968).
56. Southern Cooperative Series Bulletin 148, Selected coastal plain soil properties, Gainesville, Fla. (1970).
57. Southern Cooperative Series Bulletin 157, Soils of the Hayesville, Cecil and Pacolet

Series in the Southern Appalachian and Piedmont regions of the United States, Raleigh, N.C. (1971).
58. Southern Regional Bulletin 61, Certain properties of selected southeastern United States soils, Blacksburg, Va. (1959).
59. St. Arnaud, R. J. and M. M. Mortland, Characteristics of the clay fractions in a chernozemic to podzolic sequence of soils in Saskatchewan, *Can. J. Soil Sci.,* **43,** 336 (1961).
60. Stevens, J. H. and M. J. Wilson, Alpine Podzol soils on the Ben Lawers massif, Perthshine, *J. Soil Sci.,* **21,** 85–95 (1970).
61. Stumm, W. and J. J. Morgan, *Aquatic Chemistry,* Wiley-Interscience, New York (1970), p. 195.
62. Tamura, R., M. L. Jackson, and G. D. Sherman, Mineral content of a latosolic brown forest soil and a humic ferruginous latosol of Hawaii, *Soil Sci. Soc. Am. Proc.,* **19,** 435–439 (1955).
63. Wascher, H. L., B. W. Ray, J. D. Alexander, J. B. Fehrenbacher, A. H. Beavers, and R. L. Jones, Loess soils of northwest Illinois, Illinois Agricultural Experimental Station Bulletin 739 (1971).
64. Weaver, C. E., The clay petrology of sediments. In *Clays and Clay Minerals: Proceedings of the Sixth National Conference on Clays and Clay Minerals* (Berkeley, Calif., 1957) Pergamon Press, New York, pp. 154–187 (1958).
65. White, J. L., G. W. Bailey, and J. W. Anderson, The influence of parent material and topography on soil genesis in the midwest, Indiana Agricultural Experimental Station Research Bulletin 693 (1960).
66. Yassoglou, N. J. and E. P. Whiteside, Morphology and genesis of some soils containing fragipans in northern Michigan, *Soil Sci. Soc. Am. Proc.,* **24,** 396–407 (1960).

# 9 Chemical analyses of soils and soil fractions

**THE MINERAL PORTION OF SOILS**

All the great pedologists, as well as geologists interested in rock weathering, have used "complete" chemical analyses to good effect. It is very hard today to improve on the treatment of such analyses given by Merrill (27), Marbut (23), and Harrison (116), to name but three.

Where comparisons are between widely different climatic regions or between contrasting horizons of a given soil profile, the broad features are immediately discernible. Marbut's comparison of the Cecil soil of Georgia with the Miami of Ohio and the Barnes of Saskatchewan is a classic of its kind (23, pp. 37–40).

From the middle of the nineteenth century onward, the methods of analysis often included determinations on material soluble in boiling hydrochloric acid, as well as on the insoluble residue. Clearly the acid-soluble constituents were regarded as more important agriculturally and pedologically than the insoluble residue. Dokuchaev's memoir of 1883 on Russian Chernozem gives a number of such analyses (9). These methods have retained their usefulness over the years. They have been particularly popular in Central Europe. The immense labors of Blanck (6) on rock weathering and soil formation in various parts of the world utilized both complete rock analyses by fusion methods and determination of acid-soluble constituents.

From the analytical percentages it is easy to pass to molecular proportions, which can then be used in a variety of ways. Sensitive indicators of differences in weathering are provided by ratios such as silica to alumina, ($SiO_2/Al_2O_3$) or silica to sesquioxide $[SiO_2/(Al_2O_3 + Fe_2O_3)]$.

These are now usually thought of in relation to the ideal formula of kaolinite $Al_2O_3 \cdot 2\, SiO_2 \cdot 2H_2O$, an easily recognized weathering product that stands between the hydrous micas and smectites on the high silica side and allophane and gibbsite on the low silica side. As used earlier by Van Bemmelen (3,4), especially in his studies of lateritic soils of the Dutch East Indies, the weathered portion of soils could be divided into a reactive colloidal silicate A, soluble in boiling HCl, of $SiO_2/Al_2O_3$ ratio between 3 and 6, and a silicate B insoluble in HCl, but decomposed by concentrated $H_2SO_4$ of $SiO_2/Al_2O_3$ ratio between 2 and 3. Later investigations by Gedroiz (12) showed that silicate A contained more than that part of the mineral soil responsible for cationic exchange reactions. Considerable amounts of acid-soluble material came from other mineral sources. Thus the acid treatment does not make a mineralogically clear distinction.

Many other extracting solutions have come into use for specific purposes, and some have been exceedingly useful in the study of soil profiles. The distinction between aluminosilicate iron and aluminum and these elements in the free hydroxide form has been attempted through several extracting agents. Tamm's acid oxalate was used to good effect by Mattson and associates (25) in their detailed study of hydrologic soil series in Sweden. Two such series forming respectively Podzol and Brown Earth catenas provided a very significant series of comparisons of closely adjacent profiles on slopes rising from wet ground. The Podzol catena ("Unden") showed a remarkable distribution of extractable iron at the drier end with two zones of maximum accumulation (Figure 53). The upper zone is the normal Podzol B horizon. The lower is ascribed to the oxidation of ferrous to ferric iron where ground water comes into contact with air. At the wet end of the catena, iron is highest at the surface, corresponding to the high water table and contact with the atmosphere. Here aluminum is also highest, but it does not show the second maximum at the dry end of the catena.

The combination of mechanical analysis of soils with chemical determinations on the separates obtained soon focused attention on the clay fraction. Its chemical composition was seen to be related to exchange properties, hence was used more and more in characterization. Thus the $SiO_2/Al_2O_3$ and $SiO_2/R_2O_3$ ratios of the clay fraction became the subject of active inquiry. In the U. S. Bureau of Soils the composition of the clay fraction was extensively used in the laboratory characterization of soils and soil profiles (1). Mattson (25) in his theories of weathering and profile differentiation further drew attention to these ratios. Robinson (30) insisted that the $SiO_2/R_2O_3$ ratio provided a better characteristic than $SiO_2/Al_2O_3$, and this has been generally accepted.

The advantages of using clay composition rather than whole soil have

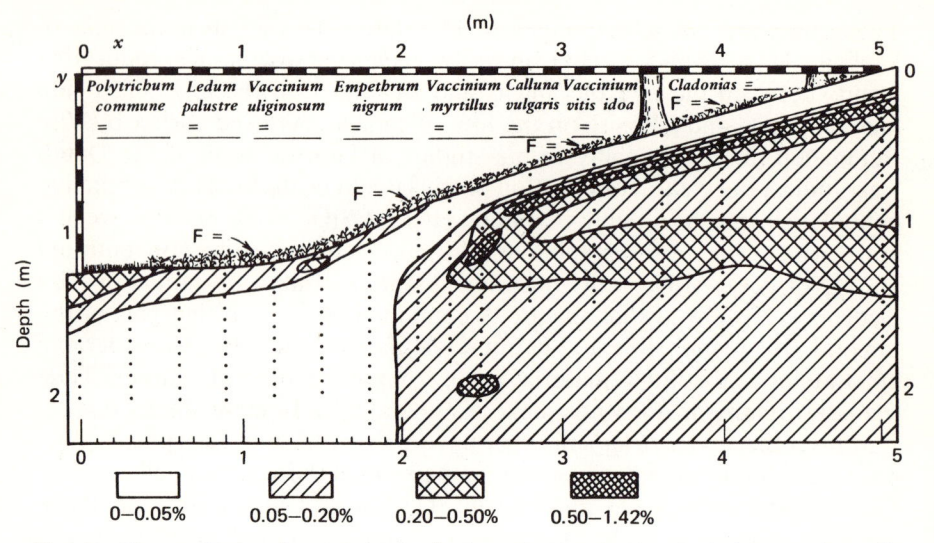

**224**   *Physical Chemistry and Mineralogy of Soils*

*Fig. 53*   Extractable iron in a catena of soils, from the "Unden" Podzol (right) to a bog soil (left). Reproduced from ref. 5 by permission, *Annals of the Agricultural College of Sweden*.

been stressed by Jenny (20) and illustrated with the comparative figures for a Podzol and Yellow-Red Podzolic soil (Figure 54). The silica-alumina ratio is plotted vertically and the depth horizontally. Variations in the clay colloid are very much less than those of the total soil. Thus the total analyses are more sensitive in these cases, but as Jenny implies, in general they are much less meaningful than the clay figures.

The relative $SiO_2/Al_2O_3$ ratio of the clay in relation to depth (setting that of the parent material at 1.00) gives characteristic relationships for different broad groupings of soils. Figures 55 and 56 show the respective patterns for Podzols and for lateritic soils as presented by Jenny. The four Podzols show a great contrast between the high relative $SiO_2/Al_2O_3$ values for the A horizons and the values less than unity for the B horizons. The four lateritic soils (this term is used by Jenny in a very broad sense) have values less than unity and show somewhat variable relationships between the A and B horizons.

In the mid-1930's it became apparent that the simple ratio $SiO_2/(Al_2O_3 + FeO_3)$ was insufficient to define the group of swelling clays or the hydrous micas. In the same way, primary rocks could not be adequately described through its use. For good correlation with mineralogy, detailed ratios involving more constituent elements were needed. The method proposed by Niggli in 1924 (28) in which the cationic elements are divided into four

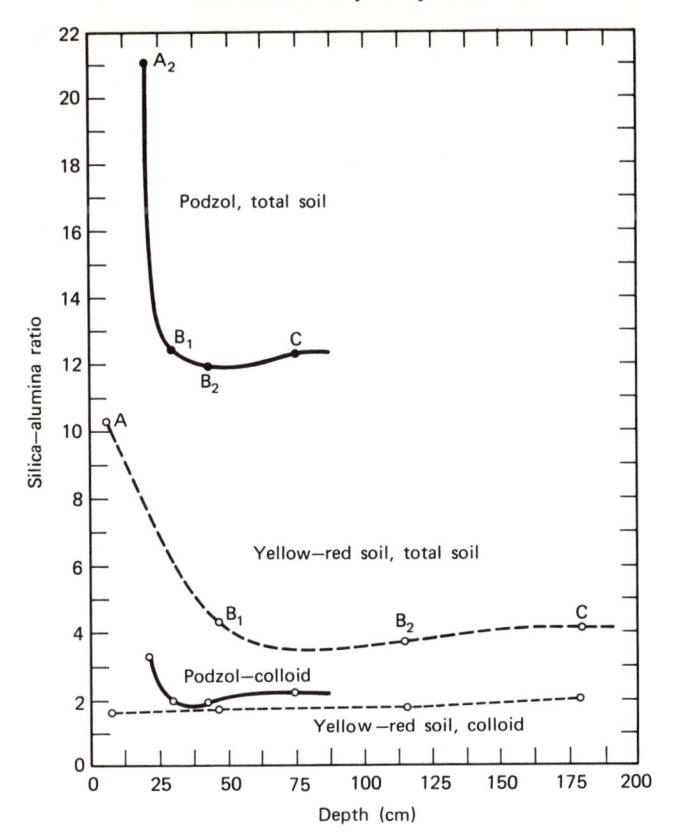

*Fig. 54* Comparison of podzolization criteria based on silica-alumina ratios of total soil mass and extracted colloid. Reproduced from ref. 20 by permission, Prof. Hans Jenny.

groups whose molecular numbers add to 100, has been widely used for geochemical and mineralogical evaluation of crystalline rocks.

A more direct translation of analytical figures into mineralogical terms has been attempted through the use of "normative" compositions calculated from the molar values of the oxides by incorporating them, in a definite order, into idealized mineral compositions. However the great variability in composition of weathering products such as hydrous micas and smectites makes the method difficult to apply to soil profiles. Where the products of weathering and pedogenesis can be identified as kaolinite or hydrous oxides the method is more straightforward.

In addition to the silica-sesquioxide ratio, used both for the whole soil and for the clay fraction, soil scientists have employed special ratios to

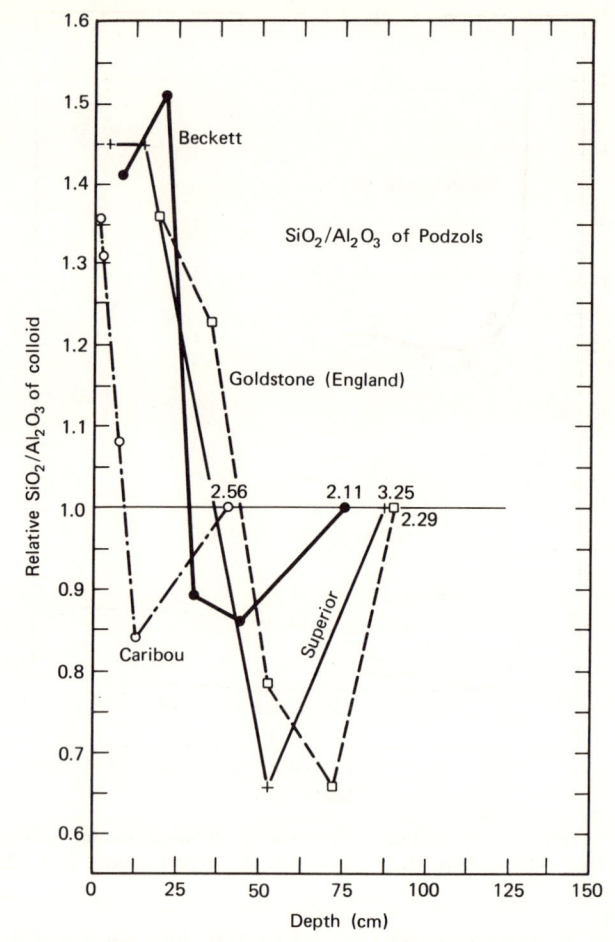

**226**   *Physical Chemistry and Mineralogy of Soils*

*Fig. 55* Characteristic patterns of Podzol profiles. The relative silica-alumina ratio of the inorganic colloidal clay is plotted as a function of depth, the C horizon being taken as unity. The actual $SiO_2/Al_2O_3$ ratios of the C horizon are given as numbers along the abscissa. Reproduced from ref. 20 (p. 160), by permission, Prof. Hans Jenny.

give a measure of weathering and leaching, aluminum being assumed constant. Harrassowitz (15) favored for this purpose the ratio $(K_2O + Na_2O + CaO)/Al_2O_3$ known as the *ba* value. Jenny (19) used two ratios, the $ba_1$ value $(K_2O + Na_2O)/Al_2O_3$ and the $ba_2$ value $(CaO + MgO)/Al_2O_3$. In respect of leaching Jenny defines the leaching factor $\beta$ as the ratio of $ba_1$ for the leached horizon to $ba_1$ for the parent material. Table 25 gives the calculation of the leaching factor for a soil derived from limestone.

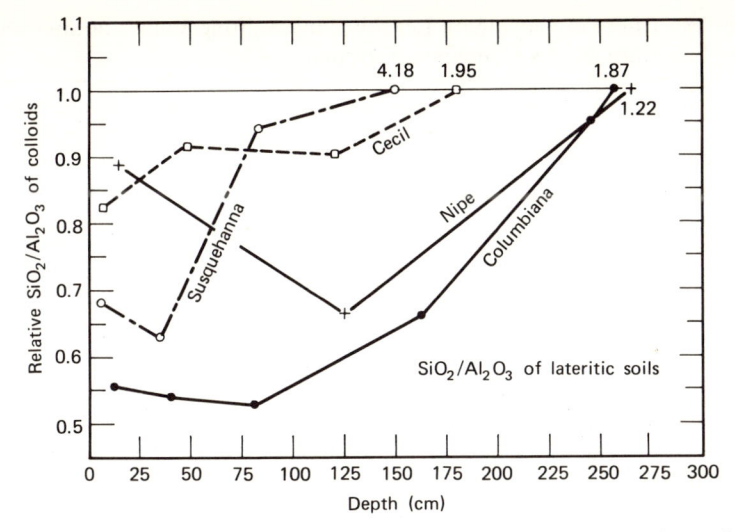

*Fig. 56* Characteristic patterns of lateritic soils. The relative silica-alumina ratio of the inorganic colloidal clay is plotted as a function of depth, the C horizon being taken as unity. The actual $SiO_2/Al_2O_3$ ratios of the C horizon are given as numbers along the abscissa. Reproduced from ref. 20 (p. 163), by permission, Prof. Hans Jenny.

Jenny then introduces another factor to distinguish between sodium and potassium. The shifting value $\mu$ is defined as (K/Na of leached horizon)/(K/Na of parent material). Since leaching causes greater losses of sodium than of potassium, the value of $\mu$ tends to be greater than unity. The contrast between podzolized soils and Chernozems, when both are formed from calcareous parent material, is well shown in Table 26.

The sequence podzolized soils–yellow-red soils–lateritic soils, formed

**TABLE 25** Calculation of Leaching Factor $\beta$ (20)

| Constituent | Rock Percent | Rock Molecular Value | Soil Percent | Soil Molecular Value |
|---|---|---|---|---|
| $Al_2O_3$ | 0.15 | 0.00147 | 7.59 | 0.0745 |
| $K_2O$ | 0.20 | 0.00213 | 1.38 | 0.0147 |
| $Na_2O$ | 0.06 | 0.00098 | 0.55 | 0.0089 |
| $ba_1$ | — | 2.11 | — | 0.32 |
| $\beta$ | — | — | — | 0.152 |

**TABLE 26** Effect of Climate on Leaching and Shifting Values of Soils Derived from Sedimentary Rocks Containing Carbonates (20)

| Soil Groups | Average Leaching Value | Average Shifting Value |
|---|---|---|
| 12 Podzolized soils (humid region) | 0.719 | 1.516 |
| 15 Chernozems (semihumid to semiarid region) | 0.981 | 1.066 |

under humid climates from igneous and metamorphic rocks, was also examined. Table 27 gives the main results. There is a large reduction in leaching values, and an increased weathering from north to south in the United States.

## THE ORGANIC PORTION OF SOILS

In the conventional total analysis of soils, carbon and nitrogen are usually reported. It is customary to multiply the carbon percentage by 1.724 to give a value for organic matter, although this figure strictly corresponds to purified humic acid from peat (58% C), not to the total organic matter of mineral soils. The nonhumified portion of the soil organic matter can vary considerably in carbon content, depending on its plant (and animal) origin, which roughly determines the proportions of carbohydrates, lignins, proteins, and fats and waxes. The humified portion shows less variability. However the three major fractions of the humified material—namely, the water-soluble fulvic acid fraction, the alkali-soluble humic acid fraction (including the hymatomelanic acid), and the insoluble humin fraction—display differences in elementary composition. Each is also somewhat variable when materials from different soils are compared.

**TABLE 27** Leaching and Shifting Values for Soils Derived from Igneous and Metamorphic Rocks (20)

| Soil Groups | Average Leaching Value | Average Shifting Value |
|---|---|---|
| 19 Podzolized soils | 0.822 | 1.805 |
| 25 Yellow-Red soils | 0.278 | 2.767 |
| 18 Lateritic soils | 0.141 | 3.206 |

From the distribution of conversion factors as determined for 63 surface soils (Figure 57), it is clear that values in the range 1.8–2.0 occur most frequently. The corresponding carbon percentages are 55.6–50.0. For well-humified material, such values correspond to mixtures of humic acids whose carbon contents range from 58 to 62%, with water-soluble fulvic acid in the range 43–52%.

Table 28 gives detailed figures for eight soils of the U.S.S.R. The fulvic acid fraction always contains less carbon, less nitrogen, and considerably more oxygen and usually gives a higher carbon-nitrogen ratio than the humic acid fraction. It also has a very considerably lower intensity of color, as given by optical density curves (Figure 58). Figure 58 also indicates the wide variation in color intensities of humic acids from different soils. Chernozems exhibit much greater color intensity than Podzols.Intensity of color is usually related to the degree of polymerization or to the proportion of aromatic units in the polymer. Intensely colored humic acids from Chernozems have higher exchange capacities and somewhat higher cationic bonding energies than the less colored humic acids from acidic peats (Marshall and Patnaik, 24).

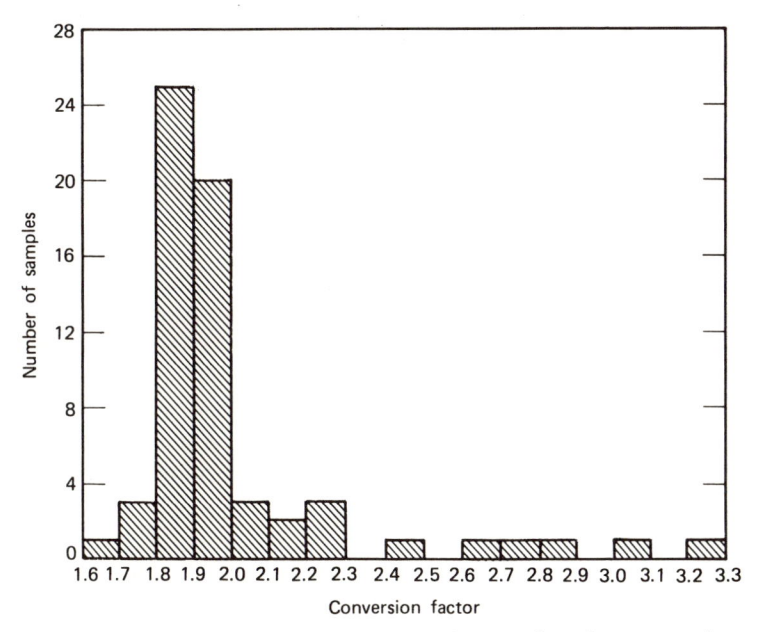

*Fig. 57* Frequency distribution of factor for converting organic carbon to organic matter in 63 surface soils. Reproduced from ref. 7 by permission of the American Society of Agronomy.

**TABLE 28** Elementary Composition of Humic and Fulvic Acids from the Main Soils in the U.S.S.R. (as a percentage of absolutely dry ash-free substance) (22)

| Soil | Acid[a] | C | H | O | N | C/H |
|---|---|---|---|---|---|---|
| Northern Podzol under forest; humus-illuvial horizon, 16–24 cm, Arkhangel'sk region | I | 58.11 | 5.37 | 32.00 | 4.52 | 10.82 |
| | II | 52.37 | 3.53 | 42.89 | 1.21 | 14.84 |
| Mountain-taiga ferruginized soil; 1–7 cm, Western Trans-Baikal | I | 58.84 | 6.02 | 29.70 | 5.44 | 9.8 |
| | II | 50.64 | 3.95 | 43.48 | 1.93 | 12.9 |
| Sod-podzolic soil; arable land, 0–20 cm, Moscow region | I | 57.63 | 5.23 | 32.33 | 4.81 | 11.02 |
| | II | 46.23 | 5.05 | 44.60 | 4.12 | 9.15 |
| Dark-gray forest soil under oak; 12–19 cm, Shipov forest, Voronezh region | I | 61.20 | 3.60 | 31.32 | 3.88 | 17.00 |
| | II | 47.46 | 3.64 | 45.87 | 3.03 | 13.04 |
| Ordinary Chernozem; arable land, 0–20 cm, Kamennaya steppe, Voronezh region | I | 62.13 | 2.91 | 31.38 | 3.58 | 21.35 |
| | II | 44.84 | 3.45 | 49.36 | 2.35 | 13.00 |
| Chestnut soil; virgin land, 0–20 cm, Valuisk experimental station, Volvograd region | I | 61.74 | 3.72 | 30.62 | 3.92 | 16.60 |
| | II | 43.19 | 3.61 | 51.43 | 1.77 | 11.96 |
| Light Serozem; arable land, 0–20 cm, Pakhta-Aral, Kazakhs S.S.R. | I | 61.94 | 3.93 | 29.46 | 4.67 | 15.76 |
| | II | 45.80 | 4.30 | 46.00 | 3.90 | 10.65 |
| Krasnozem under fern; 0–20 cm, Anaseuli, Gruzinskaya S.S.R. | I | 59.65 | 4.37 | 31.54 | 4.44 | 13.65 |
| | II | 49.82 | 3.35 | 44.33 | 2.50 | 14.87 |

[a]I, humic acids; II, fulvic acids.

Under these difficult circumstances, translation from carbon percentage to humus is inherently uncertain. To some extent it has been the subject of research into different soil groups of the world and their horizonation. One aspect of figures for total carbon derived by high temperature combustion is that elementary carbon, coal, and so on, are included. This may be important in areas subject to intermittent burning, or where coal-bearing strata are near the surface. It can be corrected for in a variety of ways (see Jackson, 17).

It is not possible to discuss soil organic matter solely in terms of carbon, hydrogen, and oxygen. Nitrogenous components are of major importance, in association with those containing sulfur and phosphorus. No methods are available for completely separating the nitrogenous from the nonnitrogenous components. Since the majority of the reactions of

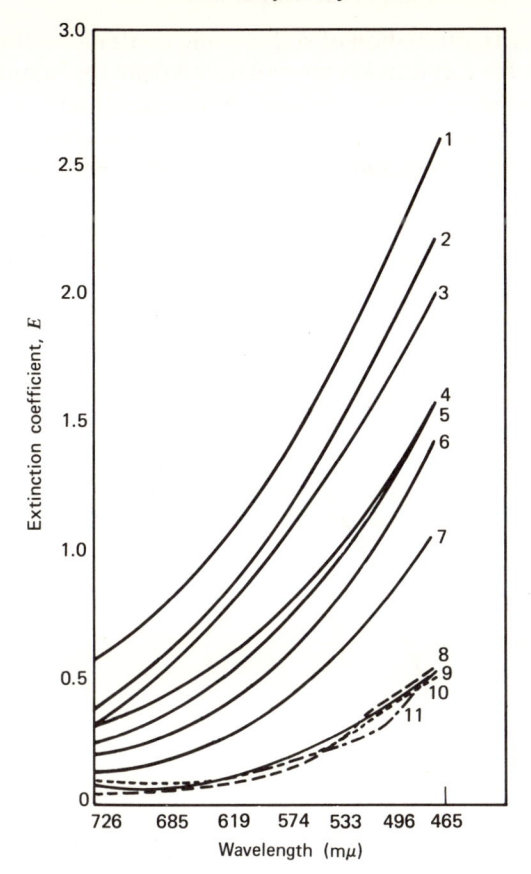

*Fig. 58* Optical density of humus substances. Humic acids: curve 1, ordinary Chernozems; curve 2, Dark-Gray Forest soil; curve 3, Chestnut soil; curve 4, Light Serozem; curve 5. Sod-Podzolic soil; curve 6, Krasnozem; curve 7, strongly podzolic soil. Fulvic acids: curve 8, Krasnozems; curve 9, Light Serozem; curve 10; strongly podzolic soil; curve 11, ordinary Chernozem. Reproduced from ref. 22 by permission of Springer-Verlag.

soil organic matter including humification are effected through biological pathways, the intimate chemical association of the nitrogen, sulfur, and phosphorus with humic matter is not surprising. Indeed the general relation of the carbon to the nitrogen finds both climatic and vegetational expression as the carbon-nitrogen ratio decreases from the Podzol zone through the Chernozem to soils of the desert.

The distribution of organic matter in soil profiles is clearly a matter of first-rate importance when pedological processes are under consideration. Russian soil scientists under the leadership of Tyurin (35) have considered the quantitative fractionation of soil organic matter as a function of depth. The water-soluble fulvic acids are separated from the humic acids in three stages corresponding to three colloid-chemical distinctions. Adding these together gives the total fulvic acids and total humic acids. Alcohol-soluble hymatomelanic acid is included with humic acid. The ratio of humic acids to fulvic acids is an important characteristic, since the fulvic fraction is the most mobile in soil profiles. Kononova (22) discusses both carbon to nitrogen and humic carbon to fulvic carbon ratios for soil groups on a worldwide basis.

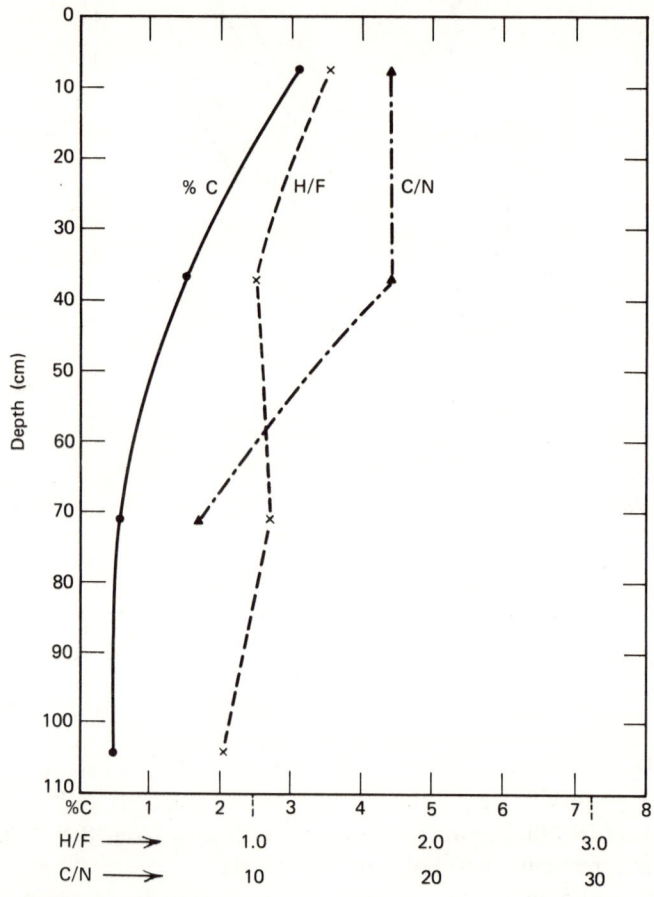

*Fig. 59* Variation with depth of total carbon, C/N ratio, and ratio of humic acids to fulvic acids (H/F) for an ordinary Chernozem soil (N. Khazakstan, Tour 5, 10th International Congress on Soil Science, profile 8).

Figures 59 to 62 give three depth functions: total carbon, carbon to nitrogen ratio, and humic acid carbon to fulvic acid carbon, for a variety of examples among the great soil groups. These utilize modern Russia data. Figures 63 to 68 give two depth functions; total carbon and carbon to nitrogen ratio, or total carbon and humic carbon to fulvic carbon from other and mainly older studies.

All these results indicate that the fulvic acid fraction is the most mobile part of the humus. This shows itself extremely strongly for low degrees of saturation with bases. Hence in Podzols and in lateritic soils the humus

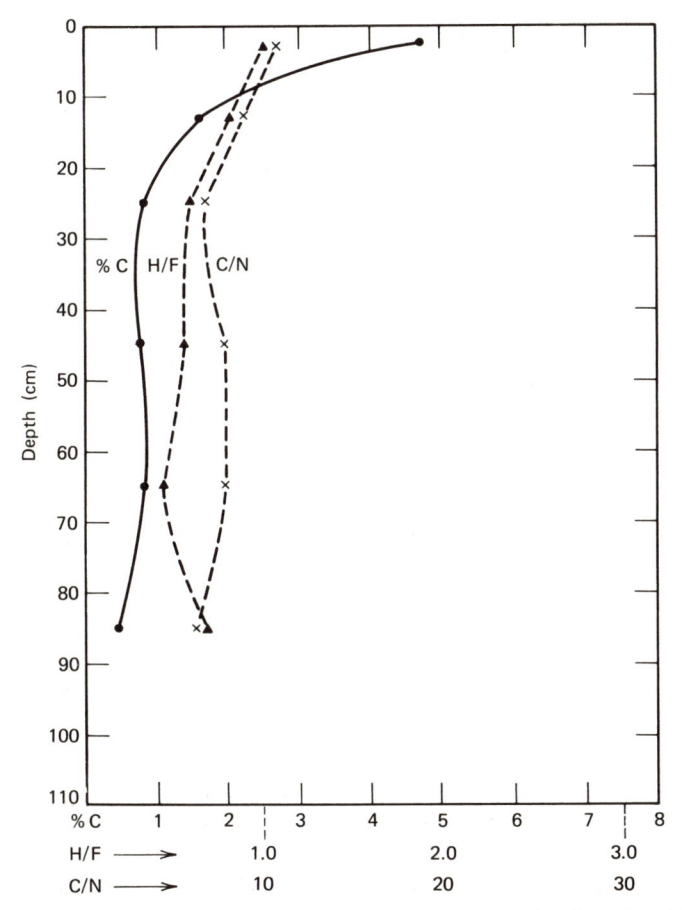

*Fig. 60* Variation with depth of total carbon, C/N ratio, and ratio of humic acids to fulvic acids (H/F) for a Cinnamon-Brown soil (Samarkand, Tour 7, 10th International Congress on Soil Science, profile 6).

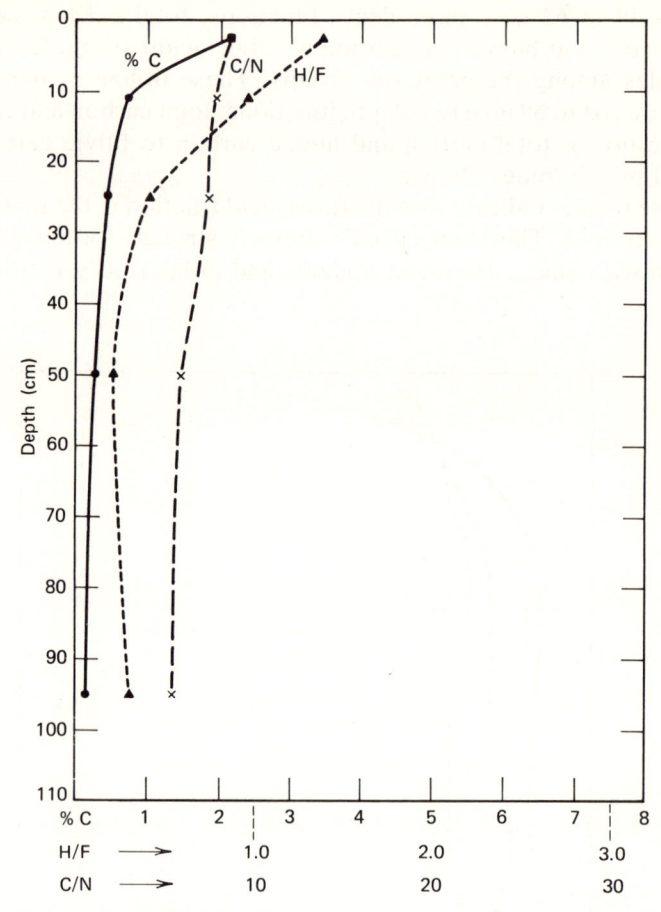

*Fig. 61* Variation with depth of total carbon, C/N ratio, and ratio of humic acids to fulvic acids (H/F) for a virgin Serozem soil (Tashkent, Tour 7, 10th International Congress on Soil Science, profile 2).

carried down pedologically consists very largely of the fulvic acid fraction. At high degrees of saturation, as in Chernozems and desert soils, this tendency is still present but the ratio of humic carbon to fulvic carbon decreases only moderately with depth.

The carbon-nitrogen ratio has been the subject of intensive investigation for many years. Soil profiles reflect the original composition of the annual increment of plant material, the biochemical factors in its transformation to humus, and finally the chemical and colloid-chemical factors governing its distribution in the soil profile. The annual increment of plant

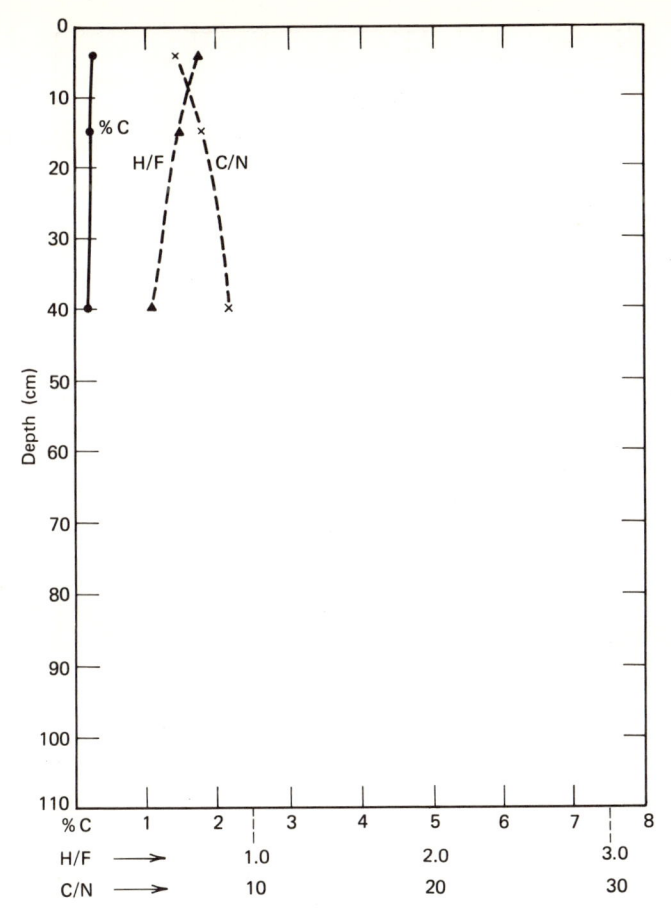

*Fig. 62* Variation with depth of total carbon, C/N ratio, and ratio of humic acids to fulvic acids (H/F) for a Gray-Brown Desert soil (Bukhara, Tour 7, 10th International Congress on Soil Science, profile 8).

material varies in its carbon-nitrogen ratio according to origin, but such values are high compared with those for the corresponding humified products. Depending on climatic and vegetational factors, and in the case of cultivated soils on management practices also, most soils attain a steady state in relatively few years (< 100), with regard to both organic matter content and carbon to nitrogen ratio.

Jenny (20) concluded that the nitrogen content of soils bore a functional relationship to temperature and moisture on a worldwide basis. However later work on tropical soils of India and Colombia demonstrated that their

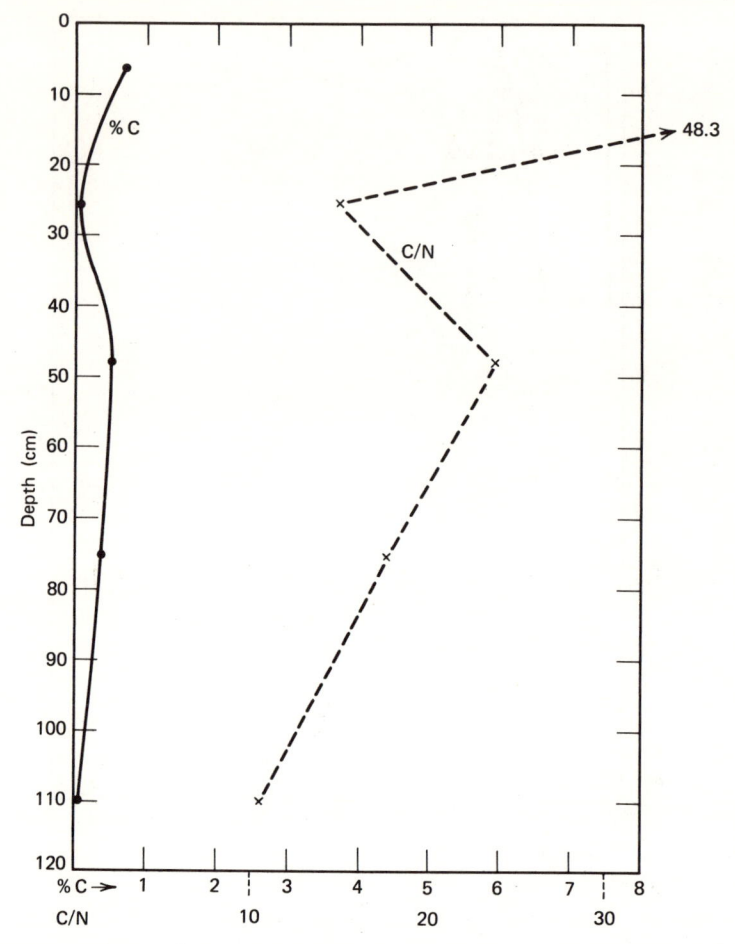

**Fig. 63** Variation with depth of total carbon and C/N ratio for a Podzol (Lakewood series, New Jersey, ref. 21).

nitrogen contents were higher than the previously defined functional relationship indicated (19). It was concluded that improved definitions of the climate and other factors were needed.

Extensive data on carbon-nitrogen ratios in soils of the United States, presented by Anderson and Byers (2), confirmed in general what had been found for Russian soils: namely, that the carbon-nitrogen ratio decreases from the moist Podzols and related soils (16–12) through the Chernozems (12–10) to the Desert soils (10–5). A number of examples illustrating profile variations are presented in Figures 59-68. In interpreting them it is

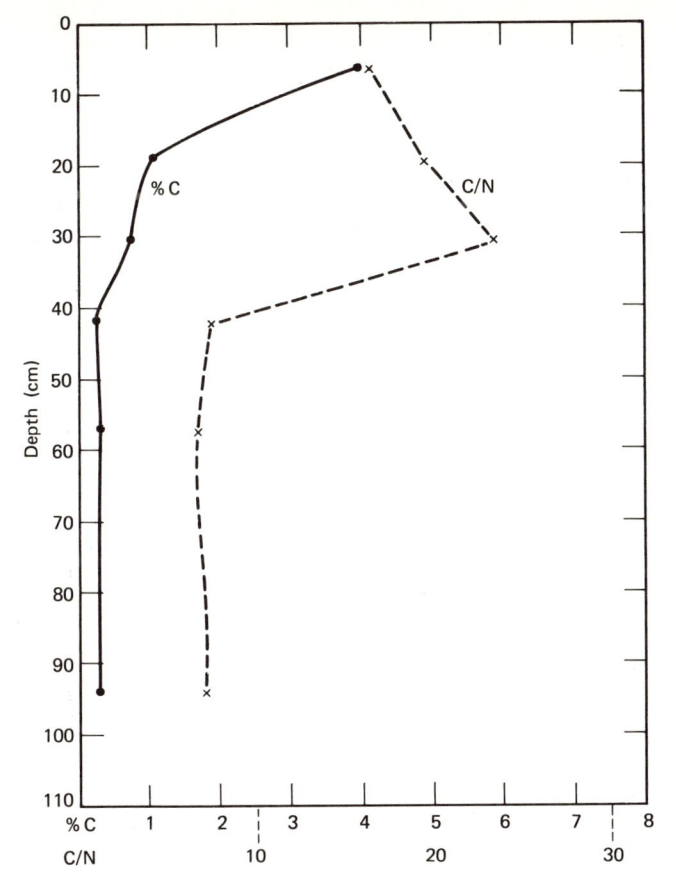

**Fig. 64** Variation with depth of total carbon and C/N ratio for a Gray-Brown Podzolic soil (Ontario loam, New York, ref. 8).

necessary to consider varying rates of humification of plant residues, together with differential movement of the humic matter formed. Table 28 indicates that in many soils, but not all, the mobile fulvic acid fraction has a higher carbon-nitrogen ratio than the less mobile humic acid. This effect would cause the carbon-nitrogen ratio to rise with depth. On the other hand, an increase in degree of humification with depth would cause a decrease in the carbon-nitrogen ratio.

The special characteristics of deep peaty soils (Histosols) formed under conditions of high water table both in high and low latitudes have recently been summarized (34). Figure 68 gives data on a Florida peat, where the major part of the organic matter is in the humin fraction. Thus variations

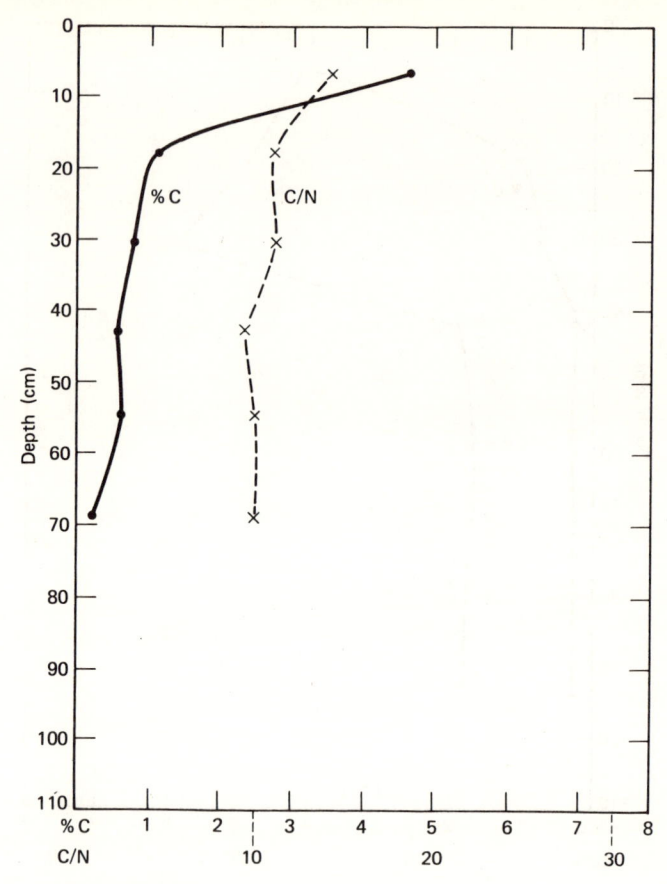

*Fig. 65* Variation with depth of total carbon and C/N ratio for a Brown Forest soil (Fenner silt loam, New York, ref. 8).

in the ratio of humic to fulvic acid affect less than half the organic matter. The carbon-nitrogen ratio is high and relatively constant with depth.

## INDIVIDUAL ELEMENTS AND COMPOUNDS AS INDICATORS OF PEDOLOGICAL PROCESSES

Most elements present in soils occur in more than one compound. Thus pedological processes can show themselves horizon by horizon in relation to total amounts of a given element and to the distribution of that element among its compounds. The latter frequently provides the more sensitive criteria of change.

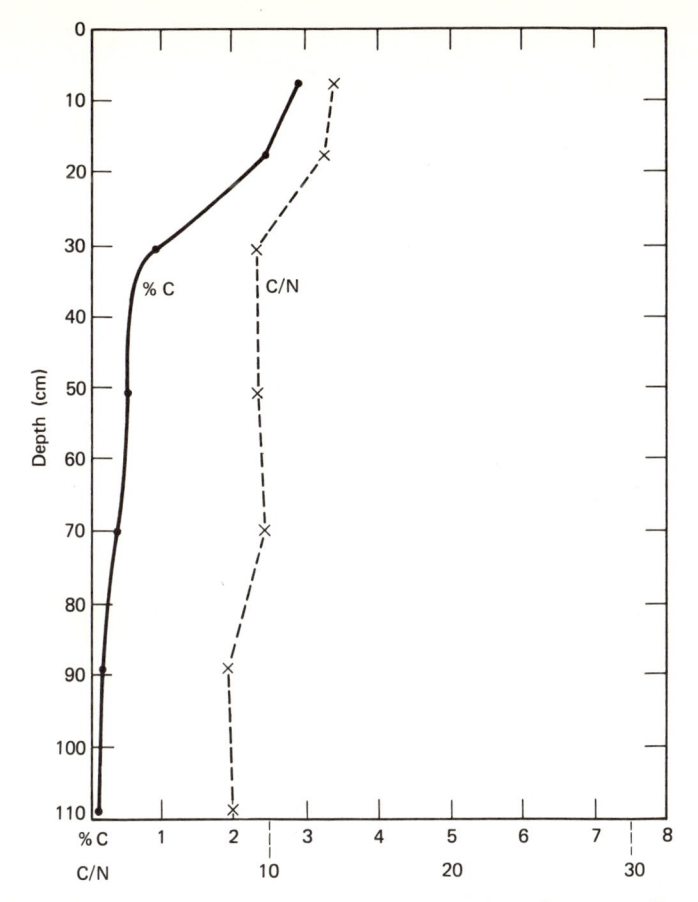

**Fig. 66** Variation with depth of total carbon and C/N ratio for a Chernozem (Barnes loam, North Dakota, ref. 26).

## Calcium

As illustration, consider the situation of calcium in the broad transect of soils extending from deserts (warm and dry conditions northward toward the Podzol region (cool and moist conditions). The parent material is taken as granitic throughout. The forms of calcium considered are (1) highly soluble salts, chloride, nitrate, and bicarbonate; (2) calcium sulfate; (3) calcium carbonate (with calcium phosphates); (4) exchangeable calcium; (5) calcium in primary crystalline silicates and aluminosilicates. It is assumed that erosion is absent and that additions of calcium in rainfall or windblown material are very small.

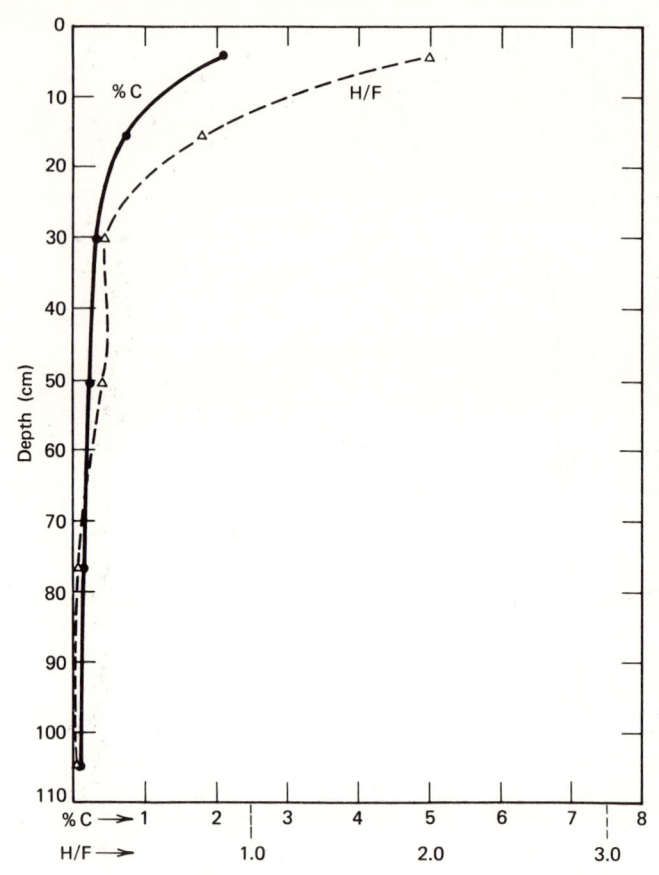

**Fig. 67** Variation with depth of total carbon and ratio of humic acids to fulvic acids (H/F) for a ferrallitic soil (Huambo, Angola, ref. 29).

Figure 69 approximates the distribution of the total calcium of a soil profile down to 120 cm under temperate conditions. The diagram is illustrative only; no actual soil data are used. If this had been done, relationships would have been more complex because the biotic factor is assumed to be constant and of small effect. In practice this would be impossible over such a transect.

The parent material is supposed to be a slightly weathered granitic rock with about 1% calcium. Calcium liberated by further weathering will have mobility in the soil profile in a decreasing order of solubility, numbers 1 to 4 as mentioned earlier.

Profile A is considered to be a relatively shallow Desert soil in which

# Chemical Analyses of Soils and Soil Fractions

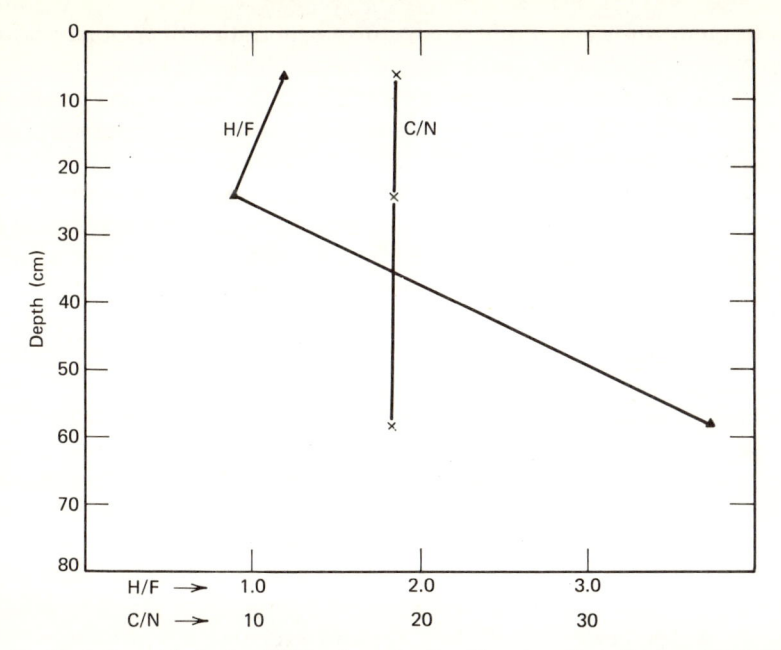

*Fig. 68* Variation with depth of C/N ratio and ratio of humic acids to fulvic acids (H/F) for a peat (Monteverde series, Florida, ref. 37).

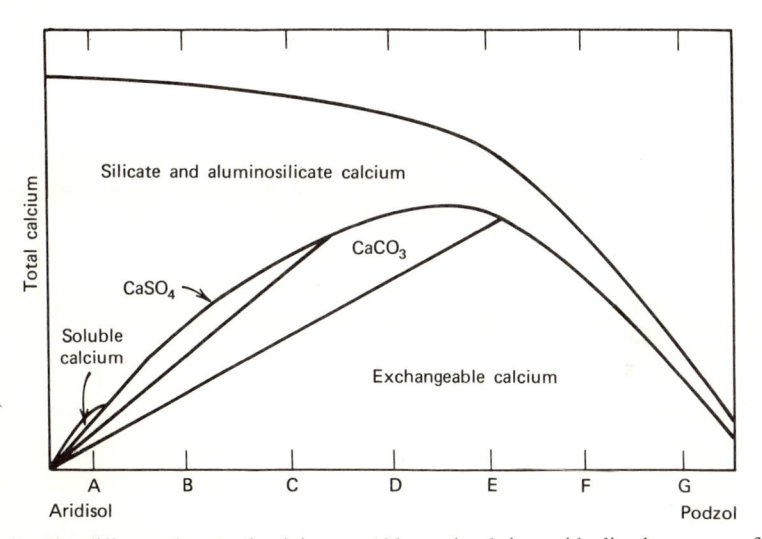

*Fig. 69* The different forms of calcium to 100 cm depth in an idealized transect of soils from Desert soils to Podzols.

the infrequent showers of rain never cause penetration by a moisture front below 60 cm, and most reach to shallower depths. During the drying period after each rain, moisture moves up to the surface by capillarity. Finally, complete drying occurs. Under such conditions soluble calcium salts, gypsum, and calcite are all somewhat concentrated in the surface layer, together with larger amounts of other salts such as sodium carbonate. However since weathering only occurs during a small fraction of the soil formation period, the total amount of potentially mobile calcium remains small. Parent material is unchanged below 60 cm. If there is eluviation of clay, the exchangeable calcium is higher in the subsoil than near the surface. There is negligible cycling of calcium from subsoil to topsoil by plants.

Profile B is similar to A except that from time to time limited amounts of water pass through the whole profile. They are sufficient to carry away soluble salts but only a little of the gypsum and calcite. Soluble salts are low throughout, otherwise the profile would be similar to A, possibly with greater movement of clay.

In profile C rainfall has become more frequent, and gypsum is lost down to 40 cm. There is a zone rising to maximum gypsum with smaller amounts at greater depths. Carbonates and exchangeable calcium extend throughout the profile in larger amount than for profiles A and B. The position and intensity of the zone of maximum gypsum largely depend on the balance between the losses by downward movement and the gains by upward capillary action. However the latter becomes less and less effective with increasing depth and is usually considered to be negligible below 1 m. The ratio of carbonate calcium to sulfate calcium is a sensitive function of depth and of the intensity of soil development processes.

In profile D increasing frequency and intensity of rainfall have produced further weathering with increases in total calcite and exchangeable calcium. Their relative distribution with depth is almost constant except for the upper horizons, where gains of carbonates have not kept pace with losses. Correspondingly, a zone of accumulation exists below this.

In profile E there are no free carbonates down to 60 cm. The surface layers are beginning to show losses of exchangeable calcium by weathering, but the total for the profile remains near the maximum. From here on the ratio of exchangeable calcium to mineral calcium provides the criterion of soil development. The distribution of exchangeable calcium with depth is now controlled by the distribution of organic matter and clay.

In profiles F and G acidic conditions cause increased weathering of mineral silicates and also increased losses of exchangeable calcium. We have increasingly podzolic conditions with illuviation of clay. Thus an $A_2$ horizon very low in exchangeable calcium becomes an important feature. In general one could say that from A to DE, increased weathering produces increased clay (calcium saturated). From E to G increased weathering corresponds to decreased saturation of the exchange complex with calcium.

*Other Cationic Elements*

The tremendous difference in behavior between calcium and sodium in soil profiles can readily be understood by constructing a diagram similar to Figure 69 for sodium. Highly soluble salts now include the carbonate and sulfate. Exchangeable sodium is lost much more readily by weathering than is exchangeable calcium. Thus the changes illustrated in Figure 69 occur over a much smaller range of rainfall whenever sodium dominates the profile.

Magnesium may be expected to lie between sodium and calcium in response to the factors illustrated in Figure 69. Its sulfate is highly soluble, but the carbonate is similar to calcium carbonate. Exchangeable magnesium is held with only a slightly different range of bonding energies than calcium.

Superficially, potassium might be equated to sodium, since its solubility relationships are similar. However potassium is much more readily taken up by fixation into secondary clay structures than is calcium, magnesium, or sodium. Thus its overall mobility is reduced. Offsetting this factor to some extent is the high preferential uptake by plants, with subsequent release through decomposition into the exchangeable form in the uppermost horizons.

*Phosphorus*

Considerable work has been done on the distribution of total phosphorus in soil profiles, particularly in the midwest (32). In some cases available phosphorus has also been measured (33) and in others a partition of the phosphorus into organic phosphorus, apatite, and secondary inorganic phosphate (occluded and nonoccluded) has been attempted.

A generalized diagram expressing forms of phosphorus in terms of soil development from basalts in New Zealand has been presented by Williams and Walker (36, Figure 70). Organic phosphorus and nonoccluded secondary inorganic phosphorus (sometimes categorized as available phosphorus) show maxima for a moderate degree of development. Phosphate in the apatite form decreases at an increasing rate as development advances. Occluded phosphate (in concretionary material) increases with and provides a good measure of advancing development.

The distribution of phosphorus as a function of depth and of soil development was first studied by Godfrey and Riecken (14) on loessial

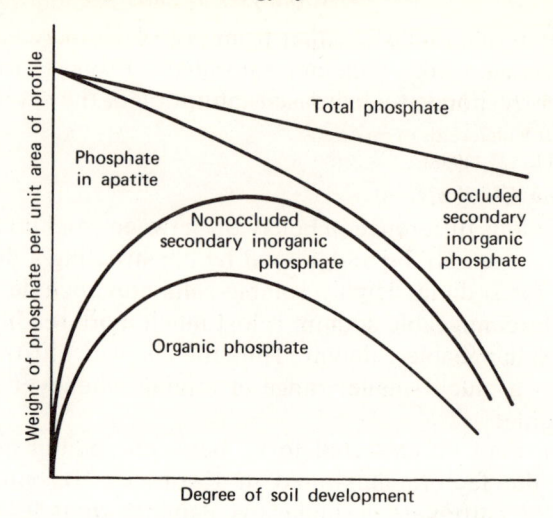

*Fig. 70* Forms of phosphate in relation to soil development. Reproduced from ref. 36 by permission of the Williams and Wilkins Co., Baltimore.

soils of Iowa and Missouri. They did not find definite accumulation in heavy B horizons of claypan soils, whereas Smeck and Runge (33), using a sequence of Illinois soils, provide clear evidence of this. They found two zones of accumulation—one in the B horizon, where it could be ascribed to formation of iron phosphates and concretions, and one at the contact with carbonates, where calcium phosphates would precipitate. These authors point out that phosphorus in soil profiles is strongly affected by landscape features, since surface erosion and horizontal movement of soil water carry considerable amounts of phosphate to low-lying sites.

## REFERENCES

1. Alexander, L. T. and H. G. Byers, U.S. Department of Agriculture Technical Bulletin 317 (1932).
2. Anderson, M. S. and H. G. Byers, The carbon-nitrogen ratio in relation to soil classification, *Soil Sci.*, **38**, 121 (1934).
3. Bemmelen, J. M. Van, Die Absorptionsverbindungen der Ackererde, *Landwirtsch. Versuchssta.*, **21**, 135 (1877); **23**, 265–303 (1878).
4. Bemmelen, J. M. Van, Die Zusammensetzung der Ackererde. . . , *Landwirtsch. Versuchssta.*, **37**, 347–373 (1890).
5. Bengtsson, B., N. Karlsson, and S. Mattson, The pedography of hydrologic soil series. IV. The distribution of Si, Al, Fe, Ti, Mn, Ca and Mg in the Unden podzol and the Dala brown earth series, *Ann. Agr. Coll. Swed.*, **11**, 172 (1943).

6. Blanck, E., *Einführung in die genetische Bodenlehre*, Vandenhock and Ruprecht, Göttingen (1949).
7. Broadbent, F. E., The soil organic fraction, *Advan. Agron.* **5,** 153–183 (1953).
8. Cline, M. G., Major kinds of profiles and their relationships in New York, *Soil Sci. Soc. Am. Proc.* **17,** 123 (1953).
9. Dokuchaev, V. V., *Russian Chernozem,* Israel Program for Scientific Translations, Jerusalem (1967).
10. Drew, J. V., Ed., Selected Papers in Soil Formation and Classification, Soil Science Society of America special publication series no. 1, Madison, Wisc. (1967), pp. 18–28.
11. Drew, J. V., Ed., Selected Papers in Soil Formation and Classification, Soil Science Society of America special publication series no. 1, Madison, Wisc. (1967), pp. 83–94.
12. Gedroiz, K. K., Contributions to our knowledge of the absorptive capacity of soils, *Zh. Opit. Agron.,* **19,** 269–322 (1918); **20,** 31–58 (1919).
13. Gieseking, J. E., Ed., *Soil Components,* Vol. I, *Organic Components,* Springer, New York (1975).
14. Godfrey, C. L. and F. F. Riecken, Distribution of phosphorus in some genetically related loess-derived soils, *Soil Sci. Soc. Am. Proc.,* **18,** 80–84 (1954).
15. Harrassowitz, H., Laterite, *Fortschr. Geol. Paleontol.* **4,** 253 (1926).
16. Harrison, J. B., *The katamorphism of igneous rocks under humid tropical conditions,* Imperial Bureau of Soil Science, Harpenden, England (1934).
17. Jackson, M. L., *Soil Chemical Analysis,* Prentice-Hall, Englewood Cliffs, N.J. (1958).
18. Jenny, H., Behavior of potassium and sodium during the process of soil formation, University of Missouri, Agricultural Experiment Station Research Bulletin 162 (1931).
19. Jenny, H., Bödenstickstoff und seine Abhängigkeit von Zustandsfaktoren, *Z. Pflanzenernaehr. Düng. Boden.,* **109,** 97 (1965).
20. Jenny, H., *Factors of Soil Formation,* McGraw-Hill, New York (1941).
21. Joffe, J. S., *Pedology,* Pedology Publications, New Brunswick, N.J. (1949).
22. Kononova, M. M., Humus of virgin and cultivated soils. In *Soil Components,* Vol. I, *Organic Components,* John E. Gieseking, Ed. Springer, New York (1975), Chap. 8, pp. 475–520.
23. Marbut, C. F., *Soils: Their genesis and classification,* Lectures delivered 1928, published by the Soil Science Society of America, Madison, Wisc. (1951).
24. Marshall, C. E. and N. Patnaik, Ionization of soils and soil colloids. IV. Humic and hymatomelanic acids and their salts, *Soil Sci.,* **75,** 153 (1953).
25. Mattson, S. and Y. Gustavson, The electrochemistry of soil formation. I. The gel and sol complex, *Ann. Agr. Coll. Swed.,* **4,** 1 (1; 37).
26. McClelland, J. E. et al., Chernozems and associated soils of Eastern North Dakota: Some properties and topographic relationships, *Soil Sci. Soc. Am. Proc.,* **23,** 51 (1959).
27. Merrill, G. W., *Rocks, Rock Weathering and Soils,* Macmillan, New York (1906).
28. Niggli, P., *Gesteins und Mineralprovinzen,* Vol. 1, Borntraeger, Berlin (1923).
29. Ricardo, R. P., Composition de la matière organique de quelques sols ferrallitiques typiques, *Trans. 9th Int. Congr. Soil Sci.,* **III,** 257–263 (1968).
30. Robinson, G. W., *Soils,* 2nd ed., Murby, London (1936).
31. Robinson, W. O. and R. S. Holmes, The chemical composition of soil colloids, U.S. Department of Agriculture Bulletin 1311 (1924).

32. Smeck, N. E., Phosphorus: An indicator of pedogenetic weathering processes, *Soil Sci.*, **115**, 199–206 (1973).
33. Smeck, N. E. and E. C. A. Runge, Phosphorus availability and redistribution in relation to profile development in an Illinois landscape segment, *Soil Sci. Soc. Am. Proc.*, **35**, 952–959 (1971).
34. Stelly, M., Ed., *Histosols: Their characteristics, classification and uses, Soil Science Society of America,* special publication no. 6, Madison, Wisc. (1974).
35. Tyurin, I. V., *The Organic Matter of Soils,* Leningrad (1937).
36. Williams, J. D. H. and T. W. Walker, Fractionation of New Zealand basaltic soil profiles. II. *Soil Sci.*, **107**, 213–219 (1969).
37. Zelazny, L. W. and V. W. Carlisle, Physical, chemical, elemental and oxygen-containing functional group analysis of selected Florida histosols. *Histosols, Soil Science Society of America,* Special publication no. 6, Madison, Wisc. (1974), Chap. 6, pp. 63–78.

# 10 Ionic properties of the exchange complex and the soil solution

**RELATIONSHIPS BETWEEN EXCHANGEABLE CATIONS**

It is interesting to review the conclusions reached by those who in the early part of this century attempted to assess the importance of ionic exchange in pedology. Nineteenth century investigators, from Way to Van Bemmelen, had established the exchangeability of metallic ions and ammonium in the colloidal mineral matter of soils. At this time colloidal organic matter was viewed somewhat differently. Its function as an acid in acidic peats or as a calcium salt in neutral organic soils was emphasized. Treatment of the exchange complex as an entity, with pedological emphasis on the part played by the hydrogen ion, came in the period 1900–1920 through the work of Gedroiz, Hissink, and Wiegner (28, 29). The degree of saturation with metallic cations was viewed in relation to soil acidity and also as a measure of the developmental processes that soils had undergone under high rainfall (Podzols, podzolic soils, Solods, red tropical soils, and other acidic profiles). The hydrogen ion was found to be especially effective in exchange reactions. Its effects on physical properties were connected with its powerful action as a coagulating ion, which placed it in the same class as the divalent cations rather than the monovalent. Its presence also undermined the stability of the mineral part of the exchange complex. In Podzols the latter came to have different elementary compositions in different horizons.

Toward the latter part of this period, ionic exchange reactions were firmly connected (especially through the work of Gedroiz 9, 10) with the special properties of alkali and saline soils. Here the exchange complex was greatly influenced by the sodium ion. Even in relatively small amounts on the exchange complex, sodium greatly favored dispersibility of the soil colloids and, in Solonetz soils, the production of a typical columnar structure. Based on this improved understanding, the reclamation of alkali soils was undertaken as a problem in exchange chemistry, actively pursued in the years 1920–1930 (Kelley 14, Von Sigmond; Gedroiz).

Characterization of the exchange complex by determination of the exchange capacity, the degree of saturation, and the individual exchangeable cations thus became an important part of the analysis of soils, and has remained so.

This traditional treatment of the exchange complex as a single entity, which led to such valuable pedological conclusions, is now seen to be inadequate in several respects. (1) Identifiable components of the exchange complex show different exchange properties, especially regarding the selective behavior toward particular cations and anions. Fixation reactions are strongly selective. (2) All components that have been studied individually show a range of bonding energies for ions (see Vol. I, Chapts. 4 and 6). (3) The total exchange capacity of a soil is a function of the method of determination. It is now customary to separate pH-dependent and pH-independent exchange functions, as well as to define fixation reactions for particular cations.

In relation to soil profiles, the exchange complex is a very sensitive indicator of changes caused by the operation of factors of soil formation. Differences in the amounts and proportions of exchangeable cations become apparent long before mineralogical changes can be recognized. This is because of rapid equilibration of the colloidal exchange complex with the aqueous environment. Changes in relation to depth are easily recognized in individual cations and anions, in ratios between pairs of ions and, most sensitively of all, in selectivity numbers for certain pairs of ions.

Relationships involving the exchange complex and the immediate factors of soil formation are likely to be most clearly evident in soil profiles dominated by a single reactive exchanger. Complex mixtures of colloidal minerals and of these with humus or with primary unweathered minerals may be difficult of interpretation. Thus we should look first to the Histosols, dominated by humus, to the Planosols and Prairie soils, dominated by beidellite, to certain limestone soils dominated by illite, to the Ultisols, dominated by kaolinite, and to the Oxisols, dominated by gibbsite. In each case the following factors require consideration: accession of cations from outside sources, cycling of cations by biological means, the special properties of the dominant exchanger with regard to cationic bonding and fixation, and the internal moisture regime, horizon by horizon. The presence of humus in the upper horizons of mineral soils needs special consideration. Its exchange capacity weight for weight is much higher than that of the clays, but its bonding energy for the common exchangeable cations is lower.

In the consideration of acidic soils, we must take into account the exchange and fixation properties of the hydronium ion $H_3O^+$, of the aluminum ion $Al^{3+}$, of various hydroxylated aluminum ions, and of complexed aluminum, together with the possible presence of free hydroxides of aluminum and iron as separate phases, as surface layers, or as molecular interlayers, along with clay minerals themselves. This complexity makes clear-cut analytical distinctions almost impossible, because of overlapping mechanisms by which aluminum can be released. In general, acidic mineral soils are aluminum-hydrogen systems to a greater or lesser extent. Increasing weathering reduces exchangeable calcium, magnesium, potassium, and sodium and increases the aluminum-hydrogen function. At the same time the nature of the exchange complex itself may change; interlayering of hydroxide layers with 2:1 clay layers occurs, leading to chloritic-type material and eventually to 1:1 clays. These factors clearly come into play in the study of the broad climatic sequence from Mollisols through Alfisols to Ultisols. In any given soil profile the identifiable mineralogical differences between horizons are often slight, but the relation of the H-Al function to the exchange complex as a whole may be very marked.

The expression of this function in relation to depth has taken a number of forms. The "exchange acidity" or "unsaturation" is usually expressed as exchangeable hydrogen plus exchangeable aluminum, to be contrasted with the sum of exchangeable calcium, magnesium, potassium, and sodium. The total exchange capacity can be plotted as a function of depth and can be separated for each horizon into these two main parts. This apparently simple method has been found to be subject to numerous limitations, many of which, as already mentioned, center around the complex role of aluminum. In addition to being exchangeable in the ordinary sense, the presence of aluminum in clays (as inter-layers) and in organic matter (as complexes) has the effect of rendering exchange sites inoperative. However under the right conditions it can be released from these forms. Thus the choice of salts and buffer systems for measuring H + Al is difficult. Few studies of soil profiles have taken account of these conditions. Comparisons on a worldwide basis are hindered by the varied combinations of reagents used. Empirically, however, some knowledge of particular profiles and restricted areas has accumulated.

The difference in exchangeability between potassium and sodium on the one hand and the divalent cations on the other led Jenny (12) to the potassium-sodium theorem. Leaching causes preferential losses of monovalent ions over divalent and at the same time greater losses of sodium than potassium. The consequence for profiles in which exchangeable divalent cations predominate over monovalent (Jenny's class I) is that the horizon with the smallest sum of potassium and sodium should have a wider ratio of potassium to sodium than the horizon with the greatest sum of these elements. Limestone soils should have a low content of monovalent cations, hence relatively high values of K/Na in the exchange complex. This is often found to be the case, but another factor should also be considered. Limestones commonly bear clay minerals of the illite-glauconite group which contain potassium. The reaction of these with water to produce kaolinite liberates potassium ions, which are mainly held by the exchange complex. Thus the general effect of the presence of these potassium minerals is the same as an enhancement of the bonding energy of potassium relative to sodium.

Jenny also briefly considered soils in which monovalent cations were dominant in the soil solution (class II). The saline and alkali soils naturally come into this category, but so also do certain soils of the humid tropics in which the total exchangeable bases are very low. He recognized that predictions were rendered difficult by lack of information on the competitive situation between potassium, calcium, and sodium. It is now known (see Vol. I, Chap. 1) that monovalent-divalent exchanges are affected by total volume, which means that in soil profiles, variations in moisture content greatly affect the equilibrium between soil solution and exchange complex. In relation to soil-water ratio the $\sqrt{\text{divalent}}$/monovalent activity ratio tends to remain constant in the soil solution rather than the ratio of divalent to monovalent.

### *Single Cation Functions*

In addition to studying variations in the amounts of single exchangeable cations with depth in soil profiles, one can investigate cationic bonding through measurements of single cation activities. The conditions for valid results are discussed in Volume I, Chapters 1 and 6, where a number of bonding energy curves derived from titrations of acidic clays with bases are presented. The application of this method to a soil profile was made by Brydon and Marshall (6). Figure 71 gives $\Delta F$ values for potassium and for calcium, from cation activity measurements on fine clay fractions. Exchange capacity and exchangeable calcium and magnesium of the whole soil (Hagerstown series in Missouri) are also given. This is an Alfisol from dolomitic limestone. The percentage saturation of the clay chosen for the

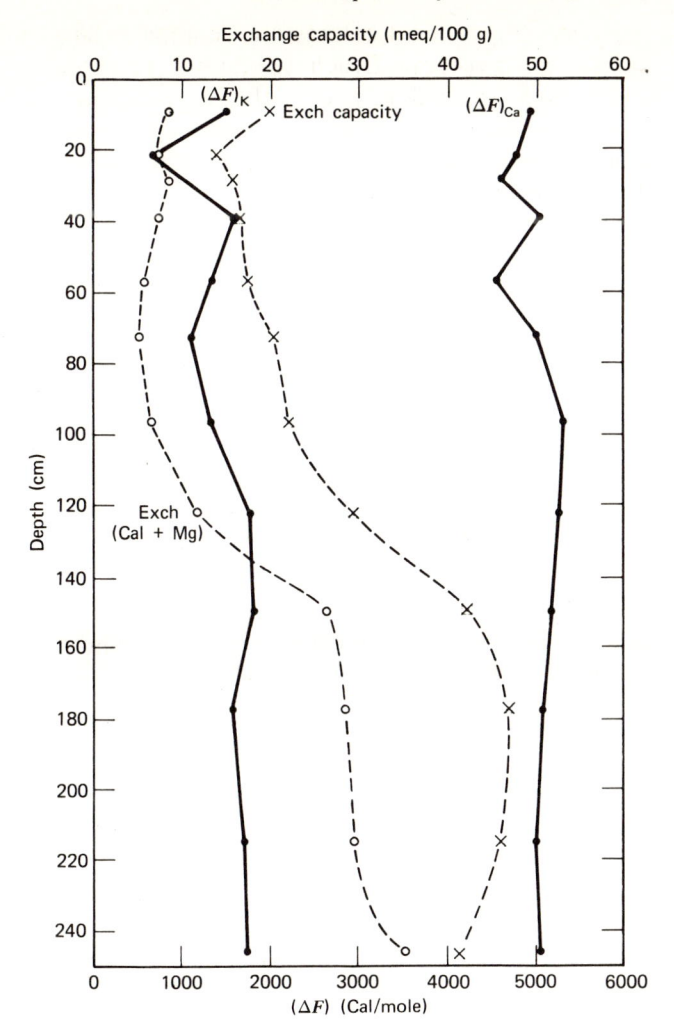

*Fig. 71* Electrochemical and exchange data for the Hagerstown (limestone soil) profile in Missouri. Exchange values are for the whole soil. Cationic bonding energies are for the clay fraction (6).

$\Delta F$ values was 60%, which is in the region of the maximum on the individual detailed bonding energy curves.

This clay is very consistent in its properties down the profile, even though the total clay increases from 20% at the surface to more than 80% at 150 cm depth. X-Ray results indicate that it is an iron-rich illite, that is, a member of the illite-glauconite series with some expanding layers, mixed with kaolinite in moderately small proportion. The bonding energy for calcium is unusually high, which helps to explain the pedological uniformity, since the rate of loss of calcium and magnesium by weathering would be greatly reduced. This is an extremely stable 2 : 1 clay mineral.

## Detailed Study of a Missouri Soil

Before reviewing examples of different great soil groups, we present the case of a Missouri soil studied in detail. On this Prairie soil, the Mexico silt loam (an Alfisol formerly classified as the rolling phase of the Putnam soil), a series of lysimeters, and associated plots have been studied over the past 10 years (25,15). Exchange capacity, exchangeable cations (Ca, Mg, K, Na), and cations in the equilibrated soil solution were determined for soil samples down the profile. The ratios of exchangeable potassium to exchangeable sodium and exchangeable calcium to exchangeable magnesium were determined and the corresponding selectivity numbers calculated using the ratios in solution as follows: selectivity number $_{Na}^{K}K_s$ = (solution $K^+$/solution $Na^+$) ÷ (exchangeable K/exchangeable Na). As pointed out in volume I, this is equivalent to the ratio $f_K/f_{Na}$, where $f$ is the activity coefficient for a given cation on the clay phase. It can also be expressed as a difference in free bonding energy for the two cations concerned: $(\Delta F)_K - (\Delta F)_{Na} = -1364 \log {}^{Na}_{K}K_s$.

These results (Figures 72 and 73) reveal a series of depth functions for different exchange properties. The total exchange capacity shows little change to 20 cm, then increases rapidly to 35 cm, remains almost constant to 55 cm, and slowly decreases to 80 cm. Exchangeable calcium and magnesium show very similar relationships. Potassium displays a definite decrease to 25 cm, then an increase to 35 cm, constancy to 55 cm, and a decrease to 80 cm. Exchangeable sodium is entirely different—constant to 15 cm, then increasing with depth to 80 cm. In consequence the ratio of exchangeable potassium to exchangeable sodium begins at the surface with a ratio around 4, rapidly decreases to a value less than unity at 25 cm, then slowly decreases to 65 cm. More sensitive expressions of the potassium-sodium relationships are the selectivity numbers $_{Na}^{K}K_s$ and the mean free bonding energy differences $(\Delta F)_K - (\Delta F)_{Na}$, which both show a strong maximum in the region 35–45 cm. The rise in $(\Delta F)_K - (\Delta F)_{Na}$ from about 890 cal./mole near the surface to 3800 cal. at 35 cm is extraordinarily high.

The recycling of potassium by plants through accumulation from the subsoil and release at and near the surface would lead to this kind of result. The subsoil would become depleted in potassium of relatively low free bonding energy, and this would increase in the surface layers. Sodium is not subject to this kind of discrimination. Since, however, it is

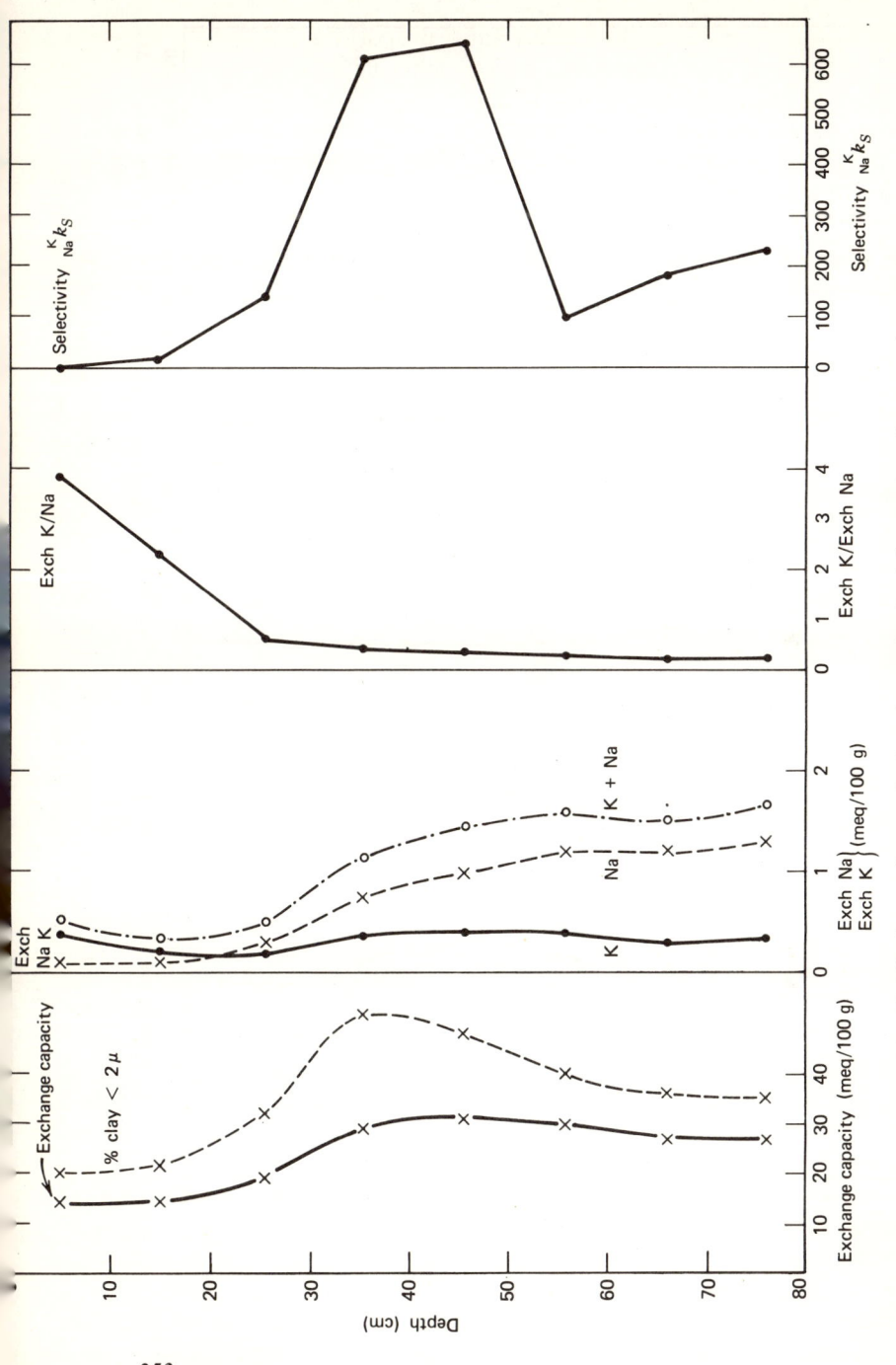

*Fig. 72* Exchange data, cationic ratios, and selectivity numbers for potassium-sodium in the Mexico silt loam profile (Missouri) (15, 19, 25).

253

*Fig. 73* Exchange data, cationic ratios, selectivity numbers, and free bonding energy differences for calcium-magnesium also potassium-sodium in the Mexico silt loam profile (Missouri) (15, 19, 25).

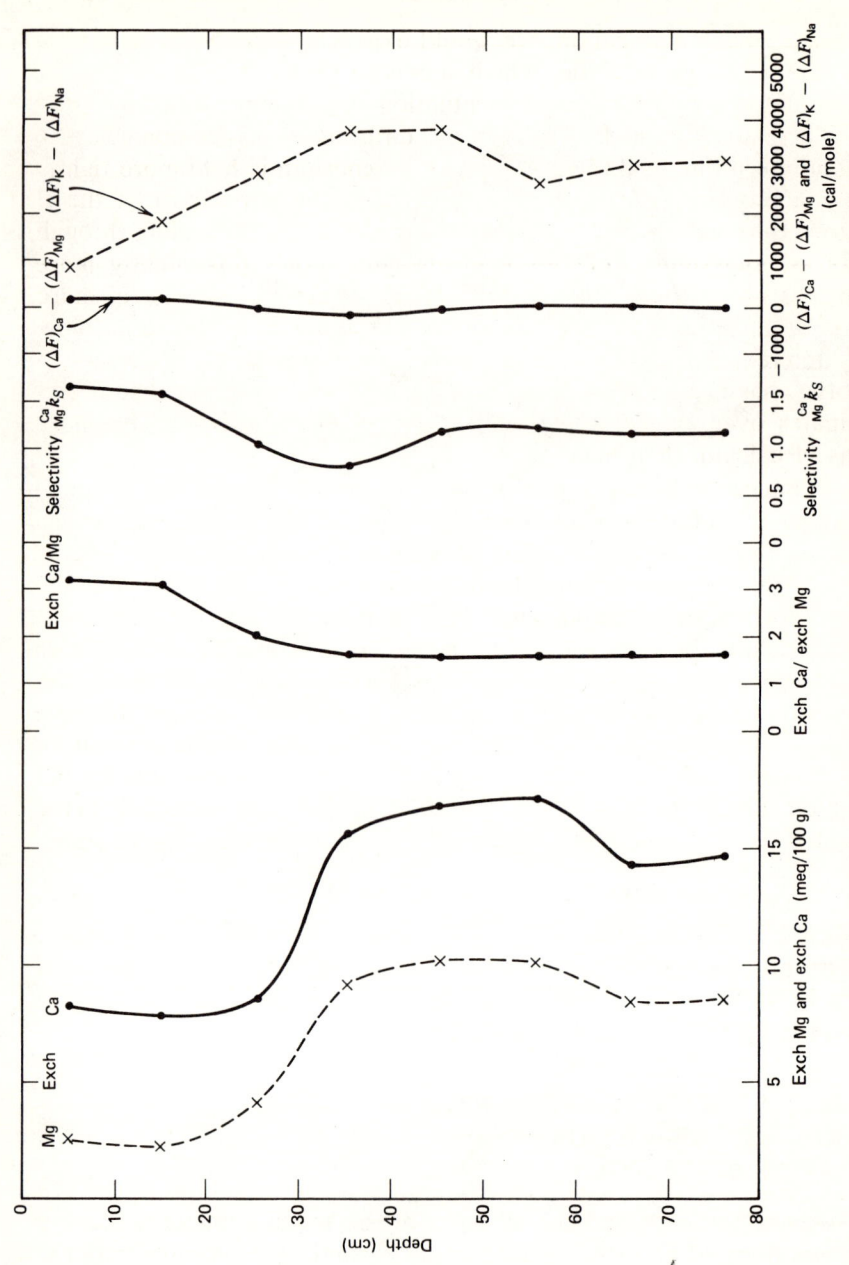

the most loosely held cation, we would expect it to increase somewhat with depth through leaching, which seems to be the case.

The relative calcium-magnesium situation also changes with depth, but through relatively small free energy differences as compared with potassium-sodium. The region in which magnesium is held more tightly than calcium occurs around 35 cm depth, in the upper part of the B horizon. Both cations are subject to the same kind of recycling through plants as potassium. Apparently magnesium shows this slightly more strongly than calcium. On an atomic basis some plants take up more magnesium than calcium. This is not uncommon with grasses (e.g., bluegrass and redtop) grown on the somewhat acidic soils of the midwest. On the other hand, coniferous trees provide needles in which calcium predominates over magnesium. Most cultivated crops also take up more atoms of calcium than magnesium.

The changes with depth just discussed are not independent of small mineralogical differences between horizons. These show themselves mainly in the coarse clay fraction ($2-0.2\mu$), the fine clay ($<0.2\mu$) being very uniformly beidellite throughout. In the heavy B horizon the proportion of coarse clay is significantly less than in the A and C horizons. Variation in the composition of the coarse clay is shown by a mixed layer smectite-mica that is highest in the A horizon and by the beidellite component that is lowest there. Illitic mica is relatively constant throughout, as are small amounts of kaolinite and quartz. The strong potassium-fixing tendency of this beidellitic clay is undoubtedly responsible for the very high values of $(\Delta F)_K - (\Delta F)_{Na}$ found throughout, but especially in the subsoil. Thus it probably accentuates the effect of the recycling of potassium discussed above.

In Figure 72 the gradual fall in the ratio of potassium to sodium below 25 cm may be an example of Jenny's sodium-potassium theorem, since concurrently the sum of these elements rises with depth. The very rapid fall in K/Na from the surface to 25 cm is best ascribed to recycling of potassium by prairie grasses, since the sum changes little.

## EXCHANGEABLE CATIONS IN RELATION TO SOIL CLASSIFICATION AND CHARACTERISTICS

Studies of soil profiles from all parts of the world commonly include determinations of cation exchange capacity and exchangeable cations. Unfortunately sodium is sometimes omitted, making it impossible to calculate potassium-sodium ratios. Only in rare instances are data on the composition of the soil solution also given. Few selectivities are available.

### Alfisols and Mollisols

The example just discussed was an Alfisol from loess. An extensive comparison of Alfisols and Mollisols from Peorian loess in Illinois is available in the work of Wascher et al. (27). These authors calculated the exchangeable calcium to exchangeable magnesium ratios, because in an earlier study by Muckenhirn of loessial Gray-Brown Podzolic soils (now classified as Alfisols), this ratio was found to afford a measure of soil development. It decreases as the latter increases. This generalization is extended by Wascher et al. to the Mollisols. An example of a well-developed Mollisol from this study, the Ipava silt loam, is given in Figure 74, which includes also the ratio of exchangeable potassium to sodium K/Na calculated from their data. The general trends with depth of K/Na and Ca/Mg are similar to those of the Mexico soil, but appreciably higher; the ratios do not fall below unity. The Ipava soil is considered to show less development than the Mexico; its general mineralogy is similar.

### Ultisols

Fairly extensive data are available on soils of the southeastern United States formerly classified as Red-Yellow Podzolic soils (24). They are dominated mineralogically by kaolinite, often with hematite, vermiculites, chlorites, and mica intermediates, and sometimes gibbsite. The exchange capacity and degree of saturation are low, corresponding to the mineralogical characteristics and high degree of weathering. Many are from granitic rocks or their outwash and in consequence are extremely low (<1 meq/100 g) in calcium and magnesium. Exchangeable aluminum is often higher than calcium and magnesium in profiles in which the sum of these elements is very low. Figures 75 and 76 give examples of Cecil and Davidson profiles, respectively, the former from granite and the latter from diorite gneiss.

### Oxisols

The Oxisols comprise the most extreme examples of tropical weathering and are dominated mineralogically by gibbsite and oxides and hydroxides of iron and titanium. Consequently the low cation exchange capacity varies greatly with pH, and anion exchange is an important characteristic. The name "Laterite" is now used more restrictively than in the past, and it refers to tropical Oxisols that possess a plinthite layer (i.e., a dense layer that hardens on exposure). This corresponds to Buchanan's original definition of Laterite.

The Nipe clay in Puerto Rico, which is an old Oxisol derived from serpentinite, has been extensively studied. A compilation of modern data on this soil by Beinroth (3) indicates that except in the surface layer the

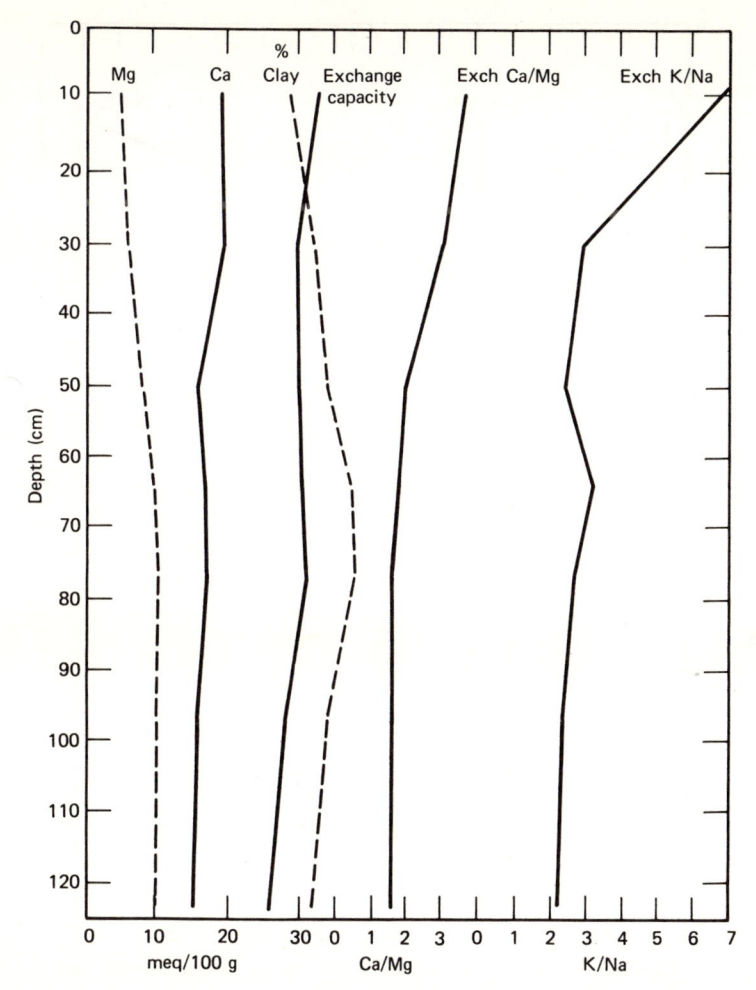

*Fig. 74* Exchange data, clay percentages, and cationic ratios for the Ipava profile, a Mollisol from Illinois (27).

exchangeable mono- and divalent cations are extremely low (i.e., no more than 0.1 meq/100 g total bases below 28 cm). Sulfate displaced by phosphate was low in the surface horizon, then increased rapidly with depth up to 1234 ppm sulfur at 120–160 cm. This corresponds to about 8 meq exchangeable sulfate per 100 g. The soil contains both gibbsite and kaolinite; the former increases from about 25% in the surface horizon to 60% at 120–160 cm.

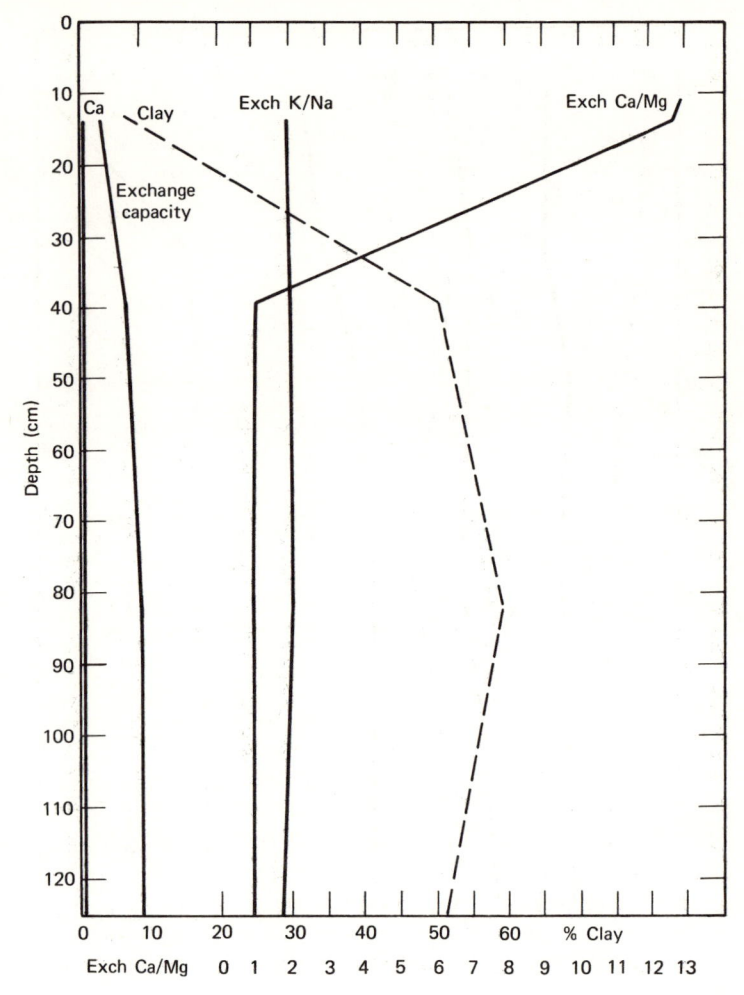

**Fig. 75** Clay percentages and exchange data for the Cecil sandy loam, South Carolina, from granitic rock (24).

## Histosols

Ratios of exchangeable cations are available in a few instances. Mattson, Sandberg, and Terning (17) investigated the ratio of exchangeable calcium to exchangeable magnesium in relation to depth for a drained bog (Ramna) in southern Sweden. From the surface to 30 cm there was a rapid fall from 0.99 to 0.39, then some fluctuation in the range 0.26–0.53 at greater depths. The plant cover gave ratios 1.2 to 1.8 for different species. Thus there is evidence for the cycling of calcium through vegetation. The

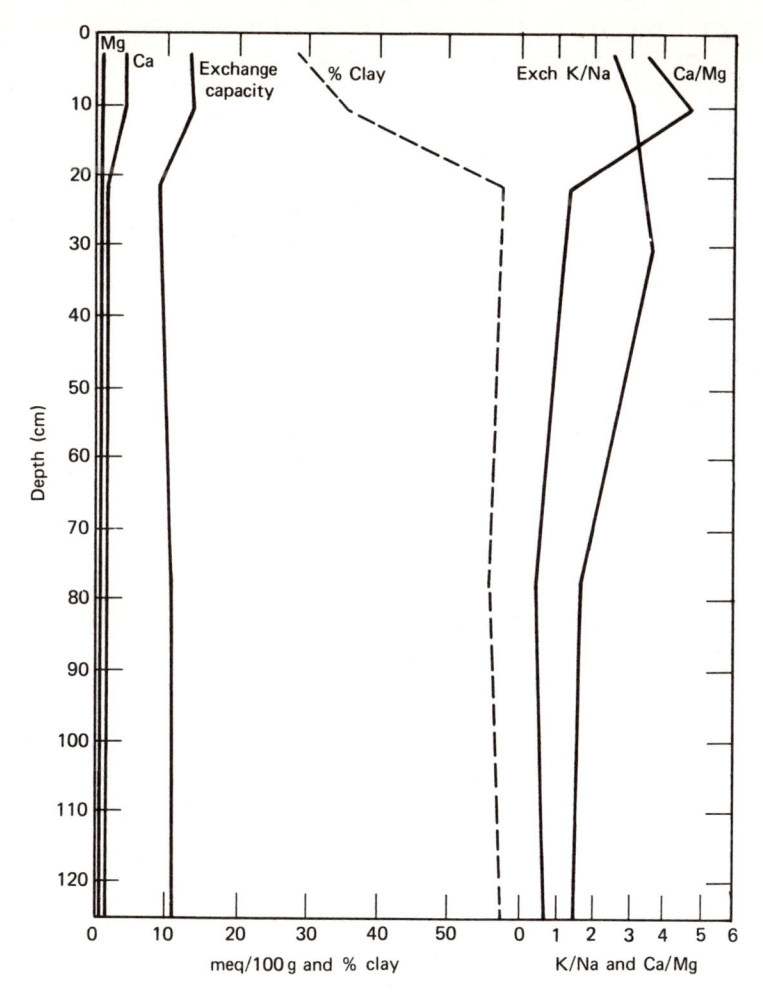

*Fig. 76* Clay percentages and exchange data for the Davidson clay loam, Georgia, from diorite gneiss (24).

relative dominance of magnesium in the profile was ascribed to the influence of oceanic salts brought down in rainwater, whose calcium to magnesium ratio is about 0.2.

## Podzols and Brown Earths

The work of Mattson and associates on catenas of Podzols and Brown Earths mentioned in Chapter 9 affords also information on exchangeable calcium, magnesium, and potassium (4, 16). The Podzol catena

**260**   *Physical Chemistry and Mineralogy of Soils*

("Unden") was from parent material very low in exchangeable bases. Respecting the change from typical Podzol at the drier upper end of the slope to semibog at the lower end, all three cations showed similar relationships. For the Podzol, the cation content near the surface was moderate and fell to extremely low values in the leached horizon. At the wet end relatively high concentration of all three cations were present at the surface down to about 40 cm, with subsequent decrease with depth. The concentration in the organic matter under wet conditions was very marked; for instance, the 0–10 cm layer of the Podzol contained 1.34 m eq/100 g of exchangeable calcium, whereas the bog, 5 m down the slope, contained 12.95 meq in this surface layer. Figure 77 is the diagram for potassium, showing a clear minimum in the bleached horizon.

The corresponding data for the Brown Earth catena ("Dala") reveal very similar relationships except that here the parent material was high in exchangeable bases. The relative concentration of calcium, magnesium, and potassium in the bog soil as compared with the drier brown earth was again great.

*Selectivity Curves*

A soil sample can be characterized for any pair of exchangeable cations by the curve relating selectivity (or a derived quantity) to cationic compo-

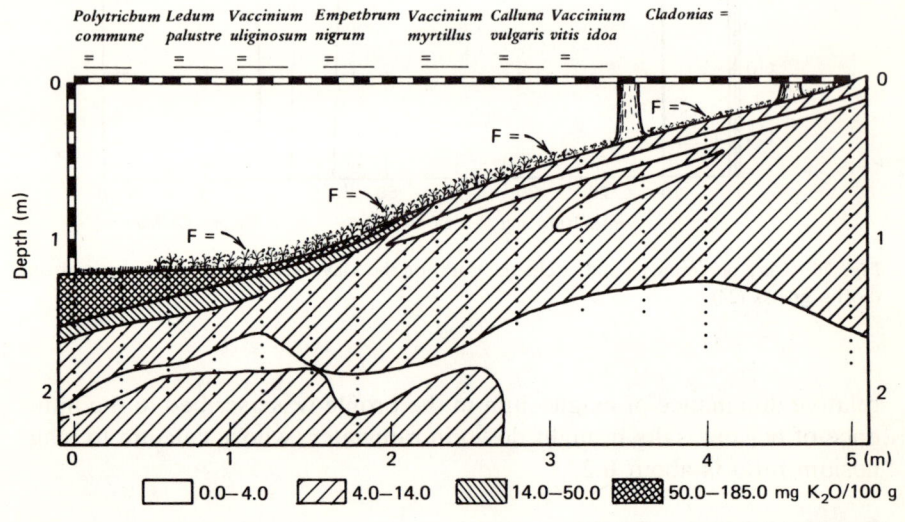

**Fig. 77**   Distribution of chloracetic acid–soluble $K_2O$ in the Unden Podzol series (catena) in Sweden. Reproduced from ref. 16 by permission of the *Annals of the Agricultural College, Sweden.*

sition. Pairs of ions having the same valency are usually favored because of the simple character of the exchange equations. For the simplest monofunctional exchangers, the curve should consist of a straight line parallel to the compositional axis and displaced from it by the magnitude of $K_S$, the selectivity number. All clay minerals thus far studied have complex curves, and the same is true of soil samples. Curves obtained for different horizons of a soil profile are found to differ significantly. A study by Roth (22) in the author's laboratory using three soils of relatively uniform mineralogical character (Mexico, beidellite; Sharkey, montmorillonite; Hagerstown, illite-glauconite) afforded good examples of this. The cation pairs were as follows: Na-K, Li-K NH$_4$-K, and Mg-Ca.

Figure 78 gives the K-Na and Ca-Mg curves for different horizons of the Mexico soil in Missouri. It is advantageous to use the free bonding energy differences instead of selectivity numbers because in this way values of $K_S$ less than unity are not crowded together in the range 0–1, but are spread out as negative free energy differences. The Na-K systems of Figure 78a actually show considerably smaller free bonding energy differences than the corresponding Li-K systems, which might prove more suitable for pedological comparisons.

Both sets of curves display relative uniformity in the middle ranges of composition but very significant bonding energy changes for small proportions of either cation. They show also that the surface soil samples are significantly different from the rest. In the K-Na series small proportions of potassium show low values of $\Delta F_K - \Delta F_{Na}$ for the surface sample but high values for all deeper layers. In the Ca-Mg series small proportions of magnesium show an upward trend in $\Delta F_{Ca} - \Delta F_{Mg}$ for the surface sample but a downward trend to negative values for deeper layers. These differences may reflect a greater proportion of the 14Å component in the clay of the surface horizon as well as a much greater content of organic matter.

The K-Na curves of Figure 78a (soil saturated with Na + K) display smaller $(\Delta F)_K - (\Delta F)_{Na}$ values and much smaller variations with depth than the individual selectivity values calculated for the natural soil samples (Figures 72 and 73). The natural soils contain only 2–5% of Na + K in the exchange complex. Their K-Na selectivities are sensitively affected by the Yarusov effect (Vol. I, Chap. 1), which influences Na much more strongly than K, and by differing pedological regimes at different depths, especially by the differential uptake of potassium relative to sodium.

Selectivity has been used in a somewhat different manner by Murdock and Rich (20). Single exchanges of rubidium and strontium against sodium were conducted, followed by exchange with magnesium or with ammonium. Three soil profiles (Ramon, Nason, and Tatum, all Alfisols)

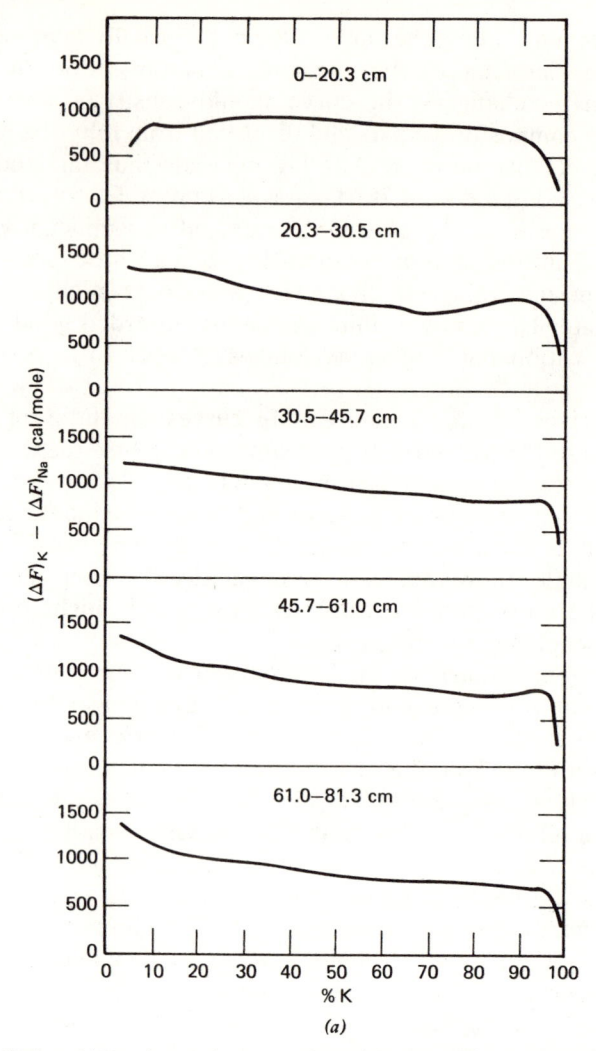

**Fig. 78** (*a*) $(\Delta F)_K - (\Delta F)_{Na}$ in relation to exchangeable potassium and sodium content for various horizons of the Mexico soil (beidellite). (*b*) $(\Delta F)_{Ca}-(\Delta F)_{Mg}$ in relation to exchangeable calcium and magnesium for the same soil (22).

dominated by vermiculite and mica were studied. The percentage rubidium or strontium released by magnesium or ammonium was plotted as a function of depth. Rubidium was more strongly released by ammonium than by magnesium, but strontium showed the reverse relationship. The results were interpreted in relation to X-ray studies of the clay from the different horizons. The presence of hydroxylated aluminum layers in the Tatum clay reduced the fixation of rubidium very strongly.

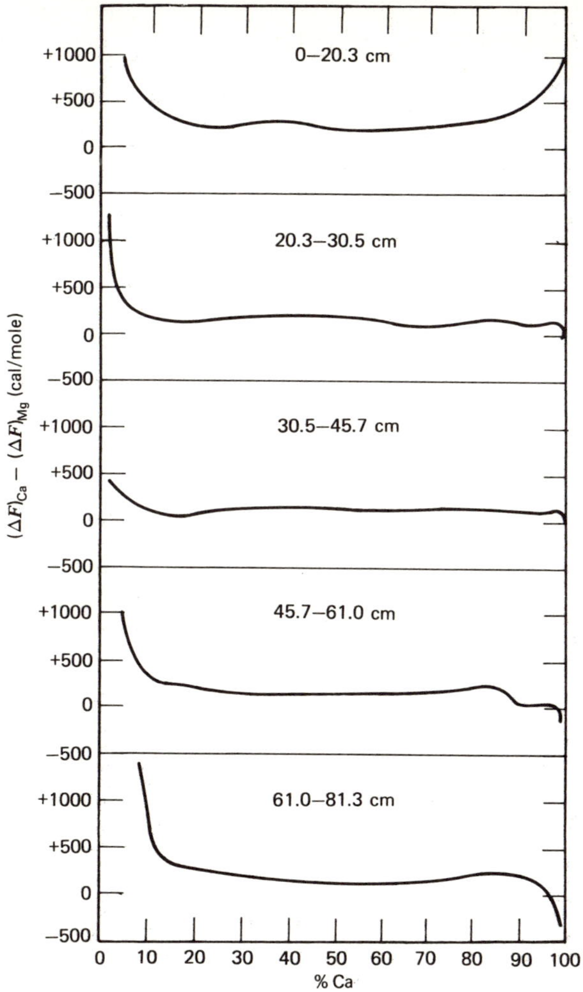
0–20.3 cm
20.3–30.5 cm
30.5–45.7 cm
45.7–61.0 cm
61.0–81.3 cm
+1000
+500
0
–500
(ΔF)Ca – (ΔF)Mg (cal/mole)
0
10
20
30
40
50
60
70
80
90
100
% Ca

(*b*)

## THE SOIL PROFILE IN RELATION TO THE SOIL SOLUTION

The differences that reveal themselves in the selectivities of cationic pairs naturally affect the equilibria between soil solution and the solid phases of soil profiles. Thus diagrams of the Garrels-Christ type should show variations with depth affecting the $cation^{+}/H^{+}$ or $\sqrt{cation^{2+}}/H^{+}$ values. There may also be variations in the silicic acid activity. These are discussed below in relation to the Mexico soil profile (15, 19, 25). This was investigated both in its natural condition and after large additions of carbonates or bicarbonates of calcium, sodium, potassium, and magnesium in certain layers. The amounts so used brought the respective soil layers close to saturation. The original soil showed only very slight variation with depth of the pH of the extract. Thus for the natural soil, variations along the vertical axis of the Garrels-Christ diagrams can be ascribed chiefly to the metallic cations concerned; but where additions were made, the pH was raised appreciably.

In Figures 79 to 82, the beidellite-kaolinite boundary is shown by two lines, calculated, respectively, for magnesium and sodium beidellites (Helgeson et al.). Potassium and calcium lines lie between. The numbers 1–15 refer to consecutive depth samples as in Figures 27 and 28.

Figures 79 and 80 give Garrels-Christ diagrams for calcium, potas-

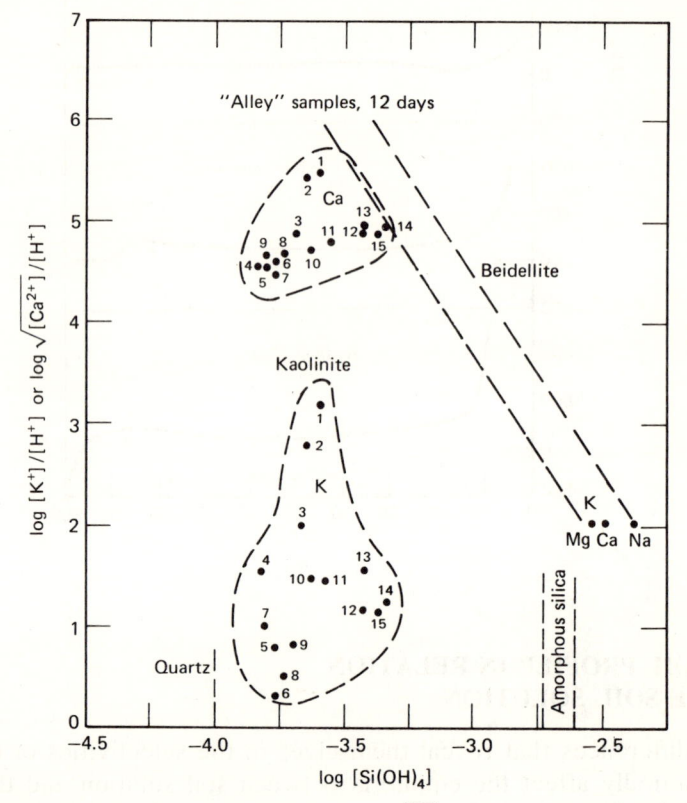

*Fig. 79*  Individual values of log [K]/[H] and log [$\sqrt{Ca}$]/[H] as functions of log [Si(OH)$_4$] for "alley" samples of the Mexico soil profile (12 days equilibration). Samples are numbered consecutively 1–15 according to depth (15 = 76–81 cm).

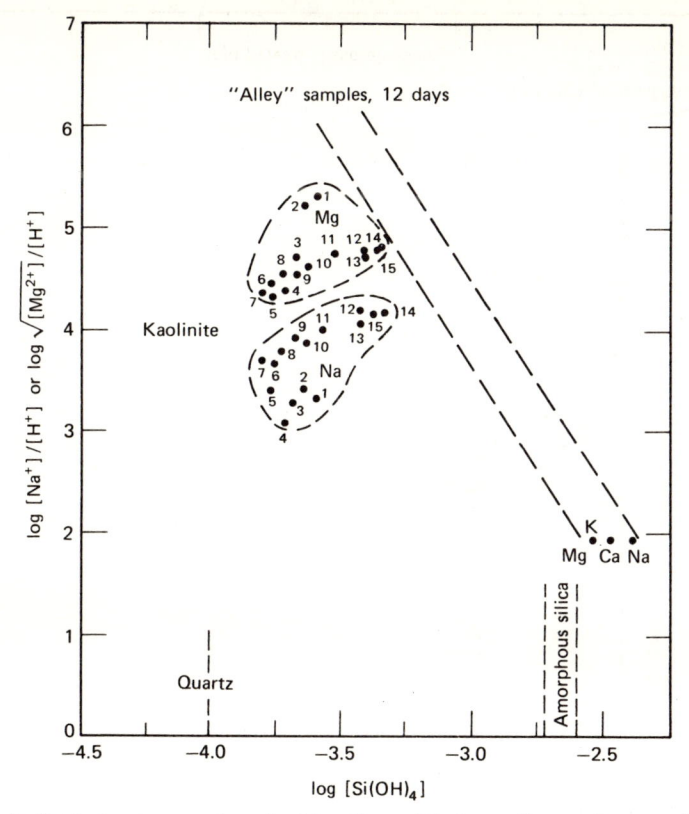

**Fig. 80** Individual values as described for Figure 79, for sodium and magnesium (15).

sium, magnesium, and sodium at various depths in the Mexico profile. For the log (cation$^+$/H$^+$) or log ($\sqrt{\text{cation}^{2+}}$/H$^+$) values, the changes with depth follow the same order for potassium, magnesium, and calcium. The surface values are highest; there is a fall to the lowest values for samples 4–7 (depths 20–41 cm) and a rise to intermediate values for samples 10–15 (depths 51–81 cm). However the potassium values are spread over a much greater vertical distance than those for calcium and magnesium. This reflects the very low values for potassium in solution in the subsoil. Sodium is quite different. The surface layers show the lowest values, which rise slowly with increasing depth.

The soluble silica also varies down the profile as shown by values along the abscissa. The surface layers are intermediate, the subsoil shows low silica, and the layers below 61 cm high silica. The total range is from about $10^{-3.8}$ to $10^{-3.4}$ $M$.

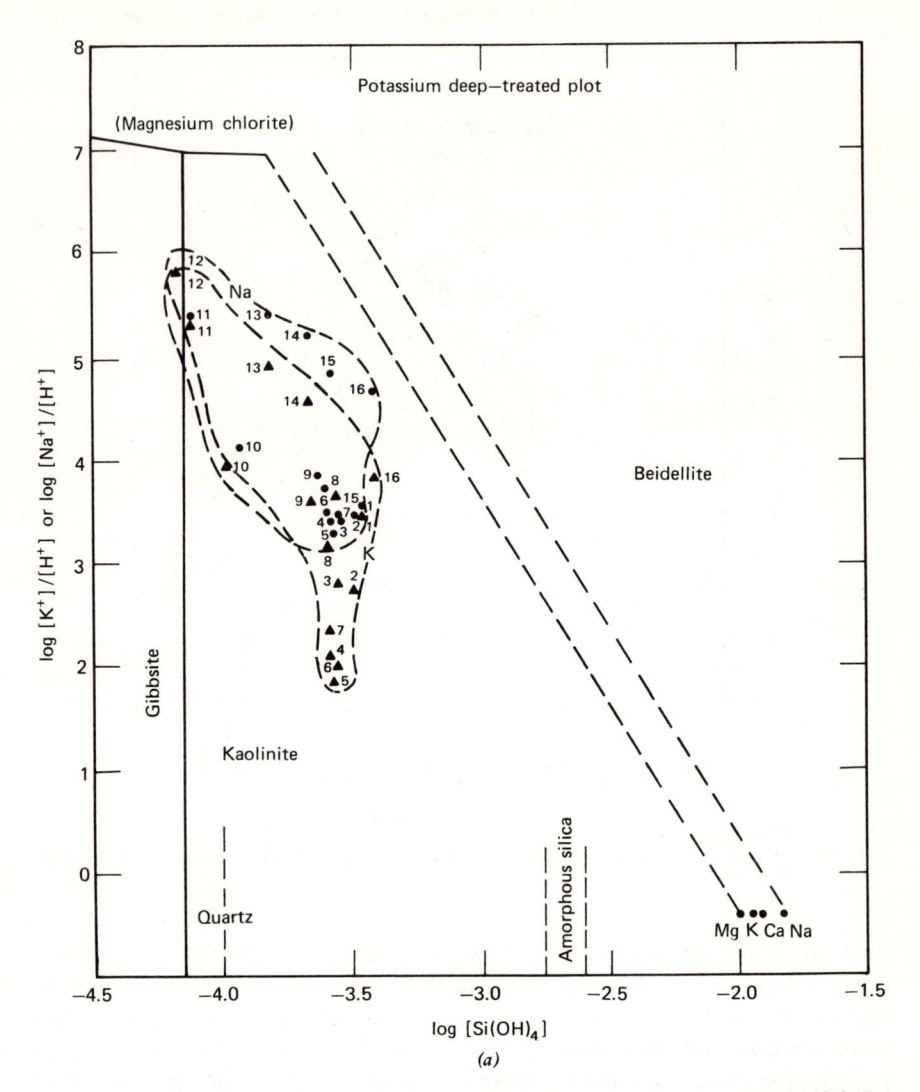

***Fig. 81*** (a) Individual values of log [K]/[H] and log [Na]/[H] as functions of log [Si(OH)₄] for Mexico soil, treated with potassium bicarbonate in the layer 41–61 cm, samples 9–12 (19). (b) Individual values for magnesium, as in (a) (19).

266

## Ionic Properties of the Exchange Complex 267

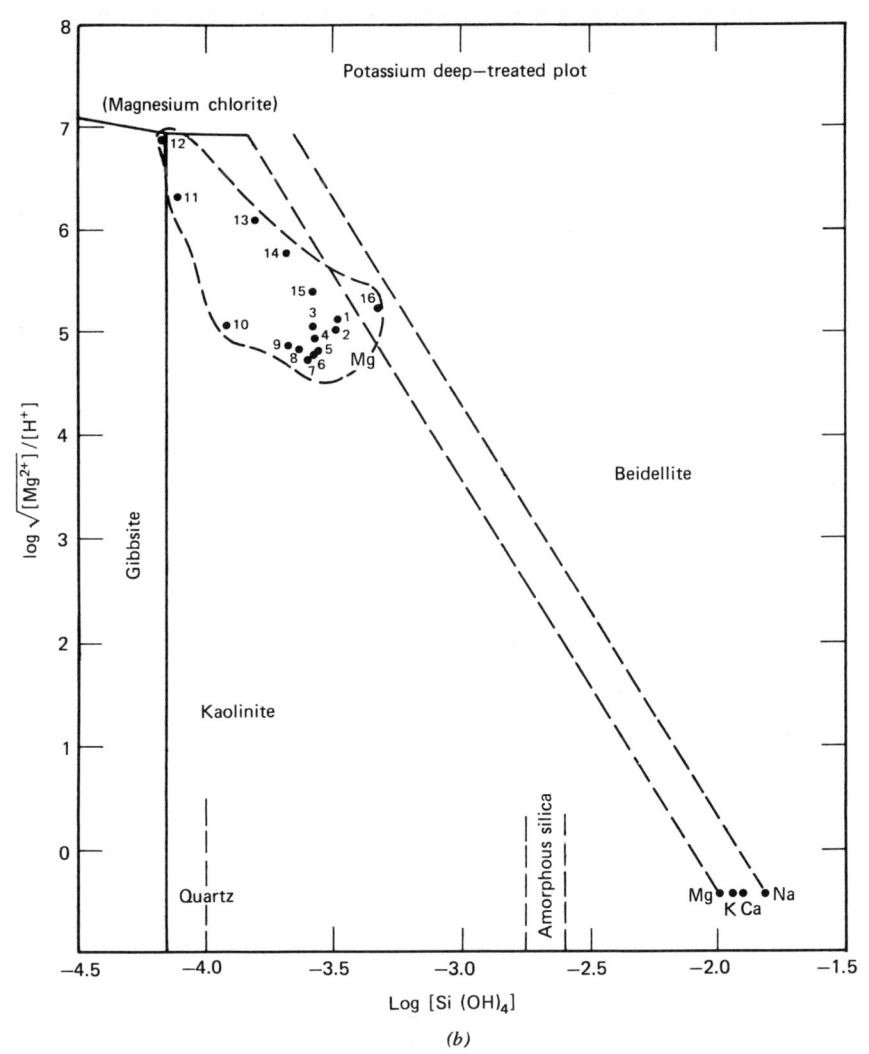

[Figure (b): Potassium deep-treated plot. Axes: log √[Mg²⁺]/[H⁺] versus Log [Si(OH)₄]. Stability fields shown for Magnesium chlorite, Gibbsite, Kaolinite, Beidellite, with Quartz and Amorphous silica reference lines. Data points 1–16 cluster in the upper region near the Mg label; Mg, K, Ca, Na markers appear at the lower right.]

(b)

The variations in $\sqrt{Mg/H}$ and $\sqrt{Ca/H}$ and in silica work together to bring the values for the lowest horizons sampled very close to the beidellite-kaolinite boundary. The sodium values are slightly further away in the kaolinite area, and those for potassium are very far from this boundary.

The relationship of these results to the natural moisture regime of the soil is as follows. The solution was extracted under a pressure of 10 atm starting with a saturated sample; thus its composition is an average for

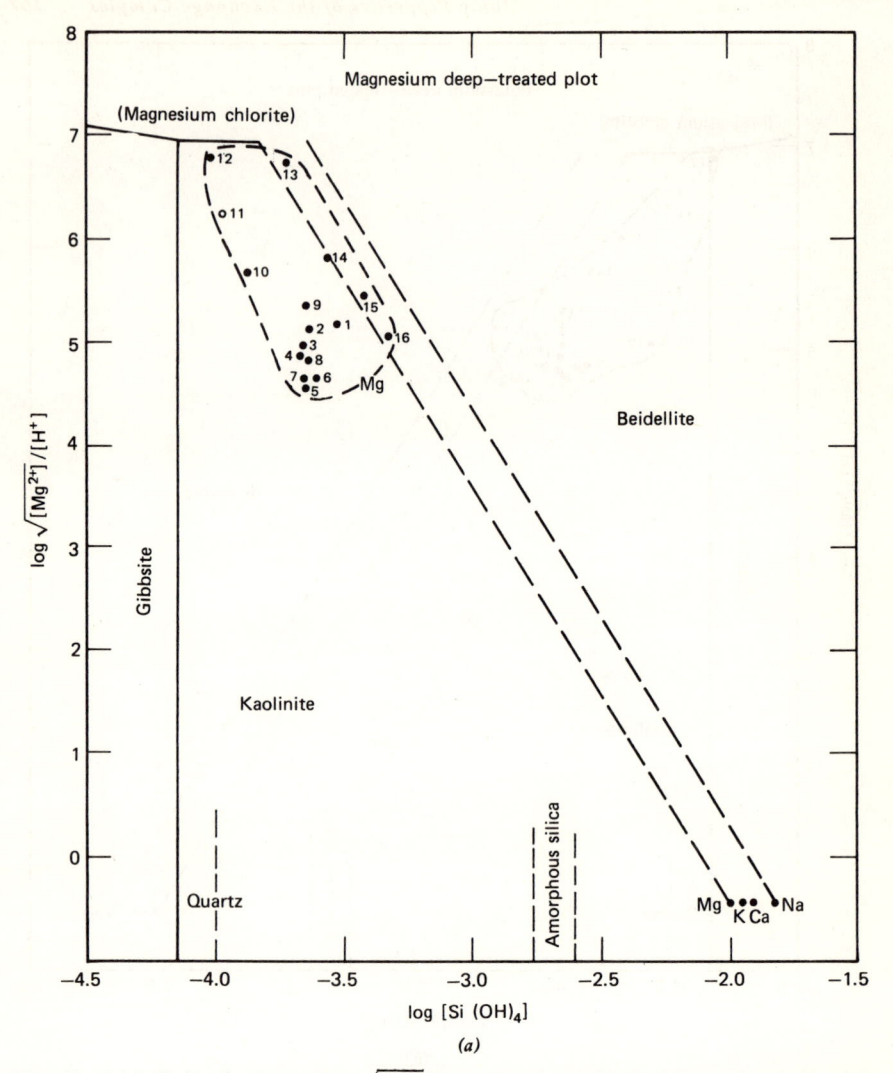

*Fig. 82* (*a*) Individual values of log $\sqrt{[Mg]}/[H]$ as functions of log $[Si(OH)_4]$ for Mexico soil treated with magnesium carbonate in the layer 41–61 cm, samples 9–12 (19). (*b*) Individual values for sodium and potassium as in (*a*) (19).

268

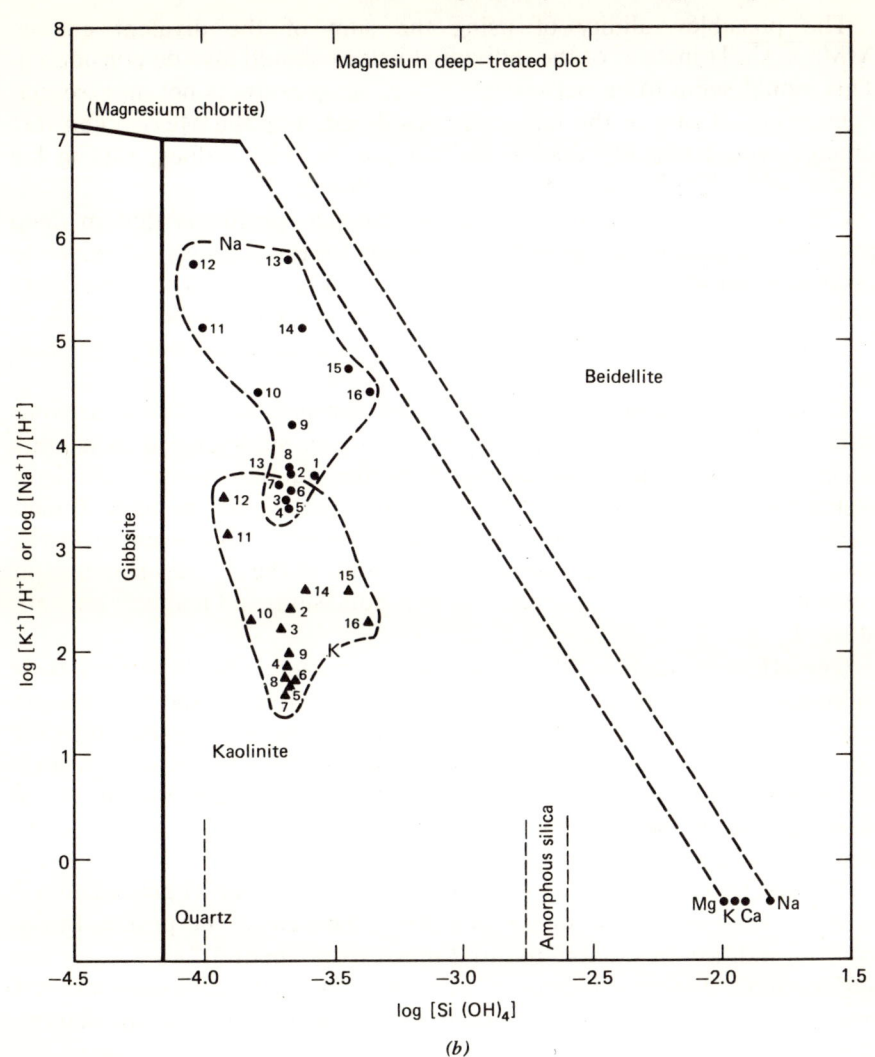

moisture tensions from near zero to this value. However during most of the year the tension varies between 0.3 and 15 atm. A smaller volume of water would be extracted over this range than over the actual range employed. Under prevailing natural moisture cycles, the solution phase and the colloidal phase would be more concentrated and the log [cation$^+$]/[H$^+$] or log [$\sqrt{\text{cation}^+}$/[H$^+$] ratio would be somewhat higher than our values. Large differences would not be expected, either on a Donnan or a Gouy model.

The possible validity of using the sum of the divalent cations $\sqrt{Mg + Ca}/H$ instead of the individual values should also be considered. This would seem to be permissible where magnesium is not an essential constituent of one of the minerals considered. For the Mexico soil this change would roughly double the calcium activity values, raising log $[\sqrt{cation^{2+}}]/[H^+]$ by around 0.15, a small change.

Thus the conclusion regarding the Mexico profile is that in deep horizons, conditions are generally favorable for the synthesis of beidellite or at least for the coexistence of beidellite and kaolinite. The subsoil and topsoil would appear to favor the beidellite→kaolinite reaction most of the time, although according to the X-ray evidence, it has not progressed far.

Under these circumstances it is of particular interest to determine what effect additions of carbonates or bicarbonates sufficient to bring this acidic soil to saturation would have. Experimental plots were available in which such additions had been made (*a*) to the 10–30 cm layer (6 meq carbonates or bicarbonates per 100 g soil), (*b*) to the 40–60 cm layer (12 meq carbonates or bicarbonates per 100 g soil). In the *a* series sodium and calcium additions were investigated; in *b* potassium and magnesium were studied.

One effect of such additions is obvious, namely, that wherever their influence is directly exerted, there will be a rise in the appropriate cation-hydrogen ratio. However there will also be rises in the ratios for the nonadded cations; because the increased anion activity brings with it an increase in the activities in solution of all exchangeable cations; and also because of the increase in pH. The combined effects were clearly demonstrated in both series.

The addition of sodium bicarbonate in the 10–30 cm layer caused after 7 years a very general rise in log $[Na^+]/[H^+]$ throughout the profile. In the zone of addition this amounted to 2 units, from about 3.7 to 5.7; for the surface soil it was raised about 1.7–1.8 units; for the layers below 40 cm it rose about 0.5 unit. Thus the mobility of sodium was clearly demonstrated.

It was otherwise with calcium. There was a rise in log $[\sqrt{Ca}]/[H]$ of about 0.5 unit in the 10–30 cm layer of the addition, little change in the surface layer, a rise of about 0.4 unit in the 30–51 cm layer, and little change below this. Calcium was strikingly less mobile than sodium over the 7 year period of the investigation.

Potassium was investigated for the large addition at 40–61 cm. Its direct effect in raising log $[K]/[H]$ was naturally very large, because the values for the control plot were remarkably low. Thus the rise in log $[K]/[H]$ for this layer was from 0.5–1.3 up to 3.5–5.8 (Figure 81*a*). Its indirect effect on magnesium was almost equal to the direct effect of magnesium in the layer of addition (Figure 81b). Adjacent layers were also affected: upward to 30 cm and downward to 80 cm depth.

Additions of magnesium had less effect on log $[\sqrt{Mg}]/[H]$ than those of potassium on log [K]/[H]. There was no effect above 40 cm. The 61–71 cm layer was distinctly enhanced, but below 71 cm there was no change (Figure 82a). Thus magnesium moves down very sluggishly. Its indirect effect in increasing log [K]/[H] was well marked (Figure 82b).

The indirect effects of potassium and magnesium on calcium and sodium in the profile were also evaluated. The great quantitative similarity in $\sqrt{Me^{2+}}/H^+$ between calcium and magnesium was maintained throughout, in indirect as well as in direct effects. Thus in Figure 82a the magnesium area can be replaced approximately by the calcium area. In this profile, however, the behavior of sodium differs from that of potassium. In the control plot the values of log [Na]/[H] increased slowly with depth throughout. Additions of potassium and magnesium in the 40–61 cm layer caused increases in log [Na]/[H] of 1–2 units, with smaller increases in the 61–82 cm layers. These additions of potassium and magnesium raised both the pH and the sodium in the extracted solution. The same was true of calcium. No appreciable change in exchangeable sodium or calcium was found. Hence the indirect effects must operate through Donnan-type equilibria to raise the degree of dissociation on the colloidal phase, that is, to lower the mean free bonding energy for all cations. This would be expected from the general reduction in cationic bonding energies near saturation as found in titration curves by the present author and his co-workers (see Vol. I, Chap. 6).

Variations in soluble silica, although much less than those of exchangeable cations, are significant in relation to the establishment of solution-mineral equilibria (Figure 83). In the zone of additions and a little beyond, there is a definite decrease in soluble silica, more marked for potassium than magnesium. In the Garrels-Christ diagram, this acts to move points to the left, that is, away from the beidellite-kaolinite boundary. However on this logarithmic scale the upward movement caused by increases in cation-hydrogen ratios is much greater.

Variations in soluble aluminum are smaller in relative magnitude, and uncertainty in the analytical figures is greater than for silica. However the lower part of the zone of additions gives higher values for soluble aluminum than the rest of the profile. This is in agreement with the action of calcium hydroxide, which lowers the silicon and slightly raises the aluminum in solution. This increase was ascribed to the production of a monovalent anion $Al_6(OH)_{15}O_2^-$ or $5Al(OH)_3 \cdot AlO_2^-$ · in a restricted pH range, 6.8–8.4 (15).

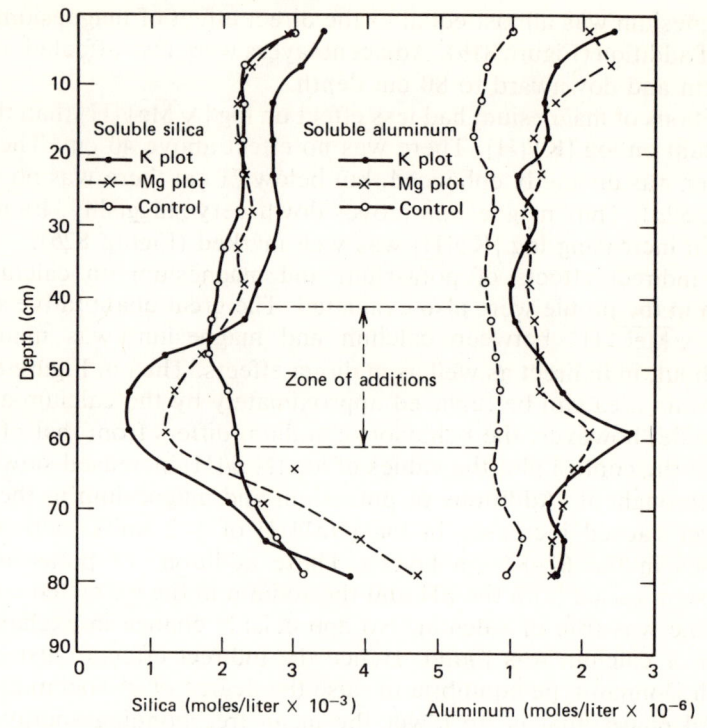

**272**  *Physical Chemistry and Mineralogy of Soils*

*Fig. 83*  Soluble silica and aluminum as functions of depth for the Mexico soil (19). (Control plot and plots with potassium and magnesium treatments in 41–61 cm layers.)

### Other Ionic Ratios

It is possible to express ionic ratios in the equilibrium solution, such as $H^+/\sqrt{Ca^{2+} + Mg^{2+}}$ (Schofield's lime potential) or $K^+/\sqrt{Ca^{2+} + Mg^{2+}}$, as a function of depth. This is equivalent to the consideration of one variable of the Garrels-Christ diagram or of the solution factor in a selectivity equation. Beckett's treatment of soil potassium (1, 2) in which potassium added to or lost from the system is plotted against $K^+/\sqrt{Ca^{2+} + Mg^{2+}}$ for the solution, gives a curve for each sample. In general these curves consist of two parts. For the larger additions of potassium they are often straight lines, that is, a constant fraction of the added potassium is effective in changing $K^+/\sqrt{Ca^{2+} + Mg^{2+}}$. The slope of this line therefore represents the potential buffering capacity for potassium ($PBC_K$). For small additions or actual depletions, there usually results a curve having increasing steepness with increasing depletion. This is a desorption isotherm for potassium at constant $\sqrt{Ca^{2+} + Mg^{2+}}$. Some soils show no straight line portions. A buffering capacity can then be defined as the slope at zero potassium addition $(PBC_K)_E$.

Although some attempts have been made to characterize surface soils through these quantities, there is little detailed information on profiles. Fernandez et al. have studied soils on the volcanic island of Tenerife (8). They found that $K^+/\sqrt{Ca^{2+} + Mg^{2+}}$ usually decreased with depth, whereas $(PBC_K)_E$ increased.

## OXIDATION-REDUCTION CONDITIONS IN SOIL PROFILES

The second most labile feature of soil profiles is the oxidation-reduction regime, the most labile being the moisture regime. Soils are consumers of molecular oxygen, which constitutes an almost constant proportion of the free atmosphere. The variability in effective oxygen pressure in soils is governed by two groups of factors—the nature and speeds of the oxygen-consuming reactions, and the effectiveness of diffusion in restoring the conditions of the free atmosphere. The second factor is, in turn, a sensitive function of the moisture regime itself. As pointed out in Chapter 2, diffusion in the vapor phase is over $10^4$ times faster than diffusion in the liquid phase. The progressive filling of soil pores with water finally changes the steady state conditions by several orders of magnitude. Thus oxidation-reduction protentials cover a wide range, both in respect of seasonal changes and of variation with depth at a given time.

As was brought out in Chapter 2, the definition of the chemical status involves both the oxidation-reduction potential itself $(E_H)$ and the pH. The two together define a point on the Pourbaix type of diagram, whose position can be related to various possible reactions. As a rule, the various practical difficulties in the measurements—poisoning of the platinum electrodes, the difficulty of securing representative contact, and so forth—are only partially overcome. Nevertheless, results that are reliable on a relative basis rather than absolutely often can be used as guidelines in the discussion of such important pedological effects as iron and manganese movement or the production of concretions.

An example illustrating variations in the Mexico soil with depth and with the season (1974) is given in Figure 84. The mean pH value was around 6.0, and all $E_H$ values were corrected to this. On the right the regions corresponding to $Fe^{2+}$, $Fe(OH)_3$, $Mn^{2+}$, and $MnO_2$ taken from the Pourbaix diagram (Chapter 4, Figure 18) are shown. For the production of iron concretions, the presence and mobility of $Fe^{2+}$ must alternate with periods of stability of $Fe(OH)_3$. This was very clearly the case at 60 cm depth. At 30 cm, reducing conditions were sufficiently severe only for 5

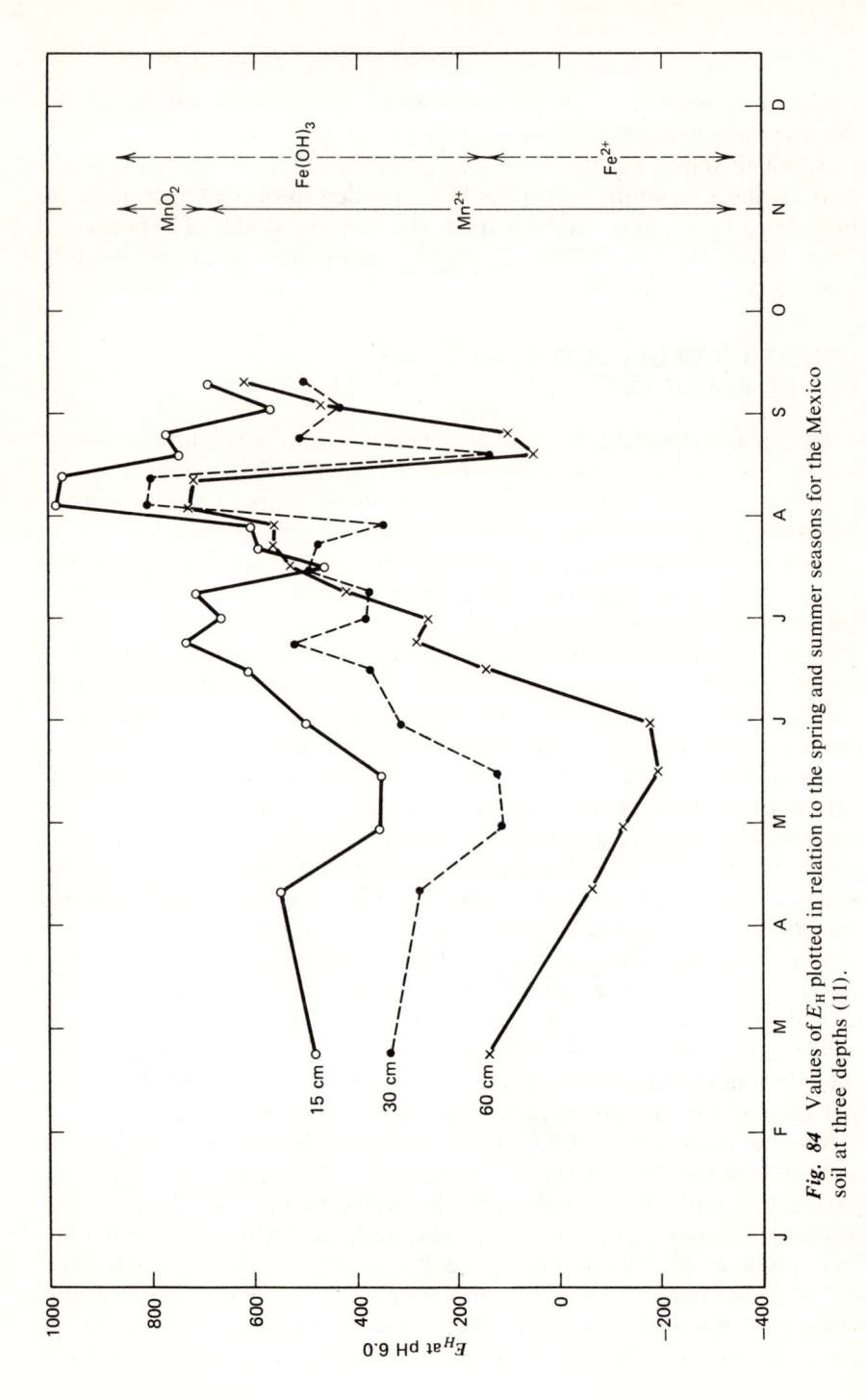

*Fig. 84* Values of $E_H$ plotted in relation to the spring and summer seasons for the Mexico soil at three depths (11).

weeks in the spring. At 15 cm, conditions stayed within the ferric hydroxide zone. The dry period in August brought all three layers close to the maximum $E_H$ value for atmospheric oxygen and $Fe(OH)_3$. Manganese was in a different situation. Only briefly in the summer were conditions anywhere favorable to the production of $MnO_2$. However a moderate rise in pH would have effectively increased such periods. Under the seasonal conditions here depicted, $Mn^{2+}$ would be present and mobile most of the year. As pointed out in Chapter 4, its reaction with $Fe(OH)_3$ may be significant, although in this soil exchangeable manganese is very low (< 0.1 meq/100 g). Hence the $Mn^{2+}$ activity cannot in practice exceed $10^{-6}$, and its quantitative role in concretion formation is correspondingly reduced.

These determinations show good agreement with qualitative field observations. The $A_p$ horizon of the Mexico soil (15 and 30 cm samples) contains less concretionary material than the B horizon (60 cm sample). The relationships in Figure 84 also suggest that the composition of concretions should vary with depth in the profile—those nearest the surface should contain more manganese than those at greater depth. Experimental evidence of this is not available. It must be emphasized that these explanations are based on a simple diagram that does not take account of the possibility of the formation of mixed precipitates of $Mn(OH)_2$ and $Fe(OH)_3$, or of actual double hydroxides (Feitknecht-type compounds).

Although concretionary material under an aerobic regime may, by field observation appear hard and apparently resistant to change, the onset of reducing conditions very rapidly removes its coherence. This was very strikingly shown by Bidwell, Gier, and Cipra (5) for pedotubules formed on grass roots. Such formations represent the most transient type of concretionary material. They showed different ratios of iron to manganese in different concentric layers, maganese being most concentrated toward the center. Schroeder and Schwertmann (23) have also emphasized the role of plant roots in the formation of concretions.

The distribution of concretions with depth was studied by Winters in Illinois (30). For well-drained soils the quantity decreased fairly regularly with depth, but impeded drainage caused accumulations at certain depths. This is understandable if their presence is a function both of the intensity and of the number of cycles of oxidation and reduction.

Very clear relationships between the distribution of concretions with depth and prevailing moisture regime have been demonstrated by Phillippe and co-workers for three soils in Kentucky (21).

Peat profiles do not show unusually low $E_H$ values as compared with mineral soils. Urquhart and Gore (26) found that poorly drained peats gave a minimum $E_H$ value at pH4 of about +50 mV for the layers around

20 cm depth, whereas well-drained peats might either give such a minimum at about +350 mV or else a gradual fall of $E_H$ with depth.

There are serious practical difficulties in the measurement of $E_H$ and pH in soil profiles by the insertion of electrodes. The heterogeneity of the soil *in situ* is one drawback; but by multiple measurements its effect on the mean value can be determined. Changes in the platinum or other metallic electrode used for $E_H$ are common with long residence in the soil. Even in laboratory measurements, close attention to the cleaning of these electrodes is essential for accuracy. If electrodes are to be repeatedly cleaned and reinserted, it is difficult to effect this without locally changing the air-water conditions.

## REFERENCES

1. Beckett, P. H. T., Potassium-calcium exchange equilibria in soils: specific adsorption sites for potassium, *Soil Sci.*, **97**, 376 (1964).
2. Beckett, P. H. T. and M. H. M. Nafady, A study on soil series: Their correlation with intensity and capacity properties of soil potassium, *J. Soil Sci.*, **19**, 216 (1968).
3. Beinroth, F. H., Occurrence of Oxisols in Puerto Rico, mimeographed tabulations (1968).
4. Bengtsson, B., N. Karlsson, and S. Mattson, The pedography of hydrologic soil series. IV. The distribution of Si, Al, Fe, Ti, Mn, Ca and Mg in the Unden podzol and the Dala brown earth series, *Ann. Agr. Coll. Swed.*, **11**, 172 (1943).
5. Bidwell, O. W., D. A. Gier, and J. E. Cipra, Ferromanganese pedotubules on roots of *bromus inermis* and *andropogon gerardu*, *Trans. 9th Int. Cong. Soil Sci.*, **IV**, 683–692 (1968).
6. Brydon, J. E. and C. E. Marshall, Mineralogy and chemistry of the Hagerstown soil in Missouri, University of Missouri Agricultural Experiment Station Research Bulletin 655 (1958).
7. Collins, J. F. and S. W. Buol, Effects of fluctuations in the $E_h$-pH environment on iron and/or manganese equilibria, *Soil Sci.*, **110**, 111–118 (1970).
8. Fernandez, C. E., M. J. Hernandez, and P. A. Borges, The $Q/I$ relationship for potassium in soils of different types of the island of Tenerife, *Potash Review*, Subject 4, No. 4, 1–8 (1975).
9. Gedroiz, K. K., *Genetic Soil Classification Based on the Absorbed Soil Cations*, 2nd ed. (1927). Translated by the Israel Program for Scientific Translation, Jerusalem (1966).
10. Gedroiz, K. K., Die Lehre vom Adsorptionsvermogen der Boden (translated from Russian by H. Kuron), *Kolloidchem. Beih.*, **33**, 317–459 (1931).
11. Hess, R. E., Arsenic equilibria in Missouri soils, Ph.D. thesis, University of Missouri, Columbia (1975).
12. Jenny, H., Behavior of potassium and sodium during the process of soil formation, University of Missouri Agricultural Experiment Station Research Bulletin 162 (1931).
13. Jenny, H., *Factors of Soil Formation*, McGraw-Hill, New York (1941).
14. Kelley, W. P., *Cation Exchange in Soils*, Reinhold, New York (1948).

15. Marshall, C. E., M. Y. Chowdhury, and W. J. Upchurch, Lysimetric and chemical investigations of pedological changes. Part 2. Equilibration of profile samples with aqueous solutions, *Soil Sci.,* **116,** 336–358 (1973).
16. Mattson, S., The pedography of hydrologic soil series. V. The distribution of K and P and the Ca/K ratios in relation to the Donnan equilibrium. *Ann. Agr. Coll. Swed.,* **12,** 119 (1944).
17. Mattson, S., G. Sandberg, and P. E. Terning, Electrochemistry of soil formation. VI. Atmospheric salts in relation to soil and peat formation and plant composition. *Ann. Agr. Coll. Swed.,* **12,** 101 (1944).
18. McCracken, R. J. et al., Certain properties of selected southeastern United States soils, Virginia Agriculture Department Experimental Station, Southern Regional Bulletin 61 (1959).
19. Misra, U. K., Mineral equilibria in a soil system under natural and modified conditions as shown by the physical chemistry of the aqueous phase, Ph.D. thesis, University of Missouri, Columbia (1973).
20. Murdock, L. W. and C. J. Rich, Ion selectivity in three soil profiles as influenced by mineralogical characteristics, *Soil Sci. Soc. Am. Proc.,* **36,** 167–171 (1972).
21. Phillippe, W. R., R. L. Blevins, R. I. Barnhisel, and H. H. Bailey, Distribution of concretions from selected soils of the inner bluegrass region of Kentucky, *Soil Sci. Soc. Am. Proc.,* **36,** 171 (1972).
22. Roth, J. R., A study of the cation exchange relationships in three Missouri soils, M. S. thesis, University of Missouri, Columbia (1965).
23. Schroeder, D. and U. Schwertmann, Zur Enstehung von Eisenkonkretionen in Böden, *Naturwissenschaften,* **42,** 255–256 (1955).
24. Southern Cooperative Series, Certain properties of selected southeastern United States soils and mineralogical procedures for their study, Virginia Agricultural Experimental Station Bulletin 61 (1959).
25. Upchurch, W. J., M. Y. Chowdhury, and C. E. Marshall, Lysimetric and chemical investigations of pedological changes. Part 1. Lysimeters and their drainage waters, *Soil Sci.,* **116,** 266 (1973).
26. Urquhart, C. and A. J. P. Gore, The redox characteristics of four peat profiles, *Soil Biol. Biochem.,* **5,** 659–672 (1973).
27. Wascher, H. L. et al., Loess soils of northwest Illinois, University of Illinois College of Agriculture Bulletin 739 (1971).
28. Wiegner, G., Zum Basenanstausch in der Ackererde, *Z. Landwirtsch.,* **60,** 11–150 (1912).
29. Wiegner, G., *Boden und Bodenbildung in kolloidchemischer Betrachtung,* Springer, Leipzig (1921).
30. Winters, E., Ferromanganiferous concretions from some podzolic soils, *Soil Sci.,* **46,** 33–40 (1938).
31. Yamane, I., $E_H$–pH diagrams of iron systems in relation to flooded soils, *Tohoku Univ. Report Inst. Agr. Res.,* **21,** 39–63 (1970).
32. Yamane, I., $E_H$–pH diagrams of manganese systems in relation to flooded soils, *Tohoku Univ. Report Inst. Agr. Res.,* **24,** 1–15 (1973).

# *11* Obstacles and vistas

Although as the curves presented in Chapter 10 show, various functions involving exchangeable cations are now used in the characterization of soil profiles, the main advance over the past 50 years has been the use of thermodynamic and quasi-thermodynamic functions involving the solution phase in equilibrium with the soil. We still treat the exchange complex as an entity, just as Gedroiz, Hissink, Wiegner, and others did in the 1920's. Even the separation of the pH-dependent from the pH-independent function has scarcely been applied in profile studies. The mineralogical information available to us has obvious implications for the exchange chemistry of soil profiles; so also has improved characterization of the organic matter. Yet we are not making the appropriate transitions from this knowledge of soil constituents to knowledge of chemical behavior, or vice versa, on any large scale. When shall we end this period of groping and fumbling? Clearly one prerequisite is to understand the obstacles, which are now reasonably apparent.

## DEFINITION OF SOLID PHASES

One set of obstacles arises from the inherent difficulty in characterizing each solid phase thermodynamically, since the standard free energy is a function of particle size, particle shape, and various kinds of departure from ideal crystallinity and composition. A small difference in standard free energy corresponds to a large difference in the concentration of an equilibrium solution. Ionic exchange reactions operate through small free energy differences; hence they are extremely sensitive to all departures from ideal macrocrystallinity. Electron microscopy in its several forms is of great value in the external characterization of finely divided single minerals. Scanning instruments now enable us to view fresh surfaces of soils with the minimum of preparation. Hence progress seems possible, beginning, as usual, with selected simple and favorable cases and gradually extending to more complex ones as techniques improve and experience accumulates.

## EXCHANGE REACTIONS

It must be admitted that application of the theory of ionic exchange reactions has reached an impasse. As far as soils and soil profiles are concerned, almost everyone seems to be satisfied with equations based on grossly oversimplified models. Over the middle range of ionic compositions, these relations appear to be adequate; but for cations in small proportion such as potassium and ammonium, it is already apparent that their behavior is governed by at least two types of site. In general, functions of cations taken singly are more sensitive than exchange reactions. Thus single ion activity measurements leading to cationic bonding energy curves are particularly well suited to characterization. Equilibration studies of single cations carried out with the addition of radioactive isotopes are also extremely valuable, especially in establishing the "labile pool" in exchange reactions.

## ELECTROKINETICS

The movement and arrest of colloidal constituents in soil profiles still remains as a central problem, to which electrokinetic methods can be applied. Very little has been done since Mattson's heroic efforts in the 1920's and 1930's. The outer parts of the electrical double layer are clearly critical. Against the natural heterogeneity of soils, we must set the tendency of the colloidal constituents to form coatings on the most varied surfaces. Thus it may well be possible to make progress, even with such a feebly discriminating technique as electroendosmosis through soil plugs. In this way the zeta potential itself, or other quantities related to it, would once again become the focus of discussion. Through modern expressions of diffuse double layer theory (Verwey and Overbeek) it may then prove possible to deal with forces between particles or, alternatively, with free energy differences arising from particle-particle interaction. These more sophisticated applications of colloid-chemical theory are available, but they have found little application in pedology.

What we have are numerous descriptions from thin section studies of clays deposited on mineral surfaces and in pores. The general conclusion is that the 2 : 1 clays (smectites and illites) readily move downward in soil profiles to form well-oriented layers. The process can be regarded as an oriented coagulation or as a tactoid formation. The 1 : 1 clays can also move, and under certain circumstances they can also form oriented layers; but they tend to show less evidence of orientation than the 2 : 1 clays. This would be expected from their greater sensitivity to random coagulation.

## INDEX METHODS

The methods discussed in Chapter 5, though inherently laborious, should find increasing application. Considering the enormous expenditure of scientific effort on agricultural and other applications of soil science, surely a small proportion could consistently and advantageously be devoted to fundamental pedological questions. This is one of the few routes by which we shall ever arrive at the understanding necessary for scientific prediction. Up to the present we have largely been concerned with the apparent simplicities of these complex systems, soil profiles. Increasingly, however, we shall find ourselves immersed in their complexities. In this situation accurate information on the balance of transformations and movement will be seen as a veritable sheet-anchor.

## SOIL ORGANIC MATTER

The nature of the obstacles presented by soil organic matter needs very careful consideration. When it is divided chemically into fractions (as in Oden's scheme, Vol. I, Chap., 5), even the mildest procedures give the following result. The total cation exchange capacity or the neutralization capacity of the fractions is greater than that of the original humus by a considerable factor, usually exceeding 1.5. This has long been known. It can be explained qualitatively by the idea that natural humus is a complex polymer or mixture of polymers. Conversion of the acidic form to the salt by raising the pH effects a depolymerization (as well as a neutralization), the extent of which varies strongly with the nature of the cation and with the final pH. The difficult task is to distinguish specific depolymerization from the ordinary peptization of a colloidal gel. In soil profiles the water-soluble component fulvic acid is the most mobile faction. With calcium saturated soils such as Chernozems, this natural fractionation is much less strongly shown than in the acidic Podzols or lateritic soils. The Solonetzs with exchangeable sodium appear from limited data to be intermediate. The movement of humus is thus in part different from that of clay. Yet in mineral soils the strong tendency of humic matter to be absorbed on clay surfaces acts to prevent cleancut separations in soil profiles.

## CATIONIC EQUILIBRIA AND MOISTURE CHANGES

The dynamic nature of the monovalent-divalent cation equilibria in relation to moisture cycles, originally predicted on the basis of Donnan theory (Vol. I, Chap. 1), is now seen as a thermodynamic necessity independent of mechanisms. In the drying process the thinning of water films involves an increase in salt concentration. But alongside this increase is a change in the proportions of the cations caused by the thermodynamic condition that the activity ratio [monovalent]/√[divalent] in the solution phase should tend to remain constant. Clearly a change in the proportions of cations in the solution phase involves a complementary change for the exchange complex, since the total cations remain constant. In relation to changes centering around field moisture capacity, the volume of the aqueous phase can be assumed to be comparable to or smaller than that of the colloidal phase. Its content of electrolyte will be much smaller than the total exchangeable cations. Under these circumstances modification of the equations presented in Volume I, Chapter 1, indicates that the molar ratio of potassium to calcium in the aqueous phase is proportional to the square root of the volume per mole of calcium. But the volume per mole is proportional to the thickness of the aqueous film in contact with the clay boundary. Hence thick films (low tensions) correspond to high ratios of potassium to calcium in the solution phase and thin films (high tensions) to low ratios. Another way of looking at the same situation is in terms of Gouy theory. The outer parts of the diffuse double layer contain a higher K/Ca ratio than the inner parts.

Experimental tests of the actual relationship between moisture tensions and K/Ca ratio are possible and would be of considerable interest in soil profiles. The effects of drought on the mineral nutrition of plants can also be looked at from this point of view, especially where different layers of the profile differ in their calcium-potassium relationship.

It seems clear both from Jenny's ideas and from the data given in Chapter 10 on the Mexico soil profile that the potassium-sodium relationships of soil profiles are likely to provide very sensitive depth functions, of great value in characterization. Calcium-magnesium selectivities are normally much less variable, but there may be cases of magnesium fixation that would become evident through these determinations.

## SILICA

We know from earlier discussions of mineral equilibria in aqueous systems and of Garrels-Christ diagrams that silica as quartz participates very sluggishly. The solution of silica can be looked on as a depolymerization and hydration either from a crystalline or an amorphous polymer. The solution phase contains monomers, dimers, trimers, and so on, of silicic acid. The monomeric form can be determined colorimetrically in solution, and its corresponding activity is employed in the equations expressing the equilibria under consideration. We are still far from understanding the kinetics of silica in soil systems or the total role it may play in natural soil profiles. At various times in the past, mobile colloidal silicic acid has been regarded as a protective colloid responsible for certain features of Terra Rossa and other soils. These ideas need careful reexamination.

## ALUMINUM

The complex problems associated with aluminum have chiefly been investigated in relation to soil acidity, which is a variable of considerable significance in soil profiles. The constants connecting the three ions $Al^{3+}$ $Al(OH)^{2+}$, and $Al(OH)_2^+$, with each other and with pH are reasonably well known. However there is also evidence for $Al_6(OH)_{15}^{3+}$ and $Al_6(OH)_{15}O_2^-$, as well as the aluminate ion $AlO_2^-$ or $Al(OH)_4^-$. The activity of any of these ions in solution may be greatly affected by the complexing of aluminum by organic matter. Some progress can be made empirically by plotting pH against p(Al $_{total}$) for soil extracts. The slope of the tangent to such a curve at any chosen pH gives the mean charge of the aluminum species concerned. If a single ionic species dominates through a certain pH range, that part of the curve will be a straight line with the appropriate slope. As brought out in Chapter 5, Figure 26, the evidence for $Al_6(OH)_{15}O_2^-$ is of this kind.

In the Mexico soil profile, variation in aluminum with depth in the soil extract was somewhat less than that of silicic acid. Both showed diminishing values with depth in the A horizon, then increases through the B horizon and into the C. A survey of these relationships for other soils would be extremely interesting. Construction of the Kittrick diagram in which $pH-1/3pAl^{3+}$ or $pH-1/2pAl^{2+}$ is plotted against $pSi(OH)_4$ showed that all points for this profile were close together and lay above the kaolinite line, thus indicating supersaturation.

The situation may of course be quite otherwise in the Ultisols and Oxisols, and in all soils in which aluminum moves as an organic complex. In the latter cases clues may perhaps be found in relationships between aluminum and fulvic and humic acids.

The complexing action of humic compounds on aluminum is clearly a matter for serious attention. As mentioned in Chapter 4, Coleman et al. demonstrated the usefulness of an overall equilibrium constant for the reaction metallic ion + acidic organic matter → metallic complex + H$^+$. If both the organic matter and the metallic complex can be treated as single solid phases, the equilibrium constant is simply related to the ions in solution by $\log K_r = n \log [H^+] - \log M^{n+}$. By plotting logarithmically the experimental relationship between H$^+$ and M$^{n+}$ in true solution in equilibrium with the solid phases, $n$ is given by the slope, and $\log K_r$ by the intercept.

When the organic matter and its metallic complex are soluble, the methods used by Schnitzer for the fulvic acid complexes are applicable, subject to the difficulties in defining the aluminum ion present. Fulvic acid is treated as a soluble acid of known molecular weight and equivalence.

These treatments are of course oversimplified by the assumption that there is a simple acidic function and a single complex in each case. Indeed, they also gloss over the presence of polymers of a range of molecular weights.

Thus, for some time, further exploration of these methods will be required. We are not yet at the stage of determining standard free energies for well-defined humic compounds. Nevertheless progress in respect of soil profiles is possible along these somewhat empirical lines.

## IRON IN SOIL PROFILES

The striking colors of the oxides and hydroxides of iron, together with their extremely low solubilities, make them valued indicators of chemical change in soil profiles. The mobility of iron is affected both by complexing and by reduction. Thus we have to take into account divalent iron and its insoluble combinations with trivalent iron, namely, magnetite Fe$_3$O$_4$, and the ferroso-ferric hydroxide Fe$_3$(OH)$_8$. Together with the trivalent oxides and hydroxides, this gives a rich variety of solid phases. The conditions for their formation are known only in small part from thermodynamic data. Even here there are uncertainties. The change goethite → hematite + water does not seem to be reversible under ordinary climatic conditions. Thus hematite is likely to remain under conditions of high vapor pressure where goethite should theoretically be formed. Under such high moisture conditions oxidation of ferrous iron or of ferrous complexes gives rise to the orange lepidocrocite (Fe$_2$O$_3$·H$_2$O). Thermodynamic data are needed for this mineral. Finally there is the ubiquitous limonite, the common brown coating of sand grains, usually thought of as amorphous ferric hydroxide, which ages to microcrystalline goethite plus water.

Attention is being focused also on aluminous goethites in soils, whose thermodynamic relationships to hematite and corundum are still undefined.

Because of its sensitivity to alternate conditions of oxidation and reduction, iron, as explained in Chapter 4, provides the most striking chemical evidence of soil heterogeneity. Variations with depth in soil profiles are readily apparent to the field observer, who sees them as concretions, mottlings, and coatings. Under conditions favoring the production of smectite and illitic clays, iron can be incorporated into these structures in appreciable quantity. The relationships between free iron oxides and hydroxides and these silicate products of weathering are undefined. Thus iron offers a broad spectrum of problems in the development of pedology. In this it is closely paralleled by manganese, cobalt, nickel, and chromium, which are usually regarded as trace elements in soils.

## THE CLAY MINERALS

Chapter 8 with its account of clay minerals in soil profiles indicates many of the uncertainties that remain in quantitative estimates. Improvements are continually being made in X-ray methods, particularly in the interpretation of shoulders and broad peaks in basal spacings. These involve mixed layers, sometimes with partial segregation of two or more constituents. Differential solubility studies provide valuable supplementary information.

Where X-ray methods fail, as they do for amorphous constituents, the other possibilities include characterization by infrared spectra, by differential thermal analysis, and by differential solubility. We are acquiring considerable information on allophanes and related amorphous aluminosilicates and hydroxides. Thus the respective parts played by crystalline and amorphous constituents in soil profiles are likely to become much better defined as these investigations progress. In the long run crystalline products will emerge, but whether this is a matter of the pedological time scale $0-10^4$ years or the geological scale needs to be investigated.

The evidence on clay minerals in soil profiles clearly indicates that under characteristic climatic regimes, crystalline minerals of the clay fraction are formed within the pedological time scale. Podzols provide the remarkable example of an A horizon favoring crystalline smectites and vermiculites, over a B horizon with a high proportion of amorphous constituents.

In general the broad climatic sequences of soils are characterized by related sequences of clay minerals. This is in excellent accord with the thermodynamic requirements for change expressed in the internal chemical environment. Thus we may summarize the general situation by saying that pedological changes account for the presence of zeolites and smectites in saline alkali soils, of attapulgites and smectites in Desert soils high in magnesium, of hydrous micas and smectites in Mollisols, of smectites and kaolins in Alfisols, of kaolins in Ultisols, and of gibbsite and hematite in Oxisols. In each case parent materials may provide an original suite of clay minerals as well as a variety of primary minerals for weathering. Vegetational and topographic factors also exert their influence by changing the internal environment. In general the pedological time scale seems adequate for the recognition of the characteristic mineralogical changes.

The modes of occurrence of clay minerals and hydroxides are of great importance in determining the structural aspects of soil profiles. This expresses itself to a considerable extent in the functions of these minerals as cementing agents between larger grains. Thus the scanning electron microscope is likely to prove to be an important tool for the investigation of soil heterogeneity, supplementing the ordinary microscopic study of thin sections, which is well established.

## INSTRUMENTATION IN THE SERVICE OF PEDOLOGY

During the lifetime of the author, vast changes have taken place in experimental methods applied to problems of the soil. Early in this century methods used by physical chemists were already utilized by soil scientists. First came those originally developed for the study of electrolytes—concentration cells with potentiometers for determination of pH, and conductivity bridges for total electrolytes in soil-water systems. Modern instruments differ from the early devices chiefly in having greater versatility and convenience. The accuracy is about the same.

Up to about 1930 chemical analyses were usually performed by standard methods with little elaboration of equipment, except for colorimeters, which became more versatile as time went on. The first purely physical method was spectrographic, used for trace elements by arc and spark techniques and for major exchangeable elements by flame emission using injection of an atomized mist of a soil extract (Lundegardh method). Originally this involved three steps: production of a photographic plate in the spectrograph, development of the plate, and determination of the density of the appropriate line by a microphotometer. Soon instruments known as flame photometers were designed which operated directly without a photographic plate. These have to some extent been superseded by atomic absorption instruments, by which very low concentrations of elements in solution can be determined.

The second physical method for quantitative determination of individual elements utilizes X-ray fluorescence as applied to solid materials. The exciting beam can be focused on a restricted area of the solid being examined. Thus some aspects of soil heterogeneity can be investigated.

Much finer focusing on micro features is achieved through the electron microprobe, in which a very narrow beam of electrons serves as the exciting agent. Different elements are recognized and quantitatively determined through the energy levels and intensities of the emitted radiation. Thus features made evident by the microscope or the scanning electron microscope can be isolated and their elementary composition determined. We are therefore in a strong position to investigate soil heterogeneity in detail with regard to the mineral elements. However these methods apply to soil organic matter only in very restricted fashion, because it is composed of elements of the lowest atomic numbers: hydrogen, carbon, nitrogen, and oxygen.

The generation of radioactive isotopes by exposure to a neutron beam provides the opportunity to employ very sensitive physical methods of detection. Activation analysis is already widely used in the determination of trace elements, both in soils and in plants.

The strong absorption of high energy neutrons by water to give low energy neutrons has provided a means of determining the moisture content of soils in the field by the neutron probe. It is a remarkable development from reactor physics, brought into the service of pedology more than 20 years ago.

Many applications of the interactions of radiation and matter are available in the low energy region of the electromagnetic spectrum. Absorption spectroscopy in the infrared region has proved to be useful in the characterization of organic matter and also of various minerals, including clays. Indeed it provides a very sensitive discriminating technique when applied to the different smectite clays. This is because the exciting units are broadly structural silicate anions, hydroxyls, and water molecules.

Then there are developments of Mössbauer spectroscopy, which enable quantitative distinctions to be made in the state of iron atoms with respect to valence and coordination. These techniques have already been used for clays, and it is natural to look for their extension in the study of soil profiles where iron is of pedological importance.

Problems involving protons and water molecules in clay systems have been investigated using nuclear magnetic resonance. Electron spin resonance has been employed to characterize the situation and behavior of certain exchangeable cations such as divalent copper and trivalent iron. Thus purely physical methods are already well established.

During the earlier years of these changes, a considerable concentration of work on the clay minerals resulted in three parallel developments involving fairly sophisticated instrumentation. In X-ray diffraction by the powder method, the camera gave way to the diffractometer as a means of detection. Differential thermal analysis established itself as a technique of special value for problems of tropical soils. Third, the transmission electron microscope, and later the scanning electron microscope, proved themselves in establishing the shapes and manner of association of clay particles.

On the organic side, various forms of chromatography with more and more sophisticated methods of detection have come into use, particularly for the determination of molecular fragments from humic matter.

The modern computer can be looked on as a piece of equipment for handling data provided in diverse ways. For instance, problems in soil physics such as the movement of water and solutes in idealized and in actual soil columns can be numerically solved. Thus models can be set up and tested with much less labor than formerly. Where an identifiable group of factors operates on a soil column (or profile) the computer can carry out operations equivalent to the solution of complex simultaneous equations. This feature offers bright hope for future advances in the science of pedology.

Thus soil scientists are already using a broad range of modern methods. In many instances they have improved techniques, to make them sufficiently accurate for determinations in the complex mixtures provided by soil systems. As earlier chapters indicate, this has been notably the case with the clay minerals. We now have a broad correspondence between clays formed pedogenically and the climatic grouping of soils. The influence of parent materials is also broadly evident in the clay minerals. Topographic and vegetational factors are similarly being evaluated. Finally, the influence of time or duration will eventually find expression as distinctive rate processes come under investigation.

We are at an early stage in the inevitable progression of pedology from a qualitative to a quantitative science. One can already see how this change is accelerating through the absorption and improvement of new methods taken from the basic sciences and through the optimism and devotion of the pedologists themselves. If this book has any human theme, surely it is that they have something to be optimistic about.

# Author Index

Note: Citations from the references are denoted by **boldface** page numbers.

Aarnio, B., 88, 90, **91**
Ahmad, H., **218**
Alexander, J. D., **221**
Alexander, L. T., **220**, **244**
Anderson, J. W., 199, **221**
Anderson, M. S., 236, **244**

Bailey, G. W., 199, **221**
Bailey, H. H., **277**
Baker, J. C., 152, **168**
Baldar, N. A., **42**
Baldar, N. D., **218**
Ballagh, T. M., 166, **167**, 212, **218**
Baren, F. A. van, 19, 21, **43**, 59, 90, **92**, **220**
Barnhisel, R. I., **277**
Barshad, I., 38, **42**, 114, 121, 122, 126, 127, 129, 130, 131, 132, 133, 138, 139, 140, 144, **146**, 195, 214, 217, **218**
Barrer, R. M., 22, 45, 47, 50, 51, **54**, **55**
Baver, L. D., 59, 73, **91**, **92**, 182, **183**
Baynham, J. W., **54**
Bear, F. E., 126, **146**
Beavers, A. H., **92**, 210, **218**, **221**
Beckett, P. H. T., 272, **276**
Beckmann, G. G., 166, **167**
Beinroth, F. H., 192, **218**, 256, **276**
Bengtsoon, B., **244**, **276**
Bemmelen, J. M. van, 5, 223, **244**, 247
Bétrémieux, R., 85, **91**
Bidwell, O. W., 275, **276**
Blanck, E., 17, 18, 19, **42**, 222, **245**

Blevins, R. L., **277**
Bloomfield, C., 85, **91**, 160, **168**
Bodman, G. B., 68, 72, **91**, 142, **146**
Bolyshev, M. V., 203, **218**
Borchardt, G. A., **218**
Borges, P. A., **276**
Bowman, J., 209, **218**
Bourne, W. C., 133, 134, 140, **146**
Bowler, J. M., 166, **168**
Braak, C., 59, **91**
Brashar, B. R., 141, **148**
Brees, D. R., 74, **91**, 152, **168**
Brewer, R., 2, 3, 4, **10**, 115, 122, 123, 126, 132, 135, 139, 140, 143, **146**, 154, 158, **168**, 171, **183**
Broadbent, F. E., **245**
Brown, G., **218**
Brown, T. H., 15, **42**
Browning, D. R., **92**, 182, **184**
Brydon, J. E., 208, 210, **218**, **219**, 250, **276**
Buchanan, F., 256
Buckingham, E., 3, **10**, 12, **42**
Buol, S. W., 140, 141, **146**, **147**, **148**, **168**, 175, **183**, 205, **220**, **276**
Burst, J. E. Jr., 210, **218**
Byers, H. G., 236, **244**

Cady, J. G., **219**, **220**
Calhoun, F. G., 215, **219**
Carn, D. B., 133, 134, **146**
Carlisle, V. W., 219, **246**

289

## Author Index

Carroll, D., 118, **146**, 210, **219**
Childs, E. C., 74, **91**
Chowdhury, M. Y., **147, 148, 277**
Christ, C. L., 14, 21, 22, 25, 28, 32, 34, 37, **42,** 63, 77, 80, **91,** 100, 102, 106, 111, 114, 187, 188, 204, 206, 263, 264, 271, 272, 281
Cipra, J. E., 275, **276**
Clebsch, E. C., 209, **220**
Cline, M. G., 140, **147**
Cole, C. V., 7, **10**
Coleman, E. A., 68, 72, **91**
Coleman, N. T., 82, 83, **91, 283**
Collins, J. F., **276**
Correns, C. W., **42**
Crawford, D. V., **92**

Darwin, C., 163
Deb, C., 88, 90, **91**
Denny, P. J., 50, 51, **54**
Deryagin, B., 66, **91**
DeVries, R. C., 50, **55**
Dokuchaev, V. V., 1, 2, **10,** 204, 222, **245**
Donat, J., 72, **91**
Donnan, F. G., 7, 8, 15, 44, 61, 79, 86, 99, 111, 269, 271, 281
Dravid, R. K., 59, **93**
Drew, J. V., **245**
Drosdoff, M., 123, **148**
Druif, J. H., 118, **146**
Dryden, C., 119, **146**
Dryden, L., 119, **146**

Eberl, D., 54, **55**
Eichenberger, E., **168**
El-Nahal, M. A., **219**
Engelhardt, W. von, **42**
Evans, D. D., **146**

Fehrenbacher, J. B., **221**
Feitnecht, W., 45, **55,** 80, 275
Fernandez, C. E., 273, **276**
Fischer, W. R., 193, **219**
Fisher, R. A., 115
Fitzpatrick, E. A., 140, **146,** 171, **183**
Flach, K. W., 141, **148**
Follett-Smith, R. R., 191, **219**
Foster, Z. C., **220**
Franzmeier, D. P., 133, **147,** 208, **219**
Frei, E., 140, **147**

Fujimoto, C. K., **220**

Galen, E., 54, 55
Gard, J. A., **219**
Gardner, W. H., **91, 183**
Gardner, W. R., **91, 183**
Garrels, R. M., 13, 21, 22, 25, 28, 32, 34, 37, **42,** 63, 77, 80, **91,** 100, 102, 106, 111, 114, 187, 188, 204, 206, 263, 264, 271, 272, 281
Gedroiz, K .K., 5, 90, **91, 245,** 247, 248, **276,** 278
Gerasimov, I. P., 1, **10, 147,** 165, **168**
Gersper, P. L., 160, **168**
Gibbs, W., 65
Gier, D. A., 275, **276**
Gieseking, J. E., **245**
Gjems, O., 209, **219**
Glazovskaya, M. A., 1, **10, 147**
Glinka, K., 7, **10**
Godfrey, C. L., 243, **245**
Goh, K. M., **168**
Gold, H. A., **218**
Gore, A. J. P., 275, **277**
Gouy, G., 111, 269, 281
Graham, E. R., 8, **10,** 38, 74, **91**
Grim, R. E., 201, **218**
Grossman, R. B., **183**
Gustavoon, G., **245**

Hagin, J., **218**
Haines, W. B., 61, 62, **92**
Haldane, A. D., 143
Halevy, E., **218**
Hall, G. F., **147**
Hallsworth, E. G., **10,** 85, **92,** 143, **147,** 170, **183**
Hardy, F., 191, **219**
Harradine, E. F., 142, **146**
Harrasowitz, H., 226, 265
Harrison, J. B., 12, 15, 16, 18, 20, 31, 32, 41, **42,** 76, **92,** 114, 122, 123, 136, **147,** 156, **168,** 187, 188, 189, 190, 191, 192, 215, **219,** 222, **245**
Haseman, J. F., 114, 118, 119, 120, 121, 123, 126, 133, 138, 144, **147,** 177, **183**
Hasler, A., **168**
Hathaway, J. C., 210, **219**
Hauser, E. A., 174
Helgeson, H. C., 15, 33, **42,** 65, **92,**

150, **168**, 264
Hem, J. D., **54, 55**, 76, **92**
Hendricks, S. B., 51, **56**
Hénin, S., 5, **55**, 85, **91**
Hernandez, M. J., **276**
Herrera, R., 166, **168**
Hervel, R. C., **219**
Hess, R. E., **276**
Hissink, D. J., 5, 247, 278
Hoeks, J., **218**
Hofmann, A., 50, **56**
Hole, F. D., 146, **168**, 183, **218**
Holmes, R. S., **245**
Holowaychuk, N., 160, **168**
Horn, F. E., 152, **168**, 178, **183**
Hower, J., 54, **55**
Hubble, G. D., 166, **167**
Huertas, F., 54, **55**
Humbert, R. P., 29, **42, 92,** 118, 119, 132, 135, 138, 139, **147**, 157, **219**

Ikawa, H., **218**
Iler, R. K., 139, **147**

Jackson, M. L., 21, 38, 39, 40, **42, 43,** 54, **56,** 140, **147,** 178, **183, 218, 221,** 230, **245**
Jarusov, S. S., 261
Jeffries, C. D., 114, 115, **147,** 216, **219**
Jenne, E. A., 76, 82, **92**
Jenny, H., 2, **11,** 163, 165, **168,** 224, 226, 227, 235, **245,** 250, 255, **276,** 281
Joffe, J. S., 84, **92,** 95, **147, 245**
Johns, W. D., 201, **218**
Jones, H. T., 82, 83, **92**
Jones, R. L., **92, 218**

Karlsson, N., **244, 276**
Karsulin, M., 52, **55**
Kawaguchi, K., 83, **92**
Keen, B. A., 59, **92**
Kelley, W. P., 90, **92,** 248, **276**
Khalifa, E. M., 141, **147,** 175, **183**
Khangarot, D. S., **147**
Kirkham, D., 74, **92**
Kittrick, J. A., 33, 37, **43,** 52, 53, 54, **55,** 186, 187, 188, 189, 216, **219,** 282
Kodama, H., 83, **93,** 208, **218, 219**
Koizumi, M., 52, **55**
Konanova, M. M., 232, **245**

Kossovich, 114
Köster, H. M., 114, 120, 122, 126, **147,** 193, **219**
Krinbill, C. A., **11**
Kubiena, W. L., 2, 6, **11, 147,** 170, **183**
Kussakov, M., 66, **91**

La Iglesia, A., 52, **55**
Langmuir, D., 64
Leary, M. L., 165
Leather, J. W., 59, **92**
LeBeau, D., 174
Leeper, R. H., 15, **42**
Linares, J., 54, **55**
Lind, C. J., 54, **55**
Lindsay, W. L., 216, **220**
Lay, T., **93**
Luna, Z. C., **219**
Lundegardh, H., 285
Lutz, J. F., 173, **183**
Lynn, W. C., **183**

McCaleb, S. B., 196, **220**
McClelland, J. E., **245**
McCracken, R. J., 168, **183,** 209, **220, 277**
McDowell, L. L., 25, 43, 102, **147,** 166, **168**
Mackenzie, R. C., **219**
McLung, A. C., 82, **91**
Marbut, C. F., 6, 7, 8, **11,** 222, **245**
Marshall, C. E., **11,** 25, 29, 42, **43,** 46, 55, **92,** 102, 114, 115, 118, 119, 120, 121, 123, 126, 132, 133, 135, 138, 139, 144, **147,** 148, 157, 174, 177, **183, 184,** 210, 214, **218, 219, 220,** 229, **245,** 250, **276, 277**
Marshall, T. J., **183**
Martell, 82
Martin-Vivaldi, J. L., **55**
Martinez, E., **93**
Matelski, R. P., 134, **147**
Matsuo, Y., 83, **92**
Matthews, B. C., 144, **148**
Mattson, S., 5, **11,** 86, 87, 88, 89, 90, **92,** 223, **244, 245,** 258, 259, **276, 277,** 279
Maxwell, C., 181
Merrill, G. W., **43,** 114, 123, **148,** 222, **245**
Meyer, A., 151
Mick, A. H., 144, **148**
Mickelson, G. A., 119, 133, **148**

## Author Index

Miller, B. L., 210, 212, **220**
Milne, G., 164, **168**, 190, 191, **220**
Misra, U. K., 220, **277**
Mohr, E. C. J., 19, 21, **43**, 59, 64, 90, **92**, **220**
Moore, D. P., 82, **91**
Morgan, J. J., 213, **221**
Mortland, M. M., 133, **147**, 203, 208, **219**, **221**
Mössbauer, 286
Muckenhirn, R. J., 256
Muir, A., **220**
Mumpton, F. A., 51, **55**
Murdock, L. W., 261, **277**

Nafady, M. H. M., **276**
Nash, V. E., 25, 42, **43**, 46, **55**
Nelson, B. W., 50, **55**
Nelson, W. L., 73, **92**
Nettleton, W. D., 141, **148**
Niggli, P., 224, **245**
Nikiforoff, C. C., 123, **148**
Noll, W., 22, 33, **43**, 49, 52, **55**, 204
Norrish, K., 206, **220**

Odell, R. T., 201, **218**
Odén, S., 90, **92**, 280
Oertel, A. C., 126, **148**
Osborn, E. F., 22, 50, **56**
Ostwald, W., 38, 44
Overbeck, J. Th. G., 279

Pallmann, H., 6, **11**, 140, **148**, 150, **168**
Patnaik, N., 229, **245**
Penck, W., 151
Pettijohn, F. J., 38
Phillippe, W. R., 275, **277**
Plas, L. Van der, **218**
Pohlman, G. G., **93**, 182, **184**
Poisson, 115
Polach, H. A., 166, **168**
Ponnamperuma, F. N., 77, 79, **93**
Pourbaix, 77, 273
Powers, W. L., 74, **92**
Protz, R., 213, **220**

Raeside, J. D., 118, 119, **148**
Rafter, T. A., **168**
Rai, D., 216, **220**
Ramdas, L. A., 59, **93**

Raz, B. W., **221**
Read, A. T., 213, **220**
Redmond, C. E., 133, 140, **148**, 203, **220**
Reesman, A. L., 155
Ricards, R. P., **245**
Rich, C. J., 261, **277**
Richards, L. A., 72, **93**
Riecken, F. F., 243, **245**
Robichet, O., **55**
Robinson, G. W., 223, **245**
Robinson, W. O., **245**
Roda, A. A., 114
Rodriguez, G., 191, **219**
Romo, L. A., 51, 52, **55**
Ross, C. S., 51, **56**
Ross, G. J., **218**
Roth, J. R., 261, **277**
Roy, D. M., 45, 50, 51, **56**
Roy, R., 22, 45, 50, 51, 52, **55**, **56**
Runge, E. C. A., 166, **167**, **168**, 212, **218**, **244**, **246**
Rutherford, G. K., **148**, **183**

Sand, L. B., 50, **56**
Sandberg, G., 258, **277**
Satyanavayana, K. V. S., 193, **220**
Scharpenseel, H. W., 165, **168**
Scheurenburg, B. Van, **218**
Schnitzer, M., 82, 83, **93**, 187, **220**, 283
Schofield, R. K., 7, **11**, **43**, 272
Schroeder, D., 275, **277**
Schuylenborgh, J. van, 84, **92**, **93**, **220**
Schwertmann, U., 193, **219**, 275, **277**
Scrivner, C. L., 152, **168**, 210, 212, **220**
Shainberg, I., 58, **93**
Shanks, R. E., 209, **220**
Sherman, G. D., 39, **42**, **220**, **221**
Shimoda, S., 208, **218**
Siffert, B., 51, **56**
Sigmond, A. A. J. von, 5, 248
Sillen, 82
Simonson, R. W., 144, **148**, 195, **220**
Singer, A., 206, **220**
Sivajasingham, S., 191, **220**
Skinner, S. I. M., 82, **93**, 187, **220**
Smeck, N. E., **244**, **246**
Smith, A., 59, **93**
Smith, B. R., **148**
Smith, R. B., 205, **220**
Smith, R. M., 73, **93**, 182, **184**

Smithson, F., 38, **93**
Springer, M. E., 212
St. Arnand, R. J., 203, **221**
Stelly, M., **246**
Stephens, J. C., **168**
Stevens, J. H., 209, **221**
Stewart, E. H., **168**
Streng, 122
Strese, H., 50, **56**
Stubican, V., 52, **55**
Stumm, W., 213, **221**

Tamers, M. A., 166, **168**
Tamm, O., 223
Tamura, R., 54, 56, **221**
Taylor, A. M., 7, **11**, 43
Terning, P. E., 258, **277**
Thomas, P. K., 193, **220**
Thornthwaite, C. W., 151
Tianes, E. M., **93**
Troughton, A., 159, **169**
Turk, L. M., 134, **147**
Tuttle, O. F., 50, 52, **56**
Tyurin, I. V., **93**, 231, **246**

Upchurch, W. J., 95, 142, **148**, **277**
Uhara, G., **218**
Urquhart, C., 275, **277**

Verwey, E. J. W., 279
Vil'yams, (Williams), W. R., 6

Walker, T. W., 243, **246**
Warshaw, C. M., 49, 52, **56**
Wascher, H. L., **221**, 256, **277**
Way, J. T., 247
Weaver, C. E., 209, **221**
Westin, F. C., 145, **148**
Wey, R., 51, **56**
White, E. A. D., 50, **55**
White, J. L., 199, 216, **219**, **221**
Whiteside, E. P., 4, **11**, 120, 133, 134, 140, **146**, **147**, **148**, 155, **169**, 174, 177, **184**, 203, 208, **219**, **220**, **221**
Whittig, L. D., **42**, **218**, **219**
Wiegner, G., 5, **11**, 247, **277**, **278**
Wild, A., 126, **148**
Wilding, L. P., **147**
Willcox, J. S., 82, 83, **92**
Williams, J. D. H., 243, **246**
Williams, (Vil'yams), W. R., 6
Wilson, M. J., 209, **221**
Winters, E., 275, **277**
Wollny, E., 58

Yaalon, D. H., **168**
Yakuwa, R., **93**
Yamane, I., **277**
Yassoglo, N. J., 120, 134, **148**, 208, **221**
Yesiloy, M. S., **146**
Yoder, H. S., 46, 50, **56**

Zelazny, L. W., **246**

# Subject Index

Abrasion pH, 14
Absorption spectroscopy, 286
Accumulation, zone of, 242
Acidic soils, 249
Acid igneous rocks, 190
   silica from, 76
Acid-soluble constituents, 222
Activation analysis, 286
Activation energy, 44, 45, 150
Activation energy relationships, 52
Activity, chemical, 22
Activity coefficient, cationic, 252
Activity measurements, ionic, 279
Adelanto silt loam, 122
Adsorption, 86
   hysteresis in, 181
   of humus, 280
   of water in soils, 66
Adsorptive model, of soil water, 67
Aeration, improvement after frost, 59
Agathis australis, 160
Age, radiocarbon, 165
Aggregate analysis, 178
Aggregation, 179
   by frost action, 59
A horizons, of Podzols, 85, 89
Aikin clay loam, 142, 143
Alaska, 215
Albaqualf, Aeric, Udollic, 96
Albite, 14, 25, 26, 29, 30, 35, 36, 48, 121, 217
   as index mineral, 133
   in weathering sequence, 39

   surface weathering, 46
Albite-analcine reaction, 34
Albite-gibbsite reaction, 32
Albite-sodium beidellite reaction, 28
Alfisol, 153, 172, 197, 198, 200, 213, 249, 252, 261, 285
Alfisols, $^{14}$C age, in relation to depth, 166
   cationic depth functions, 256
Alkali soil, 33, 34, 41, 216
   alkalis in, 250
   reclamation, 248
Alkaline earths, removal, 190
   in soil water, 189
Alkalis, removal, 190
   in soil water, 189
Allophane, 26, 191, 194, 208, 215, 216, 223, 284
   from volcanic ash, 21
   by weathering, 42, 54
   in weathering sequence, 39
Allophane-like material, 208
Allophane-organic matter association, 215
Alluvium, 198
   fine, 207
   granitic, 206
Alps, Switzerland, 209
Alumina, hydrated silicate of, 190
   hydrous, in synthesis, 52
Aluminate ion, 282
Aluminous chlorite, 26, 31, 37
   by weathering, 42
Aluminous goethite, 193
Aluminosilicate, composition, 25, 39

## Subject Index

Aluminum, accumulation, 207
  aluminosilicate, 223
  in Barshad's method, 129
  complexed, 187, 191, 207, 213, 249, 282
    by humus, 186
    by tree drippings, 84
  complexes, soluble, 82
  constant, 226
  in double hydroxides, 45
  exchangeable, 249
  extractable, 223
  hydrated silicates of, 16
  in hydrothermal synthesis, 51
  hydroxylated, 262
  ionized, activity, 207
  movement, 30
  oxides and hydroxides, 191, 192
  for silicon, proxying, 25
  in soil extracts, 103, 105, 106, 271
  soluble, from aluminosilicates, 44
    depth function, 272
  tetrahedral, 47
  translocated, 21
Aluminum chelates, decomposition, 84
Aluminum hydroxide, amorphous, 53
  charge, 88
  from silicate minerals, 89
  in laterization, 89
  interlayers, 199
  ion product, 24
  sols, 87
  solubility, 45
Aluminum hydroxides, 249
Aluminum ions, 23, 282
  hydroxylated, 249
  in mineral equilibria, 66
Ammonium, exchangeable, 261, 262, 279
Amorphous clays, 185
Amorphous material, 5
  Podzol B horizon, 208
Amorphous silica, 156, 186, 187, 206
Amphibole group, weathering, 118, 119
Amphiboles, in fine clay, 139
Amphibolite, 156
Amplitude, of diurnal temperature cycles, 58
  of moisture cycles, 61, 151
  of seasonal temperature cycles, 58
Analcime, 15, 26, 34, 35, 204, 205, 206
  by weathering, 41
  compositional diagram, 51
  synthesis, 34, 49, 51
Analcime-beidellite reaction, 33
Analysis, total, 4
Anaseuli, U.S.S.R., 230
Anatase, 116, 117, 194
  resistant mineral, 119
  weathering sequence, 39
Andesine, 26
  composition, 29
Andesite, 19
Andosols, 215
Angola, 240
Animals, effects on soils, 163
Anion exchange, in Oxisols, 256
Anisotropy, of soils, 2
Anorthite, 8, 14, 25, 26, 31, 32, 47, 48
  composition, 29
  surface weathering, 46
  in weathering sqeuence, 39
Anorthite-calcium beidellite reaction, 29
Anorthite-gibbsite reaction, 31
Antigorite, in weathering sequence, 39
  hydrothermal synthesis, 50
Ants, mounds in soil, 163
Apatite, 216, 243, 244
  weathering, 118
  in weathering sequence, 39
Appling soil, 195
Aragonite, in weathering sequence, 39
Argillaccons granite sand, 20
Argillic horizons, 141, 192
Aridisol, Ca distribution, 241
Arid regions, 204
Aripo fine sand, 195, 198, 199
Arizona, 141
Arkhangel, U.S.S.R., 230
Ashe stony loam, 118, 135, 157
Aspect, soil-forming factor, 164
Atmospheric weathering, 190, 215
Atomic absorption spectrometer, 286
Attapulgite, 26, 205, 214
  in desert soils, 285
Attapulgite-sepiolite minerals, 206
  in alkali soils, 76
Augite, weathering, 187
Australia, 124, 166
Authigenic fedlspars, 47
Available water, 179

Barnes soil, 133, 222, 239
　analysis, 145
　change in clay, 137, 138
　clay coatings, 140
　formation, 125
Basalt, 193, 195, 243
　clay formation from, 131
　weathering, 18
Basaltic hornblende, 116, 117, 119
Base saturation, in forest soils, 162
　in prairie soils, 162
Basic igneous rocks, 192, 196
　ground water from, 120
　silica from, 32, 76
　weathering, 15, 41
Basket Podzol, 160
Bauxite, ferruginous, 189
　iron minerals in, 192
　in lateritic earth, 188
Beckett soil, 226
Beech, 84, 160, 161
Beidellite, 27, 31, 32, 36, 37, 53, 158, 203, 209, 248, 255, 261, 262, 264, 266, 267, 268, 269, 270
　calcium, 34, 264
　in claypan soils, 177
　from diabase, 42
　from feldspar, 130
　hydrogen, 29, 36
　kaolinite from, 177
　magnesium, 264
　in Mexico profile, 96
　potassium, 28, 264
　sodium, 14, 15, 34, 35, 51, 204, 264
　by synthesis, 52
　in weathering sequence, 39
Beidellite-kaolinite boundary, 264, 267, 271
Beidellite-kaolinite reaction, 31
Beidellite-nontronite, 158, 205, 217
Beidellitic clay, 174
Belle Fourche montmorillonite, 52
Bennet sand, 67
Benzene, adsorption hysteresis, 62
Bethel soil, Ohio, 125, 131, 133
B horizon, 89, 143
Bicarbonate, equilibria, 15
Bicarbonates, for neutralization, 270

Biohydrologic factor, 160
Biomass, root, 159
Biotic factor, 2, 3, 6, 240
Biotite, particle size distribution, 139
　weathering, 135
Birefringent clay skins, 192
Black-alkali soils, 90
Black clays, subtropical, South Africa, 131
Black earth, 21
Black limestone soils, 214
Bleicherde, 18
Bluegrass, composition, 255
Blue Lake, soil, 208
Boehmite, 53
　by synthesis, 49, 50
　in weathering sequence, 39
Bog soil, 224
Bonding energy, calcium, 252
　cationic, 108, 149, 250
　difference, 252
　　depth function, 254
　exchange cations, 271
　humus, 249
　magnesium, 252
　potassium, 153, 210, 250
　sodium, 153
Bonding energy curves, 250, 279
Boone fine sandy loam, heavy minerals, 115, 116
Boron, in tourmaline, 122
Bozeman soil, 125, 133
　change in clay, 137, 138
　oriented clay, 140
Brittle micas, 30
Brown Earths, $^{14}$C age, in relation to depth, 166
　catenas of, 223, 259, 260
　clay formation, 131
Brown Forest soils, 84, 161, 197, 203, 209, 238
　under beech, 161
Brown Podzolic soils, clay movement, 140
Brunigra, 200
Brunizen, 161, 200, 202
Buffer, carbonate, 15
　equation, 14
　silicate, 15
Buffering, 14
Buffering capacity, 273
　potential, 272

## Subject Index

Buffer systems, 249
Bukhara, U.S.S.R., 235

Caddo soil, 176
Calcareous black soils, 214
Calcareous soils, carbonates in, 213
Calcareous till, changes in clay content, 121
Calic feldpar, 31
Calcisol, 206
Calcite, 7, 242
  concretions, 75
  index, 213, 214
  in Indian laterite, 193
  in weathering sequence, 39
  weathering, 213
Calcium, 250, 271
  additions to soil, 270
  in bog soil, 260
  bonding energy, 252
  cycling of, 242, 255, 258
  distribution with depth, 239
  enrichment in termite mounds, 163
  exchangeable, 241, 242, 249, 250, 252, 259, 260, 262, 263
    depth function, 259
    mineral ratio, 242
    in profiles, 239
  in leaf fall, 161
  loss by weathering, 252
  losses in solution, 129
  mobility in soil, 290
  movement in Mexico soil, 108, 109, 110
  in profiles, 243
  in smectites, 41
  silicate and aluminosilicate, 239, 241
  in soil extracts, 103, 104, 264
    depth function, 107, 108
  in soil solution, 186
  soluble, 241
  in Ultisols, 256
  uptake by grasses, 255
  by weathering, 240
  in weathering, 17
  in zeolites, 186
Calcium activity, 270

Calcium beidellite, 34, 264
Calcium bicarbonate, 75, 239
Calcium carbonate, 34, 205, 241, 264
  in coarse clay, 140
  equilibria of, 14
  formation, 15
  hydrolysis constant, 7
  in profiles, 239
Calcium carbonate-calcium sulphate ratio, 242
Calcium clay, zeta potential, 173
Calcium feldspars, 199
Calcium hydroxide, 271
Calcium ion, activity, 7
Calcium montmorillonite, 16
Calcium nitrate, in profiles, 239
Calcium phosphates, 239, 244
Calcium salts, soluble, 242
Calcium sulphate, 241
  in profiles, 204, 239
Calcium-magnesium, free energy differences, 255
Calcium-magnesium ratio, 256
  depth function, 257, 258, 259
  of rain water, 259
California, 124, 132, 141, 142, 217
Canopy effects, 158, 160
Capacity factors, 38
Capillarity, upward movement by, 153
  in water movement, 66, 67
Capillary conductivity, 72
Capillary pores, 158
Carbohydrates, 159, 228
Carbon, humus, 229
  soil, 165, 228
    as pedogenic entity, 165
  total, 233, 234, 235, 236, 237, 238, 239, 240
Carbon dioxide, 7, 12, 102, 190
Carbon 14, 165, 166, 212
Carbon-nitrogen ratio, 229, 231, 232, 233, 235, 236, 237, 238, 239, 241
Carbonate-bicarbonate buffer, 20

Carbonates, 170, 213, 244, 270
  alkaline earth, 12
  buffering action, 210
  depth function, 204
  in desert soils, 207
  dissolution, 209
  equilibria, 15
  pedogenic, 165, 166, 167
  secondary, 12
  by weathering, 19
Carbonic acid-carbonate equilibrium, 15, 186
Caribou soil, 226
Catalyst, water as, 45
Cataphoresis, of suspended particles, 174
Catena, 163, 223, 224, 259
Cation activity, 216, 250
Cation equilibria, 281
Cation exchange, of surface layers, 13
  pH dependent, 248
  reactions, 223
Cation exchange capacity, 255, 280
Cationic bonding energy, 149, 248, 250, 251
Cationic depth functions, 253, 257, 258
Cations, loss by weathering, 46
Cecil soil, 129, 141, 175, 195, 196, 222, 227, 256, 258
Cementation, degree of, 176
Chalcedony, 29, 41, 131, 159, 217
  by weathering, 42, 76, 120, 133, 135
Chalk, 210
Chamosite, by hydrothermal synthesis, 50
Charcoal, soil, 165
Charge, negative, 13
Chelation, 4, 83, 88
Chemical analysis, 149
Chemical environment, in Podzols, 85
  for rock breakdown, 12, 17
  of soil water, 57, 58
Chemical potential, of hydroxyl, 45
Chernozem, 37, 88, 133, 140, 200, 203, 228, 229, 230, 231, 232, 234, 280
  aggregation, 178, 179
  analysis, 145
  $^{14}C$ age in relation to depth, 166
  change in clay, 137, 138
  C/N ratio, 236
  fulvic acid, 231
  humic acid, 231
  movement of humus in, 90
  soil formation, 125, 131, 134, 227
Chert, 101, 209, 213
Chester soil, 129
Chestnut soils, 200, 230, 231
Chino silty clay loam, 72, 73
Chlorite, 194, 203, 204, 207, 208, 209, 210, 213, 217, 249
  aluminous, 26, 31, 37, 42
  interstratified, 203, 204
  magnesium, in weathering sequence, 39
Chlorites, compositional region, 46
  synthesis, 45
  in Ultisols, 256
  in weathering sequence, 39
Chromium, 284
  in montmorillonites, 45
Chromatography, 75, 287
Chromosequence, 208
Cinnamon-Brown soil, 207, 233
Cisne (Cowden) silt loam, 177, 202, 203
Clarence soil, 201, 202
Clay, colloidal, 5
  composition, 47, 223
  depth function of, 131, 135, 136, 157, 257, 258, 259
  disorientation, 171
  eluviation of, 140, 172
  fine, in weathering sequence, 40
  flocculated, 143
  illuviated, 140, 171, 172
  orientation, 143
  in pedological processes, 86, 130, 137
  transformation, 130
Clay minerals, 9
Clay movement, 140, 141, 143
Claypan soils, 141, 174, 177, 197, 212, 244
Clay skins, 141, 192, 195
Clay synthesis, 176
Climate, atmospheric, 19
  soil, 19
  as soil-forming factor, 1, 2
Climatic cycles, 188
Climatic factors, of soil formation and development, 59
Coachella soil, 73
Coagulation, 75, 86, 87, 247, 280
Coal fragments, in soils, 165
Coarse clay, 177, 211, 255
Coatings, 284

Cobalt, 5, 50, 284
Colloidal constituents, of soils, 5, 223, 226, 248
Colombia, 215
Colorimeters, 285
Columbiana soil, 227
Columnar structure, 90, 248
Complexing, 54, 81, 82, 160, 207
Computer, 287
Concretions, 4, 75, 193, 244, 284
    composition, 81, 82, 195, 199, 243
    formation, 77, 79, 273, 275
Conductors, capillary, 158
Congruency, compositional, 44
Coniferous trees, composition, 255
Constant volume, in rock weathering, 120, 126
Convection, of soil water, 66
Conversion factor, 299
Copper, divalent, 287
Corundum, in weathering sequence, 39, 284
Crystallinity, 64, 278
Crystallization, 171
Crystallographic form, 58
Cristobalite, 39, 139
Crogham soil, 175
Crust, surficial, 182, 190
Cryoboralf, 172
Cutan, 3, 140
Cycles, external, 151, 188
Cyclic changes, in soils, 58, 61, 143, 151
Cyclic process, in concretion formation, 81
Cyclic salts, 125
Cycling, of cations, 203, 248, 252

Dark-gray forest soil, 230, 231
Davidson soil, 173, 195, 256, 259
Debye-Hückel theory, 57
Deciduous forest, soil properties, 162
Deciduous trees, composition, 84
Degradation, of soils, 6, 146
Degrees of freedom, aluminum ion, 66
Degree of saturation, 247, 248
Depolymerization, 44, 94, 280, 282
Depositional factors, 111, 154, 164
Depth function, of clay formation, 131
Desert soils, 207, 231, 234, 236, 239, 240, 285
Desiccation, in formation of salonitz, 145
Desorption isotherm, for K, 272

Development, soil, 6, 9
Dew, in relation to rainfall, 59
Diabase, 190
    soil from, 119, 124, 132, 135, 136, 137, 138, 139, 157, 158
Diagenesis, 8, 9, 210
Dialysis, 189, 190, 191
Diaspore, 53
Dickite, 52
Differential dispersion, 173, 177
Differential solubility, 284
Differential thermal analysis (D.T.A.), 5, 185, 284, 287
Diffractometer, 287
Diffuse double layer, 279, 281
Diffusion, of gases, 3, 65, 273
    of solutes, 3, 13, 65, 74
Diopside, in weathering sequence, 39
Diorite gneiss, 259
Disorientation, of clay, 171
Dispersibility, of clay, 173, 177, 248
Dissociation, of exchange cations, 271
Dissociation constant, of water, 7, 65
Diurnal temperature changes, 58
Dolerite, weathering, 16, 39, 131, 187, 190, 210, 212, 213, 216
Dolomite index, 213, 214
Donnan effects, 7, 8, 15, 61, 79, 86, 94, 99, 102, 111, 271, 281
Double hydroxides, 45
Drainage, 20, 95, 99, 100, 131, 141, 163, 164
    as soil-forming factor, 163, 164
Draught, effects on mineral nutrition, 281
Duhem-Margulio equation, 57
Duration, as soil-forming factor, 1, 2, 40
Dust, 94
Dutch East Indies, Indonesia, 21, 223

Eagle Mountain, Guyana, 187, 188, 189
Earthworms, activity in soils, 163
Ecosystems, 161
Ectodynamomorphic soils, 7
Efflorescence, of calcium carbonate, 7, 75
Electrical double layer, 5, 75, 86, 279
Electrochemical potential, $E_H$, 274, 275
Electrode, reversible, 58
Electroendosmosis, of soil water, 174, 279
Electrokinetic properties, 88, 111, 143, 150, 172, 279

Electromagnetic spectrum, 286
Electron microscope, 3, 5, 158, 171, 185, 217, 278, 285, 286, 287
Electron spin resonance, 286
Eluviation, clay, 172, 242
   of aluminum, 160
   of iron, 160
Endodynamomorphic soils, 7
England, 210, 226
Entropy, of water, 57
Environment, in soil development, 149
Epidiorite, weathering, 20, 156, 157
Epidote, 116, 117, 118, 119, 132, 134
Equilibrium, in soil extracts, 102
Equilibrium constant, 22, 39, 80, 102
Equivalent diameter, of pores, 72
Ethylenediamine tetracetic acid, 54
Europe, northern, beech forests of, 161
Eutrorthox, 192
Eutrustox, 192
Evapotranspiration, 61, 95, 99, 152, 153, 160, 179
Everglades, 166
Exchange, small, dilute, 149
Exchangeability, 249
Exchange acidity, 249
Exchange anions, 149
Exchange capacity, 5, 157, 158, 248, 249, 250
   Chernozems, 229
   humus, 249
   variation with depth, 96, 251, 252, 253, 257, 258
   Ultisols, 256
Exchange cations, 149, 248, 255, 260, 278
   activities, 270
   bonding energy, 249
   depth function, 96
   in drainage, 95
   loss by leaching, 89
Exchange complex, 5, 149, 150, 243, 248, 250, 275
Exchange ions, ratios of, 111
Exchange reactions, ionic, 247, 278
Extinction coefficient, 231
Exudates, from roots, 159

Fabric, soil, 171, 192
Factor, biotic, 158
Factors of soil formation, 1

Fagus grandiflora, soil under, 160
Fats, 228
Fayette soil, 199, 201
Feithknecht compounds, 80, 275
Feldspars, 37, 39, 156, 190, 204, 205, 209, 211, 216, 217
   alkali, 216
   authigenic, 47
   calcium, 199
   in Chernozem, 134
   in coarse clay, 140
   composition, 25, 46, 47, 49
   depth function, Ashe soil, 135
      diabase soil, 136
   in fine clay, 139
   in limestones, 212
   particle size distribution, 139
   plagioclase, 8
   in Podzolic soils, 134
   potash, 36, 41, 199, 208
   silica from, 131
   sodium, 199
   weathering, 25, 46, 54, 118, 130, 135, 150
   weathering stage, 41
Feldspar-gibbite reactions, 31
Feldspar-kaolinite reaction, 29
Feldspar-mica equilibrium, 29
Feldspathoid, 25, 37, 47
Fenner silt loam, 238
Ferriallitic Latosols, 195
Ferriallitic soil, 240
Ferric hydroxide, 13, 77, 78, 80, 81, 82, 83, 273, 274, 275
   amorphous, 192, 283
   charge, 88
   from mineral silicates, 89
   sols, 87
   solubility, 45
   solubility product, 77
Ferric ion, 77, 78
Ferric oxalate, 83
Ferric oxide, 77
   hydrous, 4
Ferric tartrate, 83
Ferrisols, 195
Ferromagnesian minerals, 25, 205
Ferrosoferric hydroxide, 77, 78, 80, 81, 283
Ferrous carbonate, 77, 80
Ferrous hydroxide, 78, 80

## Subject Index

Ferrous ion, 78, 80, 81, 273, 274
Ferrous iron, 17, 77, 189
Ferruginous bauxite, 189
Ferruginous B horizon, 85
*Festuca rubra*, roots, 159
Field capacity, 68, 72, 178, 281
Field observations, 149
Film, aqueous, 281
  liquid on solid, 66
Filter bed, 142, 143
Fine clay, 211, 212, 250, 255
Fine sand, 115, 121
Fixation, 248, 249, 281
  potassium, 203, 243
Flame emisssion spectrography, 285
Flame photometer, 285
Flocculation, 143, 171
Florida, 166, 237, 241
Forest, in clay formation, 131
  deciduous, soil properties, 162
  ecosystem, 161
Forest soils, roots in, 159
Fragipan, 120, 134, 153, 176, 208
Free bonding energy, cationic, depth function, 98
  differences, 261
Free energy, 22, 63, 83
  differences, 261
  of formation, of complexes, 83
  in weathering, 46
Free water table, 13
Freezing point depression, of clay paste, 58
  at wilting point, 72
Frequency, of moisture cycles, 61
Fulvic acid, 187, 228, 229, 232, 280, 282
  adsorption, 83
  aluminum, 54
  $^{14}C$ age, 165
  complexes, 82, 283
  composition, 230
  mobility, 166

Gallium, as immobile indicator, 126
Garnet, 116, 117, 119, 120, 134
Garrels-Christ diagram, 264, 266, 268
Geological differences, 154
Geological factors, 153
Georgia, 222, 259
Gibbsihumox, 192
Gibbsiorthox, 192

Gibbsite, 14, 15, 16, 20, 21, 23, 25, 26, 28, 31, 32, 33, 34, 35, 39, 53, 156, 186, 189, 190, 191, 192, 193, 194, 209, 216, 223, 248, 267, 268
  from anorthite, 46
  in clay skins, 141
  crystalline, 33
  determination of, 191
  from igneous rocks, 41, 42, 187
  interlayers, 31
  from mica, 63
  in Oxisols, 256, 257, 285
  recrystallization, 192
  redeposition, 189
  sensitivity to pH, 187
  stability, 54
  standard free energy in relation to particle size, 76
  in Ultisols, 256
Gibbsite-kaolinite reaction, 76, 155, 186, 189
Gilgai, 214
Glacial deposits, 199, 201
  feldspars in, 140
Glacial moraines, soil formed on, 165
Glacial outwash, 200
Glacial soils, 199
Glacial till, 145, 202, 203, 208, 214
Glaebule, 4
Glass, silicate, 215
Glauconite, in weathering sequence, 39
Glauconites, formation, 49
Glucose, 85
Gneiss, clay formation, 131
  weathering, 18
Goethite, 65, 77, 192, 193, 216, 283
  aluminous, 193, 264
  crystallinity, 191
  dehydration, 64
  in weathering sequence, 39
Goldstone soil, 226
Gouy theory, 281
Granite, 196
  allumvium, 206
  Indian, parent material of Laterite, 120
  laterite from, 193
  red soils from, 213
  soil formation from, 118, 119, 120, 124, 131, 132, 134, 135
  Ultisols from, 256, 258

## Subject Index 303

weathering, 18, 20, 240
Granodiorite, soil formation from, 124, 132, 135, 137
Granular structure, by root effects, 159
Grass, in clay formation, 131
Gravitational constant, 68
Gravity, in water movement, 66
Gray-Brown Podzolic soils, 125, 133, 161, 177, 237, 256
   clay formation, 131
   clay movement, 140
Gray-Brown desert soils, 207
Gray limestone soils, 214
Great soil groups, 6, 40, 131
Greenville loam, 67
Growth, of soils, 6
Grumosols, 210
Grundy silt loam, 117, 118, 121, 122, 126, 133, 137, 138, 144, 153
Gruzinskaya,S.S.R., 230
Guyana, 156
Gypsum, in weathering sequence, 39, 75, 242

Hafnium, as immobile element, 129
Hagerstown soil, 210, 216, 250, 251, 261
Halloysite, 21, 26, 53, 189, 190, 191, 196, 203, 208, 209, 215, 217
   equilibration rate, 188
   stability, 53
   synthesis, 52, 54
   in weathering, 39, 54
Hanford soil, 142
Hastings soil, 131
Hathaway soil, 206
Hayesville soil, 197
Hawaii, 21, 192, 215
Heat of formation, standard, 22
Heath, 209
Heavy liquid separations, 216
Heavy minerals, resistant, 115, 122, 199, 216
Hematite, 65, 77, 192, 193, 194, 214, 283, 284
   crystallinity, 191
   from goethite, 64
   in Oxisols, 285
   in Ultisols, 256
   in weathering sequence, 39
Heterogeneity, in relation to diffusion, 75

of soil, 2, 3, 7, 9, 171, 279
soil, in $E_H$ measurements, 276
High level laterite, 188
Histosols, 166, 237, 248, 258
*Holars lanetus,* roots, 159
Horizon, illuvial, 174
   leached, 209
Horizonation, of soils, 2, 9, 10, 61, 171, 181, 182
Hornblende, 116, 117, 119, 134, 156, 217
   basaltic, 116, 117, 119
   particle size distribution, 139
   weathering, 118, 134, 135
Hornblende schist, weathering, 20, 156, 157
Houston soil, 142, 172, 210, 214
Huambo, Angola, 240
Humic acid, 228, 229, 230, 231, 232, 282
   $^{14}C$ age, 166, 167
Humic acid-fulvic acid ratio, 232, 235, 240, 241
Humic ferruginous Latosol, 194, 195
Humic Gleys, 200
Humic matter, 86, 159, 282, 283
   pedogenic entity, 165
Humification, 231, 237
Humin, 228, 237
   $^{14}C$ age, 166, 167
Humus, 207, 248, 249
   as weathering agent, 18
   colloidal, 5, 87
   complexing action of, 186
   in B horizon, 89
   mean residence time of, 165
   in pedological processes, 86
   peptization, 90
   protective colloid, 86, 87, 88, 90
Hyacinth (metamict zircon), 118
Hydration, 46, 173
Hydraulic conductivity, 181
Hydrogen, exchangeable, 249
Hydrogen ion, coagulation by, 247
Hydrol humic soils, 215
Hydrologic soil series, 223
Hydrolysis, Donnan, 7, 8, 15, 44, 102
   of metallic complexes, 84
   mineral, 7
Hydronium, 27, 249
Hydrophile colloid, 88
Hydrothermal synthesis, 21, 44, 45

Hydrous micas, 39, 158, 203, 207, 217, 223, 224, 225, 285
Hydrous oxides, 225
Hydroxide, interlayered, 249
Hydroxides, colloidal, in pedological processes, 86
Hydroxyl ions, as mineralizers, 52
Hymatomelanic acid, 166, 228, 232
Hysteresis, in capillary conductivity, 72
  in moisture cycles, 61, 62, 181

Igneous rocks, 12, 134, 175, 228
  acid, 18, 190
  basic, 18, 192, 196
Illinois, 174, 199, 201, 202, 256, 257
Illite, 26, 27, 28, 143, 158, 193, 201, 203, 204, 208, 210, 211, 212, 213, 248, 279, 284
  in claypan soils, 174
  formation, 49, 186
  iron-rich, 251
  in limestones, 155
  magnesium in, 186
  in Mexico profile, 96
  from orthoclase, 30
  potassium in, 186
  stability area, 155, 205
  by weathering, of dibase, 42
  in weathering sequence, 39, 41
Illite-glauconite series, 155, 250, 251, 261
Illite-kaolinite reaction, 30
Illite-montmorillonite, interstratified, 210
Illitic mica, 255
Illuviation, 212, 217
  clay, 97, 140, 171, 172, 173, 174, 195, 203, 242
Ilmenite, 39, 120, 188, 189, 190
Immature soils, 7
Imogolite, 215
Immobile indicator, 123, 177, 217
Incident light, 171
Inclusions, in zircon, 118
Index mineral, 114, 118, 145
Index mineral method, 130, 132, 133, 137, 144, 149, 195, 280
India, 193
Indicator, immobile, 114
Indonesia, Dutch East Indies, 21, 215
Infiltration, 68, 182, 183
Infrared spectra, 284

Inhomogeneity, in soils, 3
Intensity factors, 38, 79
Interlayers, aluminum hydroxide, 199
Internal climate, of soils, 163
International Method, for mechanical analysis, 177
Interparticle bridges, 134
Interpenetration, of phases, 2
Interstratified clays, 39, 41, 194, 208, 210, 212, 213, 249
Ionic exchange, 5, 247, 278, 279
Ionic product, 208
Ionic ratios, of soil extract, 114, 272
Ionic strength, 14, 103
Iowa, 244
Ipava soil, 256, 257
Iredell soil, 173
Iron, 85, 129, 164, 207
  accumulation, 207
  aluminosilicate, 223
  complexed, 84, 187, 191, 193, 207, 213
  concretions, 273
  extractable, 223, 224
  ferric, 207, 287
  ferrous, 39, 207
    in soil water, 189
  hydrated silicates of, 16, 190
  ionized, activity, 207
  in montmorillonites, 45, 49
  movement in soil, 273
  in soil heterogeneity, 284
  soluble complexes of, 82
  translocated, 21
Iron anion, 77
Iron chelates, decomposition, 84
Iron hydroxide, 16, 89, 189, 190, 249
Iron-manganese ratios, in pedotubules, 275
Iron materials, in bauxite, 192
Iron minerals, 190, 199
Iron oxides, hydrated, 76, 190, 191, 192, 256, 283, 284
Iron phosphates, 244
Ironstone, lateritic, 189
Isoelectric composition, 87, 88, 89
Isoelectric point, 89
Isoelectric precipitation theory, 207
Isoelectric state, 87
Isoelectric weathering, 86, 90

## Subject Index

Isotopes, radioactive, 279
Issorora, Guyana, 20, 156

Japan, 21, 215
Jefferson County, Wisconsin, 200

Kamennaya, U.S.S.R., 230
Kaolinite, 14, 15, 16, 21, 26, 28, 33, 34, 35, 53, 88, 143, 144, 156, 189, 190, 191, 196, 203, 204, 207, 208, 209, 210, 211, 212, 213, 214, 217, 223, 248, 252, 255, 264, 265, 266, 267, 268, 269, 270
  from beidellite, 177
  in Cecil soil, 175
  composition, 29
  in compositional diagram, 51
  disordered, 54
  equilibration rate, 188
  from feldspar, 41, 130
  in fine clay, 141
  from granites, 30
  in Laterite, 193
  from muscovite, 154, 155
  in Oxisols, 257
  reaction with water, 22, 23
  solubility, 52
  stability, 53, 54, 186, 205
  suspension, 143
  synthesis, 49, 52
Kaolinite-gibbsite reaction, 32
Kaolinite + halloysite, 194
Kaolinitic clays, 192
Kaolinitic earth, 21, 156
Kaolinization, 20
Kaolins, 58, 191, 193, 203, 216, 285
  composition, 89
  in lateritic earth, 188
  surface by nitrogen adsorption, 58
  synthesis, 52
Kauri pine, podzols under, 84, 160
Kazakhstan, 146, 203, 230, 232
Kentucky, 275
Kinetic factors, 65
Kot, India, 193
Krasnozem, 166, 230, 231
Kyanite, 119

Labile pool, in exchange, 279
Labradorite, 26, 29, 187

Lakewood soil, 236
La Motte sandstone, 115, 118
Landscape relief, as sub forming factor, 163
La Salle County, Illinois, 201
Laterite, 21, 120, 122, 126, 131, 187, 191, 192, 193, 256
  high level, 21, 187, 188
  low level, 21, 190
  primary, 20, 187, 190
Lateritic clay, aggregation of, 179
Lateritic conditions, 186
Lateritic earth, 21, 188, 191
Lateritic ironstone, 189
Lateritic products, 137
Lateritic soils, 166, 177, 191, 195, 199, 223, 224, 227, 228, 233, 290
Lateritization, 1, 89
Latitude, as soil-forming factor, 164
Latosols, ferruginous humic, 40, 194
  clay formation, 131
  ferriallitic, 195
Lava, Pleistocene, 192
Lawsonite, 15, 16, 26
Leaching, 3, 19, 145, 260
Leaching factor, 226, 227
Leaching value, 228
Leaf canopy, effect on soil, 84
Leaf drip, 160
Leaf fall, 158, 161
Lebanon soil, 210, 211
Leonhardite, 26, 161
Lepidocrocite, 192, 283
Lessivage, 145, 174, 175
Leucite, 26
Leucoxene, 39, 119
Lignins, 159, 228
Lime potential, 272
Limestone, 18, 166, 210
Limestone soils, 144, 154, 166, 209, 212, 226, 248, 250
Limonite, 39, 283
Loess, 97, 118, 133, 152, 153, 166, 199, 203, 207, 215
  Peorian, 201, 202, 256
Loessial soils, 137, 138, 199, 243
Lolium perenne, roots, 159
Lysimeter, 94, 95, 100, 141, 152, 252

Macrocrystallinity, 278
McVickers soil, 172

## Subject Index

Magnesium, additions to soil, 270, 271
  in bog soil, 260
  in compositional diagram, 100, 101
  decomposition of minerals containing, 39
  double hydroxides, of, 45
  exchangeable, 243, 249, 250, 252, 259, 261, 262, 263
  in leaf fall, 161
  losses, 129, 252
  mobility in soil, 271
  in montmorillonites, 45, 50
  movement in Mexico profile, 108, 109, 110
  in profiles, 243
  in rainwater, 259
  recycling, 255
  in soil extracts, 103, 104, 107, 108, 186, 204, 265, 267, 268
  in Ultisols, 256
  uptake by grasses, 255
  in weathering, 17
Magnesium beidellite, 264
Magnesium carbonate, 243, 264, 268
Magnesium chlorite, 267, 268, 269
Magnesium ion, activity, 204
Magnesium minerals, 199
Magnesium Solonetz soils, 90
Magnesium sulphate, 243
Magnetite, 120, 134, 283
Manganese, 85, 284
  in concretions, 195
  exchangeable, 80, 275
  in montmorillonite group, 50
  movement in soil, 273
  in soil solution, 164
  soluble complexes of, 82
Manganese dioxide, 78, 80, 81, 82, 273, 274, 275
Manganese equilibria, in soils, 77
Manganese sesquioxide, 78, 80, 81
Manganous carbonate, 77, 78, 80, 81
Manganous hydroxide, 275
Manganous ion, 80, 81, 273, 274, 275
Margarite, 30, 47, 49
Marion soil, 119, 121
Marlette soil, 125, 133, 134, 138
Marshall soil, 117, 153
Mass action equation, 37
Mass movement, water, 13, 19, 61, 65, 72
Mat, of roots, 159

Material balance equation, 130
Maturity, of soils, 6, 7, 38
Maui, Hawaii, 194
Maxwell's equation, 181
Mazaruni Quarry, Guyana, 20
Mean error, 115
Mechanical analyses, 144, 223
Mecklenberg County, Virginia, 196
Mediterranean region, 214
Memphis soil, 153
Menfro soil, 152, 178, 180
Mercury, for pore size determination, 181
Metahalloysite, by weathering, 54
Metamorphic rocks, 8, 9, 228
Metastable phases, in hydrothermal synthesis, 52
Mexico soil, 95, 96, 111, 153, 252, 253, 256, 261, 262, 263, 264, 268, 270, 272, 273, 274, 275, 281, 282
Miami soil, 131, 160, 222
Mica, 29, 32, 155, 190, 193, 203, 205, 209, 217, 262
  intermediates, in Ultisols, 256
  particle size distribution, 139
  weathering to gibbsite, 63
Micaceous groups, composition, 47
Mica gneiss, 197, 198
Mica-kaolinite reaction, 30
Mica-vermiculite-smectite interstratification, 208
Micas, in Chernozem, 134
  composition, 46, 49
  depth function, 136
  hydrous, 41, 191, 203, 206, 207, 217, 223, 224, 225
  indicators of weathering, 118
  miscibility of phases, 46
  potassium, 199
  potassium dioctahedral, weathering stage of, 41
  potassium-sodium, 49
  secondary, 41
  surface by nitrogen adsorption, 58
  synthesis, 45, 49, 52
Mica schist, 198
Mice, activity in soils, 163
Michigan, 120, 125, 134, 144, 145, 208
Microbiological activity, 163
Microcline, 27, 28, 32, 39, 156, 205, 212, 216

## Subject Index

Microphotometer, 285
Microscopic methods, 170
Middle West, U.S.A., 199, 200
Millville silt loam, 72, 73
Mineral equilibria, 150
Mineralizer, water as, 45
Mineralogical analysis, 149
Mineral synthesis, 8, 9, 37
Mississippi River, 152
Missouri, 120, 124, 132, 135, 152, 157, 174, 211, 217, 244, 250, 252, 253, 254, 261
Missouri Ozarks, 210
Missouri River, 115, 116, 152
Mixed layer clays, 52, 185
Moisture content, in relation to water movement, 66
Moisture cycles, 61
Moisture equivalent, 69, 70, 71
Moisture potential, gradient of, 68
Moisture regime, internal, 249, 275
Moisture tension curves, 179
Mole fraction, of water, 57
Moles, activity in soils, 163
Mollisols, 153, 166, 197, 200, 203, 205, 249, 256, 257, 285
Monazite, 119
Monofunctional exchange, 261
Monomer, silica, 282
Montana, 125
Monteverde soil, 241
Montmorillonite, 15, 26, 31, 33, 53, 88, 142, 199, 204, 205, 206, 207, 211, 214, 261
  in black earths, 21
  calcium, 16
  chemical potential, 58
  composition, 33, 49, 53
  from diabase, 29
  miscibility of phases, 46
  from orthoclase, 30
  pedogenic, 208
  potassium, 156
  reversible collapse, 45
  sodium, 174, 204
  sodium-potassium, 49
  solubility, 52
  stability, 51, 54, 186
  suspension, 143
  synthesis, 45, 49, 51, 52
  weathering stage, 39, 41
Montmorillonite group, 28, 194
Montmorillonite-fulvic acid interlayers, 83
Montmorillonite-mica, interstratified, 194
Montmorillonitic clay, 141, 201, 203, 208
Moscow, U.S.S.R., 230
Mössbauer spectroscopy, 286
Mottlings, 284
Mountain taiga ferruginized soil, 230
Muscovite, 14, 25, 26, 27, 28, 30, 31, 36, 150, 156, 212
  from orthoclase, 63
  in Podzols, 134
  primary, 41
  stability area, 205
  thickness of liquid films on, 66
  in weathering sequence, 39
Muscovite-kaolinite reaction, 30, 154
Muscovite-potassium beidellite reaction, 34

Nacrite, 52
Naiwa soil, 193, 194
Nason soil, 261
Nepheline, 26, 29, 30
Neutralization capacity, 280
Neutron probe, 286
New Brunswick, 208
New Jersey, 236
New South Wales, 132, 135
New York, 140, 175, 237, 238
New Zealand, 21, 84, 215, 243
Nickel, 284
  in montmorillonite synthesis, 50
Nipe soil, 227, 256
Nitrogen, 228
  of humus, 231
  of soils, 162, 235
Nitrogen adsorption, external surface by, 58, 62
Nontronite, 39, 190, 203
Normative composition, 193, 225
North Carolina, 196, 239
North Dakota, 125, 140
Nova Scotia, 208
N/S quotient, 151
Nuclear magnetic resonance, 286

Oconee County, South Carolina, 197
Octahedral groups, 25
Ohio, 125, 131, 160, 222

## Subject Index

Ontario soil, 140, 237
Oligoclase, 8
Olivine-hornblende, in weathering sequence, 39
Opal, 112
Opal phytoliths, 76, 186, 191
Opaque heavy minerals, 119, 134
Optical density, 229, 231
Oregon, 141
Organic matter, acidic function, 187
 clay, complex, 212
 complexing action, 187, 207, 213
 in pedological processes, 185
 reducing agent, 187
 of soils, 149, 216, 231, 232, 242, 280
 soil apparent age, 165
Oriented coagulation, 280
Oriented clay, 140, 143
Orthoclase, 14, 25, 26, 27, 36, 39, 48
 in podzolic soils, 134
 weathering, 46, 63
Orthoclase-gibbsite reaction, 32
Orthoclase-kaolinite reaction, 30
Orthoclase-muscovite reaction, 26
Oxalic acid, 54, 83, 223
 basic salts of, 83
Oxidation-reduction conditions, 183
Oxidation-reduction equilibrium, 187
Oxidation-reduction potential, 79, 273
Oxides and hydroxides, by weathering, 150
Oxisols, 40, 191, 192, 193, 248, 256, 285
 Al in, 282
Oxygen, atmospheric, 275
 diffusion, 13
 in rock breakdown, 12
Ozarks, Missouri, 210

Pakhta-Aral, U.S.S.R., 230
Paleargid, 206
Paleosol, 165
Palestine, 214
Palygorskites, 205, 206, 207
Paragonite, 30, 34, 36, 49
Parent material, 40, 153, 154
 in clay formation, 131
 consolidated, 4
 as soil forming factor, 1, 2
 unconsolidated, 4
Particle size, effect on solubility, 139
 influence on standard free energy, 278

Particle size distribution, 118, 176, 178
Particle size variation, in goethite-hematite reaction, 64
Pastures, grass roots in, 159
Peats, 229, 275, 276
 C/N ratio, 241
Pectins, from roots, 159
Pedalfer, 7
Pedocal, 7, 8
Pedogenesis, 225
Pedogenic clay movement, 97
Pedogenic factors, 134
Pedogenic origin, 145
Pedological processes, 9, 10, 37, 61, 154, 185, 238
Pedological time scale, 284
Pedology, 1, 5, 9, 45, 86
Pedotubules, 275
Peds, 178, 206
Peorian loess, 201, 202
Peptization, 75, 86, 87, 90, 173, 174, 280
Perched water table, 73, 164, 183
Percolation, of water, 61
Permeability, 181, 182
Peters Mine, Guyana, 20
pF-moisture content curves, 73, 181
pH-dependent exchange, 248, 278
Phase, unstable, 44, 45
Phase boundary, 27
Phase diagrams, 52
Phase rule, 3
Phase, interpenetration of, 2
Phosphate, 259
 in limestone soils, 213
 iron, 244
 occluded, 243
  secondary inorganic, 243
 total, in profiles, 244
Phosphorus, available, 243
 enrichment in termite mounds, 163
 of humus, 231, 243
 total, 243
Physical description, of soil, 149
Phytoliths, opal, 76, 162, 186
Piedmont region, 132
Pine, humus from, 84
Pine forest soils, 161
Plagioclase feldspars, 119, 134, 216
Planosol, 202, 248
Platinum electrodes, in $E_H$ measurement, 276

Playas, 207
Plinthite, 191, 193, 256
Plugs, soil, 279
Poa pratensis, roots, 159
Podzol catena, 223, 259, 260
Podzol intergrade soil, change in clay, 137, 138
Podzolic soils, 9, 84, 134, 227, 228
Podzolization, 1, 85, 87, 161, 207, 225
Podzols, 5, 9, 54, 88, 89, 90, 133, 134, 160, 191, 207, 208, 209, 224, 225, 226, 229, 230, 233, 260, 280, 284
 aggregation, 178, 179
 A horizon of, 87
 B horizon of, 87
 $^{14}$C age, 166
 C/H ratio, 236
 clay formation, 131
 cristobalite ine, 139
 degree of saturation, 247
 differential movement in, 165
 elementary composition, 247
 formation, 88
 horizon relationships, 84
 under Kauri pine, 84, 160
 from limestone, 213
 mechanism, of formation, 83, 84
  for mobility of iron in, 85
 silica-alumina ratio, 224
Podzol zone, 231, 239
Poiseuille equation, 72
Polymer, amorphous, 282
Polymerization, degree of, 229
Polymers, in humus, 200
Polyphenols, mobilizers of iron, 85
Pond soil, 206
Ponding, in formation of Solonetz, 145
Pore sizes, distribution, 142, 179, 181, 182
Pores, blocking of, 143
 capillary, 158
 irregular, 72
Porosity, 3, 13
Porosity factor, 73
Potassium, 250, 270, 271, 272
 area in compositional diagram, 100, 101
 in bog soil, 260
 bonding energy, 210, 250
 exchangeable, 112, 243, 249, 252, 253, 259, 260, 262, 279
 fixation, 203, 243
 in hydrous micas, 41
 in illite, 186
 in illite-glauconite group, 250
 leaching, 129, 227, 250
 in leaf fall, 161
 mobility in soil, 270
 movement in Mexico soil, 108, 109, 110
 in profiles, 243
 recycling, by plants, 252, 255
 in sericite synthesis, 49
 in soil extracts, 103, 104, 107, 108, 186, 265, 266, 267, 268, 269
 soluble, in chloracetic acid, 260
  depth function, 113
 upward movement, 153
 in weathering, 17
Potassium beidellite, 28, 264
Potassium bicarbonate, 264, 266
Potassium chloride, diffusion coefficient, 74
 displacement, 74
 in montmorillonite synthesis, 50
Potassium feldspars, 199, 208
Potassium micas, 199
Potassium montmorillonite, 156
Potassium-sodium ratio, 255, 256, 257, 258, 259
Potassium-sodium theorem, 250
Potential, chemical, 22
Potential buffering capacity, 272
Prairie ecosystem, 161
Prairie-forest boundary, 161
Prairie grasses, recycling of K, 255
Prairie soils, 36, 88, 90, 130, 131, 162, 198, 200, 248, 252
Precambrian rocks, 157
Precipitation, in clay formation, 131
 monthly, annual, 99, 151
 of phosphate, 213
Precipitation-effectiveness index, 151
Precipitation-evaporation ratio, 151
Pre-Kaolin, 54
Pressure deficiency, 62, 68
Preston clay, 67
Primary laterite, 12, 16, 156, 187, 189, 190, 191, 192
Probability tables, 115
Profile, soil, 4
Protective action, 5, 86, 87, 207, 282
Proteins, 228
Protons, 286

## Subject Index

Provenance, 154
Proxying, Al for Si, 28
Puerto Rico, 256
Putnam soil, 95, 96, 141, 177, 252
Pyrophyllite, 26, 33, 49
Pyroxene, in weathering sequence, 39, 139

Quantity factor, 80
Quartz, 16, 26, 28, 29, 35, 53, 121, 156, 188, 189, 191, 193, 194, 196, 203, 208, 210, 211, 213, 217, 255, 281
  from alkaline solutions, 76
  authigenic, 42, 120, 131, 133, 135
  in Chernozem, 134
  cracks in, 139
  depth function, 135
  diabase profile, 76
  in fine clay, 139
  as index mineral, 114, 118, 120, 133
  in lateritic earth, 188
  loss by weathering, 120, 187
  in Mexico profile, 112
  particle size distribution, 139
  in podzolic soils, 134
  secondary, 41, 158, 190
  solubility, 15, 34, 45, 76, 139, 155
  in weathering sequence, 39, 41
Quartz + chalcedony, depth function, 136
Quercetin, 54

Radioactive isotopes, 85, 165, 279, 286
Radiocarbon age, 165
Rain factor, 151
Rainfall, frequency, 95
Raman soil, 261
Ramna bog, 258
Random coagulation, 280
Recharge with water, months of, 152
Recycling of cations, by plants, 255
Red-brown earths, 126
Reddish Brown Lateritic soils, 196, 199
Red earths, South Africa, clay formation, 131
Redox potential, 79
Red Podzolic soils, 131, 199
Redtop, composition, 255
Red tropical soils, degree of saturation, 247
Red-yellow Podzolic soils, 144, 177, 195, 213, 256
Refractive indices, 158, 217
Relative humidity, of soil air, 63, 64, 193

Relief, as soil-forming factor, 1
Rendzinas, 210, 214
Resilication, 191
Resistant mineral indicator, 94, 123
Resistant mineral ratios, 116, 118
Rock analyses, complete, 222
Rock weathering, 4, 18, 222
Root action, 158
Roots, mat of, 159
Rothemsted subsoil, 7
Rubidium, exchange, 261, 262
Runoff, surface, 164
Russia, 236
Rutile, 116, 117
  inclusion in quartz, 119
  as resistant mineral, 119
  in weathering sequence, 39

St. Clair soil, 144
St. Peter's sandstone, 115
Saline soils, 250
Salinization, 1
Salt content, in relation to moisture movement, 66
Samarkand, U.S.S.R., 233
Sand fractions, 216
Sandstone, weathering, 3, 18, 209
Sandy soil, heterogeneity, 3
Sanidine, 48
Saponite, 39, 51, 52
Saprolitic rock, 154, 155
Saskatchewan, 203, 222
Saturated flow, 72
Saturation, degree of, 256
Saturation deficit, 151
Sausuritization, 119
Saylesville soil, 199, 200
Scandinavia, 209
Scanning electron microscope, 171, 175, 176, 278, 285, 287
Scotland, 109
Seasonal temperature changes, 58
Seawater, in limestone, 155
Sedimentary rocks, 3, 40, 196
Selective processes, in Podzol formation, 87
Selectivity, 97, 248, 255, 263
  ionic, 149, 153
Selectivity curves, 260
Selectivity equation, 272

Selectivity number, 111, 114, 252, 261
  Ca/Mg, 111, 181
  depth function, 97, 98, 99, 253, 254
  K/Na, 111, 281
Self-diffusion, of water, 74
Self-mulching soils, 142, 172, 210
Semiarid soils, 204, 205
Semibog profile, "Under," 260
Senescence, of soils, 6
Sepiolite, 205
Sepiolite-attapulgite group, 51
Sericite, 27, 39, 41, 49
Serozem, 230, 231, 234
Serpentines, 207, 256
Sesquioxides, 89, 145
Shales, calcareous, montmorillonitic, 142
Sheridan soil, 122, 130, 132, 133, 137, 138, 139, 144, 217
Sharkey soil, 261
Shifting value, 227, 228
Shipor, U.S.S.R., 230
Siderite, 77
Sierozems, 207
Sierra sandy loam, clay movement, 142
Silica, 191, 212
  amorphous, 16, 33, 35, 41, 53, 112, 156, 186, 187, 188, 206, 209
    solubility, 34, 52, 76, 139
  colloidal, 87
  in compositional diagram, 101
  cryptocrystalline, 209
  cycle of, 113
  depolymerization, 282
  in equilibria, 281
  as immobile indicator, 114
  loss in weathering, 16, 46, 129, 145
  mobility of, 30, 32
  in plants, 191
  secondary, in soils, 76
  soluble, 103, 104, 105, 112, 155, 189, 265, 271, 272
Silica-alumina ratio, 17, 19, 20
Silica-sesquinoxide ratio, 87, 222, 223, 224, 225, 226, 227
Silicate framework structures, 45
Silicate glass, 215
Silicic acid, 23, 34, 41, 155, 189, 190, 216
  activity, 33, 263
  chemical potential, 24
  dimer, 282

  dissociation constant, 14
  gel, moisture hysteresis, 42
  in laterization, 89
  from mineral silicates, 89
  monomeric, 23, 32, 54, 189, 204
  in pedological processes, 86
  protective colloid, 140, 282
  solubility, 188
  trimer, 282
Sillimanite, 119
Silt, in filtration, 143
Silt fractions, 115, 216
Siltpan soils, 182, 183
Skins, clay, 140
Slope, as soil-forming factor, 163, 164
Small dilute exchange, 149
Smectites, 25, 191, 208, 209, 210, 223, 225, 279, 284, 286
  in Alfisols, 285
  in alkali soils, 33, 285
  compositional range, 50
  in desert soils, 285
  dioctahedral, 34
  formation, 186
  in Mollisols, 285
  in Podzols, 186, 284
  synthesis, 54
  trioctahedral, 34, 204
  weathering stage, 41
Smectite-mica, mixed layer, 255
Smectite-vermiculate, dioctahedral, 209
Sod-podzolic soil, 230
Sodic alkali soils, 206
Sodium, 250, 270, 271
  area in compositional diagram, 100, 101
  exchangeable, 145, 249, 252, 253, 261, 262, 280
  leaching, 227, 250
  movement, 108, 109, 110, 153, 270
  in profiles, 243
  in smectites, 41
  in soil solution, 103, 107, 108, 129, 186, 204, 248, 265, 266, 267, 268, 269
  in weathering, 17
  in zeolites, 186
Sodium acetate, in hydrothermal synthesis, 51
Sodium beidellite, 14, 34, 35, 204, 264
Sodium bicarbonate, 264, 270
Sodium carbonate, 216, 242

## Subject Index

Sodium chloride, 50, 58, 204, 206
Sodium citrate, in hydrothermal synthesis, 51
Sodium feldspars, 199
Sodium montmorillonite, 174, 204
Sodium-potassium theorem, 255
Sodium sulphate, 204, 206
Soil development, 5, 124, 242, 244
Soil formation, 5, 124, 222
Soil-forming factors, 1, 2
Soil horizons, differentiation, 163
Soil solution, 39, 86, 250, 252, 278
Soil Survey Manual, 170
Solid phases, definition in soils, 278
Solods, 146, 203, 247
Solonetz soils, 90, 145, 204, 248, 290
Sols bruns acides, 209
Solubility, congruent, 22
   in relation to particle size, 139
Sonoita soil, 206
South Africa, 131, 195
South America, 215
South Australia, 126
South Carolina, 258
South Kanora, India, 193
Spectrographic methods, 285
Spodosols, 175, 207
Stability of minerals, 22
Standard cell, Barth's, 120
Standard free energy, 21, 80, 84, 278, 283
Stem flow, 160
Stilbite, in weathering sequence, 39
Stockton soil, 142, 143
Strontium, exchange, 261, 262
Structure, soil, 171
Structure capacity, 179
Suite, of minerals, 115, 116
Sulphate, in Oxisols, 254
Sulphur, of humus, 231
Superior soil, 226
Superstition sand, 72, 73
Surface area, by adsorption, 181
Surface crust, 182, 190
Surface erosion, of phosphorus, 244
Surface layers, 249
Surface reactivity, in relation to moisture movement, 66
Surface tension, of water in soils, 72
Suspension effect, 79
Susquehanna soil, silica-alumina ratio, 227

Sweden, 223, 258, 260
Swelling clays, 42, 224
Synthesis, clay, 176

Tactoid formation, 280
Takyr, 207
Talbott soil, 213
Tarkio, Missouri, 152
Tartaric acid, basic salts of, 83
Tashkent, U.S.S.R., 234
Tatum clay, 262
Tatum soil, 261
Temperate region, 20
Temperature, annual cycle, 150
   in clay formation, 131
   diurnal variation, 60, 130
   effect on rock weathering, 17
   monthly variation, 60
   in water vapor movement, 66
Temperature regimes, 59
Temperature variations, cyclic, 58
Tenerife, 273
Tennessee, 209, 212
Tension, moisture, 62, 67, 68
Termites, mounds, enrichment in nutrients, 163
Terra Rossa, 18, 210, 214, 282
Tetraethylammonium hydroxide, 52
Tetrahedral groups, 25
Texas, 142, 176, 210, 214
Thermochemical data, 32, 37
Thermodynamic functions, 278
Thin sections, of soils, 3, 140, 149, 171, 187, 192, 279
Till, calcareous, 144
Time, as soil forming factor, 1, 2
   factor in clay formation, 132
Time phase characteristics, of climatic cycles, 59
Time scale, pedological, 284, 285
Titaniferous iron, 190
Titanium, in Barshad's method, 129
   as immobile element, 120, 129
Titanium oxide, 189, 190, 256
Topography, as soil-forming factor, 1, 2, 163, 164, 285
Tourmaline, as index mineral, 115, 116, 117, 119, 122, 132, 133, 134
Trans Baikal, western U.S.S.R., 230
Translocation, of clay, 10, 158, 218

## Subject Index

of phosphate, 213
Transmitting zone, 68
Transpiration, 153, 158
Tremolite, 116, 117, 134
Trifoleum repens, toots, 159
Trinidad, north, 195, 198, 199
Triple point, 34, 79
Tritiated water, salt diffusion, 74
Tropical Red Earths, 131, 195
Tropical soils, $^{14}$C age in relation to depth, 166
  nitrogen content, 235
Tumatumari, Guyana, 188, 189, 190
Turbicutans, 192
Turbulent flow, 172
2:1 layer clays, 198
Typic Calciustoll, 206
Typic Hapladulf, 178, 180
Typic Haplargid, 206

Ultisols, 191, 192, 193, 195, 209, 213, 248, 249, 256, 282, 285
Unavailable water, 179
Unden Podzol, 223, 224, 260
United States, S.E., 177, 228, 236, 256
Unsaturated flow, 72, 182
Uplift, of sedimentary rocks, 164
Uptake, differential, 261
  preferential of K, 243
U.S.S.R., 229, 230
Uzbekistan, U.S.S.R., 207

Vapor phase movement, of water, 59
Vapor pressure, of water in soils, 63, 64
Valuisk, U.S.S.R., 230
Vegetation, factor in clay formation, 131
  as soil-forming factor, 1, 161, 285
Venezuela, 166
Vermiculite, 39, 45, 194, 208, 209, 210, 213, 262
  chloritized, 208
  in Podzols, 284
  in Ultisols, 256
Vermiculite-chlorite, 194, 209
Vertisols, 172, 210, 214
Vina loam, clay movement, 142
Virginia, 210
Voids, 171
Volcanic activity, 215
Volcanic ash, soils from, 5, 21, 214, 215

Volcanic dust, 210
Volcanic glass, 21
Volvograd, U.S.S.R., 230
Voronezh, U.S.S.R., 230

Water, in rock breakdown, 12
Water extract, of soil, 100
Waterlogged soils, 62, 77, 79
Wavelength, of moisture cycles, 151
Water table, depth of, 163, 164
  perched, 164, 183
Waxes, 228
Weathering, action of frost on, 61
  atmospheric, 15, 190, 215
  calcium distribution, 242
  degree of, 116, 200
  of rocks, 8, 9, 13, 175, 200
Weathering crust, 12
Weathering index, 38
Weathering mean, 40, 140
Weathering reactions, 22, 25
Weathering sequence, 21, 38, 39, 40, 41, 140, 178
Weathering stage, 46
Wetting front, 68
White House soil, 205
Wilting point, 72
Wisconsin, 199

X ray, methods, 5, 144, 158, 185, 193, 203, 209, 210, 216, 251, 262, 270, 284
  quantitative, 125
X ray fluorescence, 286

Yellow Podzolic soils, 131, 196, 199
Yellow-Red Podzolic soil, 224, 225, 227, 228
Yolo soil, 68, 69, 70, 142, 143

Zeolites, 15, 37, 47, 186, 285
  composition, 25
Zeolitic structures, collapse of, 45
Zeta potential, 86, 87, 173, 174, 279
Zinc, in montmorillonites, 45, 50
Zircon, index mineral, 39, 115, 116, 117, 118, 120, 121, 122, 132, 134, 138, 144, 145
  metamict (hyacinth), 118
Zirconium, in heavy minerals, 120
  as immobile indicator, 122, 129
Zircon-tourmaline ratio, 121, 145